AF352837

Advances in Binders for Construction Materials

Advances in Binders for Construction Materials

Editor

Jorge Otero

MDPI • Basel • Beijing • Wuhan • Barcelona • Belgrade • Manchester • Tokyo • Cluj • Tianjin

Editor
Jorge Otero
Mineralogy and Petrology
University of Granada
Granada
Spain

Editorial Office
MDPI
St. Alban-Anlage 66
4052 Basel, Switzerland

This is a reprint of articles from the Special Issue published online in the open access journal *Materials* (ISSN 1996-1944) (available at: www.mdpi.com/journal/materials/special_issues/Binders_Constr_Mater).

For citation purposes, cite each article independently as indicated on the article page online and as indicated below:

LastName, A.A.; LastName, B.B.; LastName, C.C. Article Title. *Journal Name* **Year**, *Volume Number*, Page Range.

ISBN 978-3-0365-6583-5 (Hbk)
ISBN 978-3-0365-6582-8 (PDF)

Cover image courtesy of Jorge Otero

© 2023 by the authors. Articles in this book are Open Access and distributed under the Creative Commons Attribution (CC BY) license, which allows users to download, copy and build upon published articles, as long as the author and publisher are properly credited, which ensures maximum dissemination and a wider impact of our publications.
The book as a whole is distributed by MDPI under the terms and conditions of the Creative Commons license CC BY-NC-ND.

Contents

About the Editor

Jorge Otero

Jorge Otero is a research fellow at the University of Granada (UGR, ESP). He holds a Ph.D. in civil engineering from Sheffield Hallam University (UK). He specializes in the characterization of masonry materials, their degradation phenomena, and the development of technologies for their conservation. He is now the PI of the Horizon2020-funded NANOMORT-project. Before joining UGR, he was a postdoc researcher at the Getty Conservation Institute (USA), an intern at the Smithsonian's Museum Conservation Institute (USA) and the Institute of Conservation (ICON, UK). He is a member of ICOMOS and coauthor of the open-access "Built Heritage Evaluation Manual"published by the Smithsonian Scholarly Press.

 *materials*

Article

In-Depth Analysis of Cement-Based Material Incorporating Metakaolin Using Individual and Ensemble Machine Learning Approaches

Abdulrahman Mohamad Radwan Bulbul [1], Kaffayatullah Khan [1,*], Afnan Nafees [2], Muhammad Nasir Amin [1], Waqas Ahmad [2], Muhammad Usman [3], Sohaib Nazar [2] and Abdullah Mohammad Abu Arab [1]

1   Department of Civil and Environmental Engineering, College of Engineering, King Faisal University, Al-Ahsa 31982, Saudi Arabia
2   Department of Civil Engineering, COMSATS University Islamabad, Abbottabad 22060, Pakistan
3   Interdisciplinary Research Center for Hydrogen and Energy Storage (IRC-HES), King Fahd University of Petroleum & Minerals (KFUPM), Dhahran 31261, Saudi Arabia
*   Correspondence: kkhan@kfu.edu.sa

Abstract: In recent decades, a variety of organizational sectors have demanded and researched green structural materials. Concrete is the most extensively used manmade material. Given the adverse environmental effect of cement manufacturing, research has focused on minimizing environmental impact and cement-based product costs. Metakaolin (MK) as an additive or partial cement replacement is a key subject of concrete research. Developing predictive machine learning (ML) models is crucial as environmental challenges rise. Since cement-based materials have few ML approaches, it is important to develop strategies to enhance their mechanical properties. This article analyses ML techniques for forecasting MK concrete compressive strength (fc'). Three different individual and ensemble ML predictive models are presented in detail, namely decision tree (DT), multilayer perceptron neural network (MLPNN), and random forest (RF), along with the most effective factors, allowing for efficient investigation and prediction of the fc' of MK concrete. The authors used a database of MK concrete mechanical features for model generalization, a key aspect of any prediction or simulation effort. The database includes 551 data points with relevant model parameters for computing MK concrete's fc'. The database contains cement, metakaolin, coarse and fine aggregate, water, silica fume, superplasticizer, and age, which affect concrete's fc' but were seldom considered critical input characteristics in the past. Finally, the performance of the models is assessed to pick and deploy the best predicted model for MK concrete mechanical characteristics. K-fold cross validation was employed to avoid overfitting issues of the models. Additionally, ML approaches were utilized to combine SHapley Additive exPlanations (SHAP) data to better understand the MK mix design non-linear behaviour and how each input parameter's weighting influences the total contribution. Results depict that DT AdaBoost and modified bagging are the best ML algorithms for predicting MK concrete fc' with $R^2 = 0.92$. Moreover, according to SHAP analysis, age impacts MK concrete fc' the most, followed by coarse aggregate and superplasticizer. Silica fume affects MK concrete's fc' least. ML algorithms estimate MK concrete's mechanical characteristics to promote sustainability.

Keywords: metakaolin; SHAP analysis; bagging; boosting; decision tree; multilayer perceptron neural network; random forest

Citation: Bulbul, A.M.R.; Khan, K.; Nafees, A.; Amin, M.N.; Ahmad, W.; Usman, M.; Nazar, S.; Arab, A.M.A. In-Depth Analysis of Cement-Based Material Incorporating Metakaolin Using Individual and Ensemble Machine Learning Approaches. *Materials* **2022**, *15*, 7764. https://doi.org/10.3390/ma15217764

Academic Editor: Jorge Otero

Received: 12 September 2022
Accepted: 27 October 2022
Published: 3 November 2022

Publisher's Note: MDPI stays neutral with regard to jurisdictional claims in published maps and institutional affiliations.

Copyright: © 2022 by the authors. Licensee MDPI, Basel, Switzerland. This article is an open access article distributed under the terms and conditions of the Creative Commons Attribution (CC BY) license (https://creativecommons.org/licenses/by/4.0/).

1. Introduction

Throughout the previous decades, there has been a strong demand and concern for investigation to develop green structural materials to meet the increasing need from public and private sectors. Concrete continues to be the most widely utilized manmade substance on the planet. Given the considerable environmental impact of cement production, research has concentrated on both reducing the impact on the environment and cost reductions for cement-based products [1–3]. The utilization of metakaolin (MK) as an additive or partial substitute for cement is a major area of research in the manufacture of concrete materials.

MK is an alternative to cement that is manufactured by calcining kaolin clays at elevated temperatures ranging from 700 °C to 900 °C. As a cement replacement in concrete structures, MK has been employed as a 10% to 50% replacement, depending on the specific application [4–7]. It has been shown that MK enhances the mechanical and durability properties when used in place of Portland cement [8–10]. The pozzolanic reaction, MK aggregate's fineness, and the accelerated cement hydration all contribute to an increase in concrete's compressive strength (fc') during the early curing phases [11]. Additionally, cement manufacture generates a substantial amount of carbon dioxide (CO_2) emissions; this new trend of replacing metakaolin for cement in concrete is part of a comprehensive approach to environmental sustainability. Addition of MK in concrete has various advantages as depicted in Figure 1.

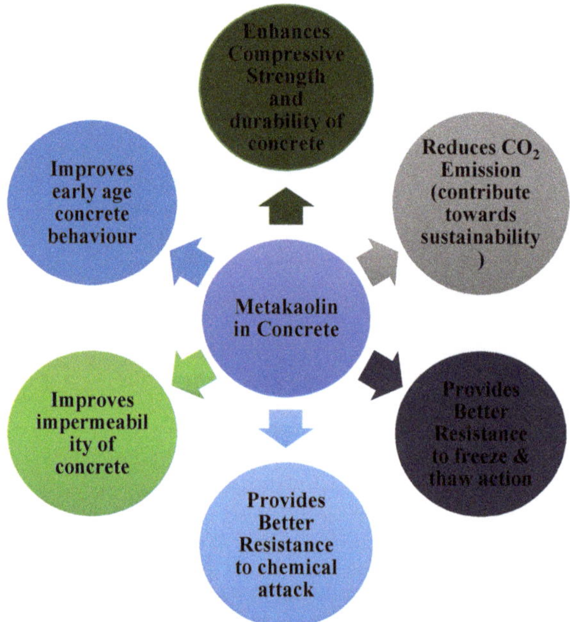

Figure 1. Advantages of metakaolin in concrete.

The cost, labour, and time consuming complexity of laboratory-based mixture optimisation might be replaced by computational modelling techniques [12]. To determine the optimum concrete mixtures, these approaches generate objective functions from the concrete components and their properties, and then use optimization techniques to determine the best concrete mixtures. Previously, goal functions for linear and nonlinear models were individually created. Due to the very nonlinear connections between concrete qualities and input parameters, the relationships of such models cannot be precisely established. Therefore, researchers are using machine learning (ML) techniques for predicting concrete properties.

Creating a concrete mix with MK in it complicates the determination of the concrete's fc' using an analytical formula, as opposed to a standard concrete, which has fewer mix parameters than cement MK specimen. This is mostly because of the enormous number of constituents and the fc' very nonlinear behaviour in regard to the mix parameters. To this purpose, when basic equations cannot directly connect the input and output values, machine learning (ML) techniques frequently give important alternatives in the context of engineering problem solving [13–22]. Owing to the intricate nonlinear interactions amongst independent and dependent variables, such techniques can be accomplished with a sufficient level of accuracy if a comprehensive library of sufficient experimental data points is accessible in the area of computational engineering structures and materials. Thus, a wide range of innovative approaches to a wide range of technological problems may be put into practice.

Until now, the literature has primarily focused on the use of ML techniques such as artificial neural network (ANN) in the field of materials science without ensemble learners [23–26]. These algorithms were utilized to predict the fc′ and elasticity modulus of materials composed of cement [26–29]. The literature has comprehensive and extensive publications on the use of ANNs in the modelling of concrete materials [30–35]. Fuzzy logic algorithms and genetic algorithms approaches have also been utilized in the recent decade in place of ANN models to describe the mechanical properties of cement-based materials [36–41].

Since cement-based materials have a limited number of ML methods, it is vital to investigate if other ML techniques may be used to improve their mechanical characteristics. Thus, the present work investigates ML approaches application for predicting the fc′ of MK concrete. Three different individual and ensemble ML predictive models are presented in detail, namely decision tree (DT), multilayer perceptron neural network (MLPNN), and random forest (RF), together with the factors that are most effective, allowing for efficient investigation and prediction of the fc′ of cement-based concrete. The authors employed a comprehensive database of MK concrete mechanical characteristics for model generalisation since it is an essential part of any prediction or simulation work. The reported database contains 551 data points with highly effective input parameters for calculating the fc′ of MK concrete. The database includes a value for cement, metakaolin, coarse and fine aggregate, water, silica fume, superplasticizer, and age, which have a considerable effect on the fc′ of concrete and have rarely been treated as vital input parameters in the past. The trained and created model has produced a holistic map of concrete fc′. Finally, the performance capabilities of the offered models are evaluated in order to select and implement the most predictive model for addressing the mechanical properties of MK concrete.

There has been a surge in increased interest in large-scale production of sustainable, low-priced, and high-performance construction materials that are also robust in adverse ecological circumstances over the previous few decades. One of the world's most common construction materials—cement-based concrete—required the incorporation of more components and additives than previously used concrete because of environmental concerns. However, the high number of mixture factors and their substantially nonlinear relationship to the mechanical characteristics of concrete, such as the fc′, challenge the analytical methods for numerically estimating the concrete fc′. To this purpose, unconventional methods become a critical instrument for resolving the afore-mentioned complicated optimisation problem. In this perspective, the most widely used ML techniques, such as, DT, MLPNN, and RF, have been suggested for estimating the fc′ of concrete, a critical parameter for the reliable design in structure. Among the proposed ML models, the optimal predictive model has shown to be extremely successful, demonstrating trustworthy projections and, most importantly, showing its highly non-linear mechanical properties.

Additionally, there seems to be a research gap in the study of MK fc′ and its influence on raw materials. It was, thus, necessary to investigate the influence of MK containing concrete's input parameters/raw components on its anticipated compressive strength using a post hoc model-agnostic approach known as SHapley Additive exPlanations (SHAP) [42,43]. Machine learning (ML) techniques were used to integrate SHAP data in order to get a better understanding of the multifarious non-linear behaviour of the MK design mix for the strength parameter and how each input parameter's weighting affects the overall contribution. ML approaches may be used to accurately forecast concrete kinds, as previously stated. The experimental setting requires a significant investment in terms of labour, time, and resources to do this. Data modelling and the discovery of interconnected independent components, as well as a rapid reduction in input matrix size are, thus, urgently required. Concrete construction materials may be accurately predicted using machine learning approaches. The use of ML methods may be justified as an alternate strategy to calculating MK fc′ in order to save on both time and money spent on experiments. We used both a stand-alone ML model and an ensemble of ML models in our investigation. Additionally, statistical tests were used to evaluate the models, and their results were compared. Later, a model with precise MK prediction was suggested based on the performance of several statistical factors. In order to get a thorough understanding of mix design in order to achieve MK concrete strength, this research also explained how input factors contributed and how ML models were integrated. Explainable ML techniques and features significance for considerable characteristics of the structure were found to be linked in the study's overall findings.

2. Data Description

Currently accessible literature has been used to get the data needed to simulate concrete's fc′ utilising MK [44–55]. The predicted output compressive strength data consists of eight input parameters, which include cement, MK, fine and coarse aggregate, water, age of concrete, superplasticizer, and silica fume. Type of cement is not considered as an input parameter, as only one type of cement (Type-I) is utilized for modelling the ML algorithms. Cement and metakaolin are two constituents that are prevalent among those selected for the database. Additionally, attempts were made to choose articles that share common components (admixtures, superplasticizers, etc.). The authors tried to choose publications based on criteria related to materials that are widely used in concrete and make important contributions to concrete's mechanical characteristics. Further, similar material in varied arrangements is required for modelling ML algorithms. Except for age in days, all characteristics are measured in kilograms/m^3. Descriptive statistics are a set of descriptive coefficients that provide a result that may be applied to the whole population or to a sample of the population. In descriptive statistics, measures of central tendency and measures of variability are used (spread). However, variance, standard deviation, maximum and minimum variables, kurtosis, and skewness are all indices of variability. Tables 1 and 2 and Figure 2 provide the variation in data used to run the models. Various information is reflected in the descriptive analysis's outcomes, which are derived from the data of all the input variables. Additionally, the table displays the ranges, maximum, and lowest values of each model variable. Nonetheless, the other parameters of the study, such as mean, mode, standard deviation, and the total of all data points for each variable, also reveal the important values. Figure 2 depicts the relative frequency distribution of each parameter utilised in the mixes. A relative frequency distribution illustrates the percentage of total observations that correspond to each value or class of values. It has tight ties to a probability distribution, which is often used in statistics. Figure 2 depicts the link of input parameters by displaying the relative frequency distribution of data items. Each chosen parameter has a significant impact on the concrete's strength characteristics. In addition, Table 1 displays the lowest and highest variable values including 551 datasets, and Table 2 provides a data analysis check with the variance, range, standard deviation, and mean.

Table 1. Metakaolin concrete compressive strength model input and output variable ranges.

Parameters Input Parameters	Acronym	Min	Max
Cement	C	30	512
Fine aggregate	FA	300	1146
Coarse aggregate	CA	0	1154
Water	W	12	220.30
Silica fume	SF	0	75
Metakaolin	MK	0	256
Superplasticizer	SP	0	43
Age	Age	1	180
Output variable			
Compressive strength	fc′	9.84	131.30

Table 2. Statistical description of Metakaolin concrete variables.

Parameters	Cement	Fine Aggregate	Coarse Aggregate	Water	Silica Fume	Metakaolin	Superplasticizer	Age (Days)
Statistical description								
Mean	396.64	737.24	853.09	161.71	4.05	45.48	2.90	37.29
Std error	3.72	7.79	15.47	1.53	0.55	2.16	0.33	1.93
Median	400	711	1037	163.40	0	40	0	28
Std. dev	87.28	182.93	363.13	36.00	13.02	50.59	7.73	45.34
Variance	7617.40	33,463.99	131,866.26	1295.65	169.49	2559.62	59.80	2056
Kurtosis	3.57	0.37	1.42	4.18	13.68	4.43	16.38	3.13
Skewness	−1.34	0.52	−1.75	−1.58	3.68	1.85	3.92	1.90
Range	482.00	846	1154	208.30	75	256	43	179
Min	30	300	0	12	0	0	0	1
Max	512	1146	1154	220.30	75	256	43	180
Sum	218,548.70	406,217	470,050.80	89,099.50	2232.15	25,060.77	1597.30	20549
Count	551	551	551	551	551	551	551	551
Training dataset								
Mean	395.18	738.49	841.41	161.06	4.09	45.50	2.66	37.85
Std error	4.19	8.86	17.83	1.74	0.63	2.39	0.35	2.20
Median	400	711	1037	163.40	0	40	0	28
Std. dev	87.97	185.84	374.05	36.54	13.24	50.03	7.38	46.20
Variance	7738.04	34,538.02	139,915.86	1334.83	175.26	2503	54.44	2134.15
Kurtosis	3.65	0.33	1.04	4.08	13.61	4.56	18.52	2.92
Skewness	−1.38	0.52	−1.66	−1.57	3.69	1.86	4.13	1.86
Range	482	846	1154	208.30	75	256	43	179
Min	30	300	0	12	0	0	0	1
Max	512	1146	1154	220.30	75	256	43	180
Sum	173,881.30	324,936.00	370,219.70	70,868.40	1798.72	20019.51	1171.52	16656
Count	440	440	440	440	440	440	440	440
Testing Dataset								
Mean	402.41	732.26	899.38	164.24	3.90	45.42	3.84	35.07
Std error	8.03	16.29	29.75	3.21	1.15	5.03	0.85	3.98
Median	400	708	1037	163.40	0	40	0	28
Std. dev	84.64	171.61	313.42	33.81	12.16	53	8.98	41.92
Variance	7163.11	29,450.55	98,231.83	1142.93	147.97	2808.83	80.63	1756.94
Kurtosis	3.30	0.59	3.85	4.79	14.25	4.19	11.32	4.29
Skewness	−1.16	0.54	−2.25	−1.62	3.64	1.85	3.31	2.06
Range	434.50	846	1149	192.40	75	256	43	179
Min	77.50	300	0	27.90	0	0	0	1
Max	512	1146	1149	220.30	75	256	43	180
Sum	44,667.40	81,281.00	99,831.10	18,231.09	433.43	5041.26	425.78	3893
Count	111	111	111	111	111	111	111	111

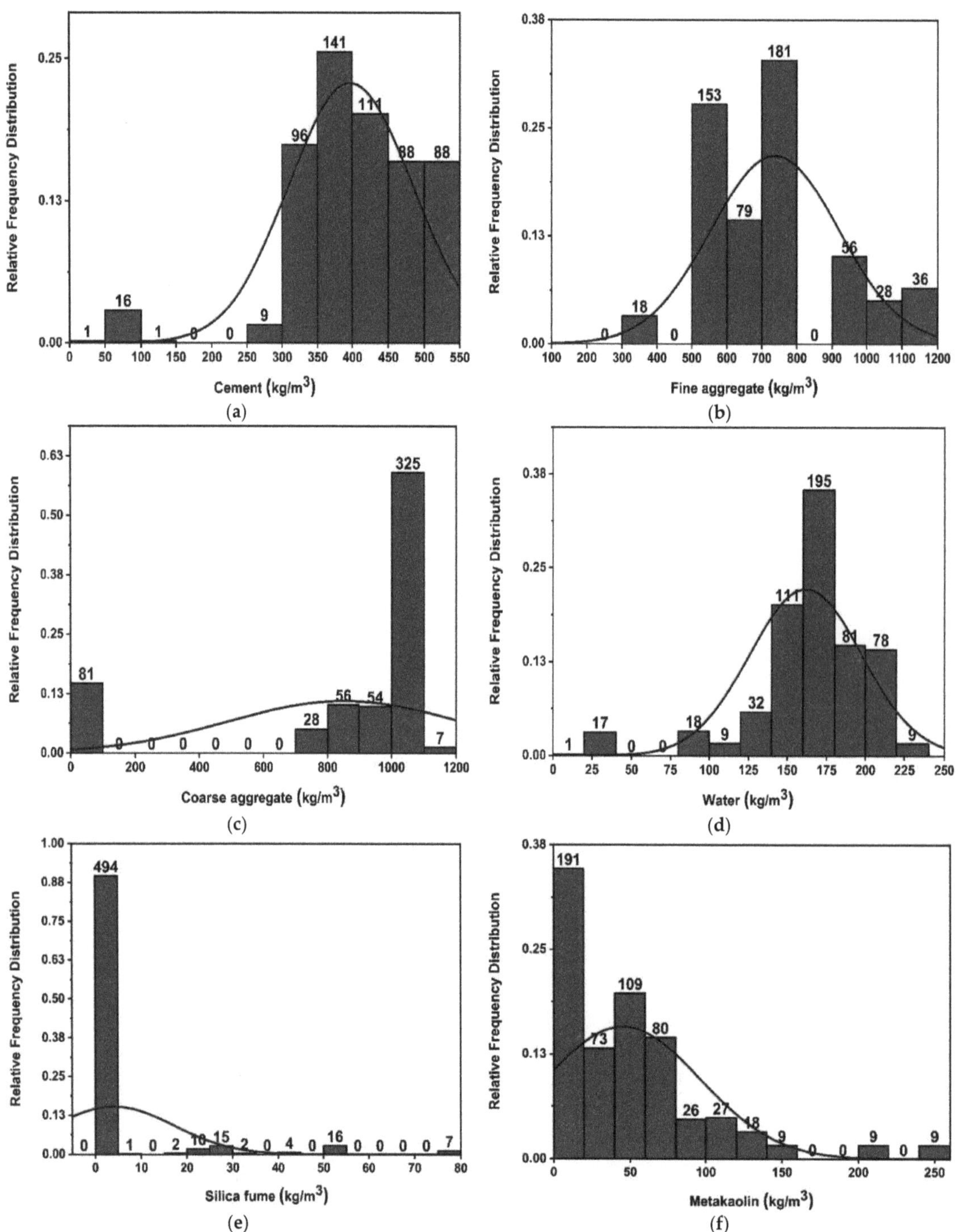

Figure 2. *Cont.*

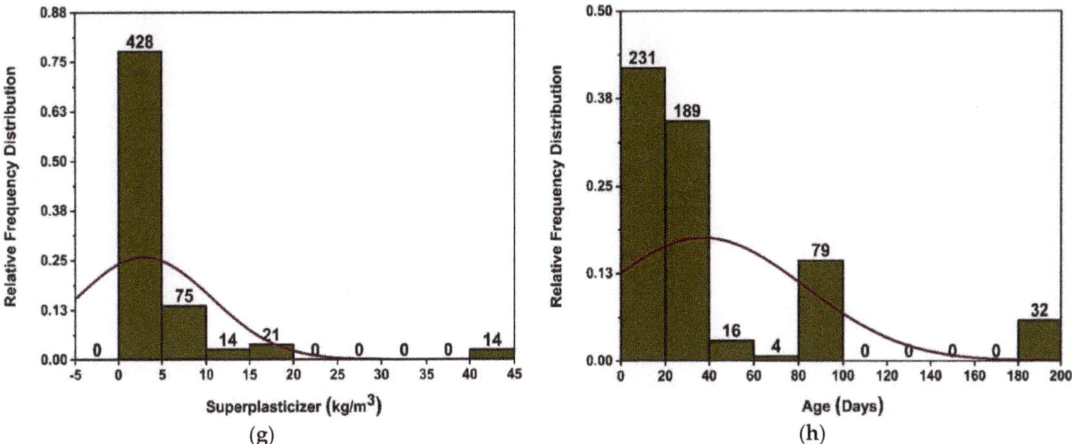

(g)

(h)

Figure 2. Compressive strength parameters' relative frequency distribution; (**a**) cement, (**b**) fine aggregate, (**c**) coarse aggregate, (**d**) water, (**e**) silica fume, (**f**) metakaolin, (**g**) superplasticizer, and (**h**) age.

3. Methodology

ML techniques are being used in a variety of fields to anticipate and understand the behaviour of materials. In this work, ML-based techniques comprising of the DT, MLPNN, and RF are employed for forecasting the fc' of MK concrete. The selection of these methods is based on their prevalence and reliability in the forecast of outcomes in comparable studies, as well as their significance as the top data mining algorithms. In addition, an ensemble algorithm was afterwards employed to simulate the concrete fc'. Figure 3 displays the technique flow chat for ensemble learning.

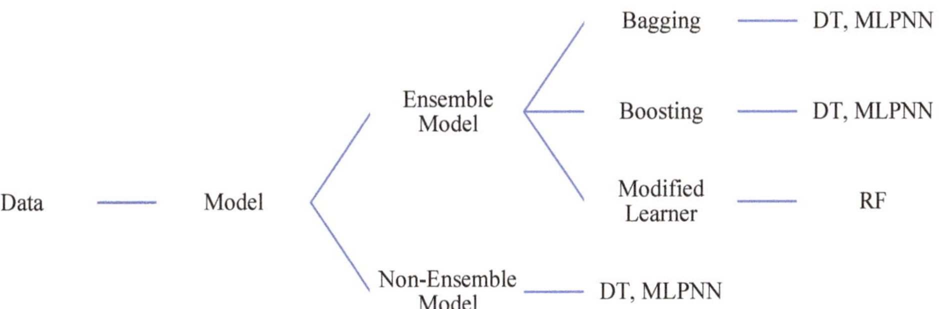

Figure 3. Flow chart of ML techniques.

3.1. Machine Learning Methods

It has been shown that artificial intelligence (AI) is a more effective modelling methodology than traditional methods. AI has a number of advantages for addressing ambiguous difficulties and is an excellent method for handling such complicated situations. It is possible to identify engineering design parameters using AI-based methods when testing is not possible, resulting in significantly reducing the workload of human testers. In addition, AI could expedite decision making, decrease error, and improve processing efficacy [56]. Recently, a rise of interest in the application of artificial intelligence to all scientific fields has been observed, sparking a range of goals and aspirations. The field of civil engineering has experienced a significant increase for utilizing different AI methods all over its numerous fields. ML, reinforcement learning, and deep learning (DL) are three AI approaches that are proving to be a new category of creative approaches to structural engineering problems.

ML is a fast-expanding area of AI that is often used in the construction sector for predicting behaviour of material. One project aims to investigate inclusion of social elements into multi criteria infrastructure assessment strategies, with inclusion of social factors into the assessment of infrastructure's long-term viability using multi-criteria assessment techniques [57]. In the framework of structural design, exhaustive study on evolutionary computation, an area of artificial intelligence, was conducted [58]. Yin et al. [59] explored AI uses in geotechnical engineering. A study was done to determine the state of high-rise building optimization. [60]. In order to synthesise concepts in the developing field of AI applications in civil engineering, this study was done. This list contains a broad variety of methods: Fuzzy systems (FS), neural networks, expert systems (ES), reasoning, categorization, and learning are only a few examples of evolutionary computing [61].

In spite of the fact that the referenced review papers discussed the use of AI in civil engineering, they mostly concentrated on the usage of old approaches and did not cover latest methods using ensemble techniques. Figure 3 shows an estimation of the fc' of MK concrete using ML approaches including DT, MLPNN, and RF. These algorithms were selected on the basis of their broad usage in relevant research and their reputation as the finest data prediction algorithms available. In addition, ensemble techniques are employed to predict the concrete modelling strength. In terms of computational speed and processing time, ML models are fairly significant. Compared to conventional models, the rate of error is almost non-existent. A comparison is made among individual and ensemble models in this research. SHAP analysis is performed to find the optimum dosage and contribution of each input parameter towards fc'. Moreover, positive and negative impacts of each input parameter and their effect on other parameters are also studied.

3.1.1. Decision Tree-Based Machine Learning

In DT, any number of nodes can be connected to any number of branches, and each node can have an infinite number of branches. Leaves are the nodes that do not have any outgoing ends, whereas inner nodes are those that do. Using an interior node for a specific event, the case utilized for classification or regression can be partitioned into multiple classes. During the learning process, the input variables play an important role. The algorithm that generates the DT from provided instances serves as the stimulant for the DT. By reducing the fitness function, the implemented algorithm calculates the optimal DT. A regression model is used because there are no classes in the dataset selected for this study, so the independent variables are used instead of the target variable. For each variable, the dataset is broken up into many subsets. The error among the anticipated and the actual values of the pre-specified relation is determined at every split point by the algorithm. The variable having least values of fitness function is selected as the split point after comparing the inaccuracies in the split point across the variable quantity. Repeatedly, this technique is carried out.

In the DT architecture, independent variables are partitioned into homogenous zones by decision rules that recursively split them [62]. DT is primarily concerned with the investigation of a system for making decisions that are suitable for predicting a result given a collection of inputs. DT is referred to as a regression tree or a classification tree, depending on whether the target variables are continuous or discrete [63]. Numerous studies have demonstrated the effectiveness of DT in a variety of real-world situations for the aim of prediction and/or categorization [64].

The primary advantage of DT is its ability to simulate complex interactions between existent variables. Through consideration of how data are distributed, DT models are capable of combining both continuous and categorical variables without making any stringent assumptions [65]. Additionally, developing a DT is straightforward, and the resulting models are easily interpretable. Additionally, when it comes to determining the relative relevance of input characteristics, DT is an excellent choice [66].

DT modelling includes two phases: tree creation and tree pruning [67]. Stage one starts with DT's root node being identified as the independent variables with the highest performance gain. Following that, according to root values, the training dataset is partitioned into subsets and sub-nodes are created. When the input variables are discrete, a sub-node of the tree is constructed for each conceivable value, whereas in some circumstances, the threshold-finding step results in the generation of two sub-nodes [68]. Following that, each sub-gain node's share is calculated, and the technique is recurring till all samples in a certain node are classified as belonging to the same class. They are then referred to as "leaf nodes," and their values are designated as the values of the classes they belong to. A flow chart of DT is shown in Figure 4.

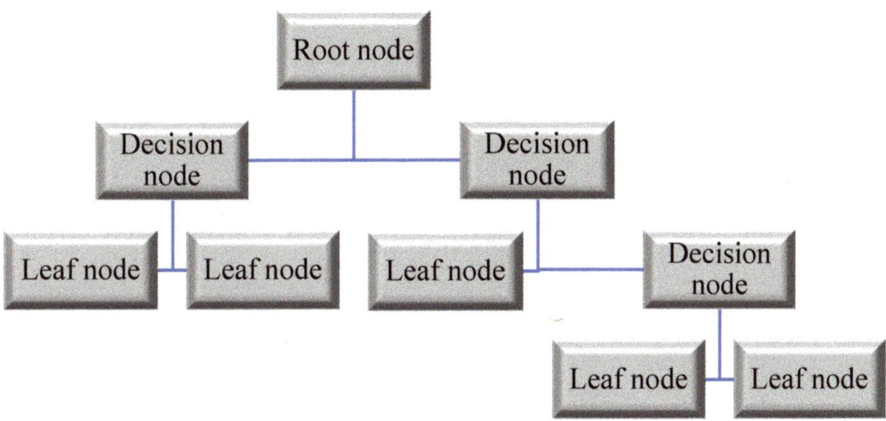

Figure 4. Flow chart of DT.

3.1.2. Artificial Neural Network-Based Machine Learning

A multilayer perceptron neural network (MLPNN), a network-based DM computing approach, is employed in this research with individual base learner and ensemble base learner methods to model and forecast MK concrete. An ANN program mimics the structure of a biological nervous system's neurons [69]. Parallel linkages provide the basis of ANNs. In order to transmit the weighted inputs from neurons, these cells use an activation function to transmit the weighted outputs. It is possible to have one or more multi-layers in these activities. Use of the multilayer perceptron network is widespread in brain activity. The perceptual response is created among the number of input parameters and the number of output parameters. There are three types of layers in a network: input, hidden, and output. Between the input and output layers, there is a hidden layer that may have a huge network of hidden layers. The perceptron can handle all of its issues with a single layer, but it is more efficient and helpful to have several hidden layers [70]. Figure 5 depicts a typical neural network design. With the exception of neurons in the input layer, all neurons in a layer perform linear addition and bias computations. Non-linear functions are then calculated in the hidden layers by neutrons A sigmoid function is a word used to describe a non-linear function [71]. This research paper models ANNs using a feed-forward multi-layer perceptron (MLPNN) network. To discover the highest-performing MLPNN, hidden layers and neuron pairs of varied numbers are meticulously selected [72]. In the hidden layer, the link between input and output variables may also be determined using a linear activation function and a nonlinear transit function. In addition, the data extracted from the published literature are separated into training and testing sets. This is conducted to reduce the influence of data overfitting, since overfitting is an intractable issue in machine learning. Randomly, 80 percent of the data is used for training the models and 20 percent for testing the trained models, as recommended by the literature [73].

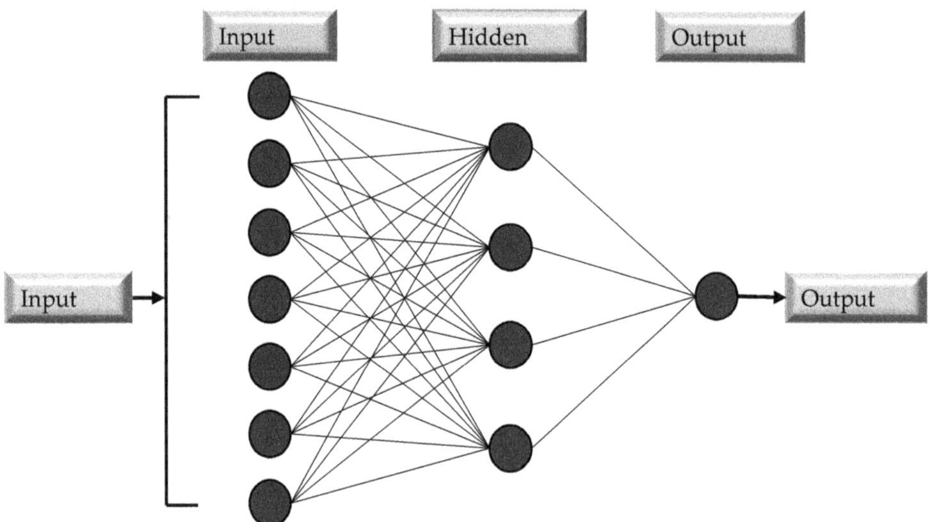

Figure 5. Flow chart of MLPNN.

3.2. Bagging and Boosting for Ensemble Approaches

ML classification and prediction accuracy may be improved using ensemble approaches. By combining and accumulating several weaker predictive models, such techniques frequently support decrease training data over-fitting problems (component sub-models). Training data may be manipulated to create various sub-model/classifier components (i.e., 1 to M) that help a learner. Furthermore, the best predictive model may be generated by merging limiting sub-models using averaging combination processes. Bootstrap resampling and benefit collection are two common methods for modelling ensembles that make use of bagging. The first training set substitute's component replicates the bagging procedure. In product models, specific data samples might appear numerous times, while others do not. An average is computed from the output of each component model. As with the bagging method, this strategy builds a cumulative model that yields several components with more accuracy as compared to the individual model. A weighted average of dependent sub-models is used to set sub-models in the final model, which is referred to as "boosting". AdaBoost [74] is a rapid ensemble learning algorithm that picks multiple classifier examples repeatedly by distributing weights adaptively across training cases. This approach linearly combines the chosen classifier instances to form an ensemble. Even when a large number of base classifier instances are included in a model, AdaBoost ensembles seldom display an overfitting issue [75]. It is possible to diminish loss function by fitting to a staged additive model. This indicates that a cost function that is not differentiable and is not smoothed has been optimized; it can best be described using an exponential loss function [75]. It is, therefore, possible to employ AdaBoost to tackle a number of classification problems with impressive results. DT, MLPNN, and RF are used in conjunction with ensemble learners to envisage the strength of consistently used concrete in this study.

3.3. Ensemble Learner's Parameter Tuning

Models of the tuning parameters that are employed in the ensemble methods might consist of (i) factors connected with the optimum model learners' number and (ii) rate of learning and other attributes that have a significant influence on ensemble algorithms. In this research, twenty sub-models were generated for each ensemble base learner. The component sub-models ranged in size from 10 to 200, and the optimum constructs were selected on the basis of the large values of determination coefficient. Figure 6 illustrates the link between performance of ensemble model and the number of component sub-models. As shown in Figure 6a,b, the assembling model with bagging and boosting yields a significant determination coefficient in the estimation terms. As demonstrated in Figure 6a, the 140th sub-model of DT with bagging as an ensemble of other sub-models provides a stronger relationship than the other sub-models of DT with bagging. Similarly, the 30th sub-model of MLPNN with bagging provides a significant higher correlation coefficient as compared to MLPNN bagging other sub-models. Similarly, as depicted in Figure 6b, the 50th DT AdaBoost and the 180th MLPNN AdaBoost sub-model provide the best results when compared to their other sub-models. Preliminary analysis indicates that the usage of ensemble modelling improves the efficiency of both models.

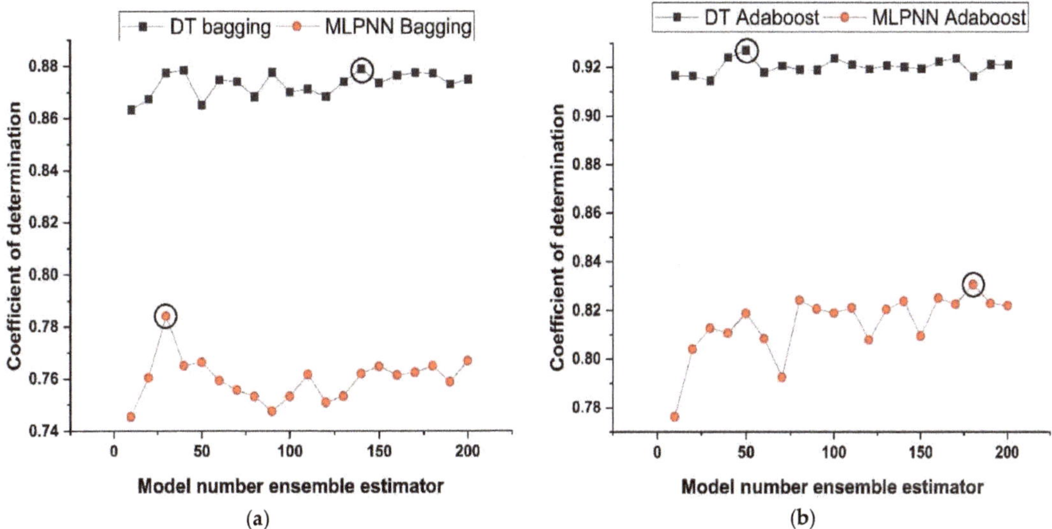

Figure 6. ML with ensemble sub-models; (**a**) Bagging; (**b**) AdaBoost.

3.4. Random Forest Regression Based Machine Learning

The RF model is a regression and classification strategy that has piqued the curiosity of a number of different researchers [76]. The main difference between DT and RF is that one tree is created in DT, but several trees are built in RF, and unlike data are randomly picked and distributed to all the trees in the forest. Each tree's data are organised into columns and rows, with a variety of column and row sizes to choose from [77]. Each stage of a tree's development is detailed below:

1. The equivalent of two-thirds of the entire dataset is chosen for each tree at random. Bagging is the term used to describe this practise. Predictor variables are picked, and node splitting is done based on the best possible node split on these variables.
2. The remaining data are used to estimate the out-of-bag error for each and every tree. To get the most accurate estimation of the out-of-bag error rate, errors from each tree are then added together.
3. Every tree in the RF algorithm provides a regression, but the model prioritizes the forest that receives the most votes over all of the individual trees in the forest. The votes might be either zeros or ones. As a prediction probability, the fraction of 1s achieved is provided.

3.5. 10 K Fold Method for Cross Validation

For training and holdout data, the k fold approach for cross validation is often employed to decrease arbitrary sampling prejudice. A stratified 10-fold cross-validation strategy was used in this work to evaluate model performance by dividing the input data into ten distinct subsets. Each of the 10 rounds of model construction and validation uses a different sample of data to test and train the model. As indicated in Figure 7, in order to validate the adequacy of the model, the test subset is utilised. The accuracy of algorithm is calculated as the mean of the 10 models' accuracy scores after 10 rounds of validation.

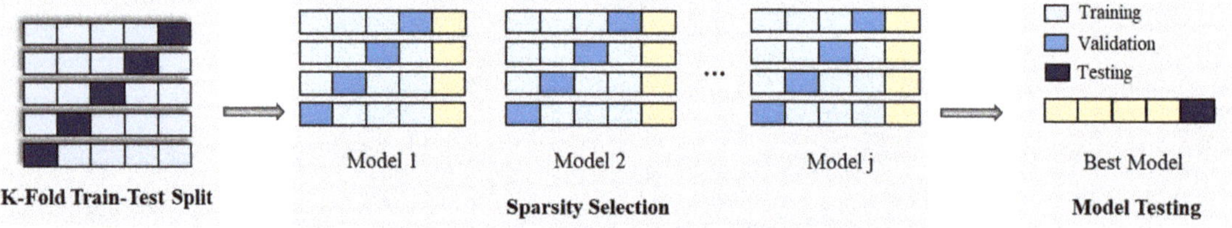

Figure 7. K-fold cross validation method [78].

3.6. Evaluation Criteria for Models

Statistical errors for example root mean squared logarithmic error (RMSLE), square value (R^2), mean absolute error root (MAE), and root mean square error (RMSE) are used to assess model performance on a training or testing set. R^2 is also called the determination coefficient and is used to evaluate a model's ability to predict. Concrete's mechanical characteristics may now be predicted with greater accuracy due to advances in artificial intelligence modelling methods. The models are assessed statistically by calculating error metrics in this research. There are a variety of measures that might help us better understand the model's inaccuracy. In addition, the model's performance may be assessed using the variance coefficient and standard deviation. According to the coefficient of determination, the model's correctness and validity may be confirmed. Models with R^2 values ranging from 0.65 to 0.75 indicate promising results, whereas models with R^2 values lower than 0.50 reveal disappointing outcomes. Equation (1) can be used to determine R^2. The units used in MAE are the same as the ones used in the output. It is possible for a model with a value of MAE that falls within a certain limit to have large errors at some points in time. In order to calculate MAE, Equation (2) is used. RMSE is the under root of the average of squared differences between estimates and measurements. Error squared is calculated by summing the squared errors. This method gives a greater weight to outliers and significant exceptions than other methods, which results in bigger squared differences in certain cases and lower squared differences in others. Using RMSE, the model's average estimation error given an input can be calculated. Improved models have fewer root mean squared errors of variation. The lower the value of RMSE, the less accurate the model is in predicting the data. Equation (3) is used to determine RMSE. Relative imprecision amongst forecasted and actual values is taken into account by RMSLE. It is the difference between the expected value and the actual value, expressed as a logarithmic scale. RMSLE is calculated using Equation (4).

$$R^2 = \frac{\sum_{i=1}^{n}\left(M_i - \overline{M_i}\right)\left(P_i - \overline{P_i}\right)}{\sqrt{\sum_{i=1}^{n}\left(M_i - \overline{M_i}\right)^2 \sum_{i=1}^{n}\left(P_i - \overline{P_i}\right)^2}} \tag{1}$$

$$MAE = \frac{1}{n}\sum_{i=1}^{n}|P_i - M_i| \tag{2}$$

$$RMSE = \sqrt[2]{\frac{\sum_{i=1}^{n}\left(P_i - M_i\right)^2}{N}} \tag{3}$$

$$RMSLE = \sqrt{\frac{1}{N}\sum_{i=1}^{N}\left(\log(yi + 1) - \log(yj + 1)\right)^2} \tag{4}$$

4. Model Result

4.1. Decision Tree Model Outcomes

As seen in Figure 8, the DT is modelled using various ensemble techniques including bagging and boosting. The actual prediction from individual base learner DT produces a high relationship with predicted values having a $R^2 = 0.868$, as seen in Figure 8a. Figure 8b depicts the error distribution of an individual DT model. Figure 8b indicates that the testing set has an average inaccuracy value of 5.79 MPa. In addition, 82.88 percent of the data exhibit error below 10 MPa, and 11.7% of the data exhibit error between 10–15 MPa. In contrast, each domain of 15–20 MPa, 20–25 MPa, and 35–40 MPa contains 1.8 percent data error, with a maximum and minimum error of 35.3 MPa and 0.085 MPa, respectively, as illustrated in Figure 8b. Individually, DT provide accurate predictions; but, if the DT is an ensemble of several methodologies, it yields a more precise outcome, as seen in Figure 8c–f. Bagging ensemble yields a conclusive and favourable result with $R^2 = 0.879$ and minimal testing data error. The data indicate an inaccuracy of 84.685% below 10 MPa, 9% between 10 and 15 MPa, and 3.6% between 15 and20 MPa. As shown in Figure 8d, only 1.8% of the data fall between 20 and 25 MPa and 0.9% between 30 and 35 MPa, with a maximum and minimum error of about 33.06 MPa 0.029 MPa, respectively. Similar to individual DT and bagging DT algorithms, boosting with AdaBoost produces models with a significant correlation. As seen in Figure 8e–f, this is because of the influence that a strong learner has on the aspect of prediction. A DT AdaBoost ensemble model has a R^2 equal to 0.924. The error distribution is minimised by applying AdaBoost with a DT, with an average error of 4.12 MPa, a maximum and minimum error of 34.578 MPa, and 0.065 MPa, respectively. Approximately 92.79%of the data is below 10 MPa, with 6.3% between 10 and 15 MPa and 0.9% between 30 and 35 MPa. Table 3 presents the statistical information pertaining to DT with bagging and boosting ensemble learners.

Table 3. DT model statistical evaluation of errors.

Statistical Analysis	DT	DT-Bagging	DT-AdaBoost
Average	5.79	7.29	7.05
Minimum	0.08	0.11	0.07
Maximum	35.3	34.74	31.31
No. of data points below 10 MPa	92	94	103
No. of data points between 10 and 20 MPa	15	14	07
No. of data points between 20 and 30 MPa	02	02	00
No. of data points between 30 and 40 MPa	02	01	01
No. of data points testing points	111	111	111
Average below 10 MPa	82.88	84.68	92.79
Average in range of 10 to 20 MPa	13.51	12.61	6.31
Average in range of 20 to 30 MPa	1.80	3.60	00
Average in range of 30 to 40 MPa	1.80	2.70	0.90

4.2. MLPNN Model Outcomes

In the field of ML and AI, neural networks fall under the rubric of supervised learning, and its implementation yields a rigid correlation between prediction and target response. As illustrated in Figure 9, MLPNN is also modelled utilising ensemble learner's methods, similar to the DT. Figure 9a depicts the actual projection of MK concrete with $R^2 = 0.724$ with its error distribution as seen in Figure 9b. MLPNN error distribution indicates that a test set has an average error of 8.70 MPa, with lowest and highest errors of 0.044 MPa and 35.15 MPa, respectively. However, MLPNN ensemble model reduces the distribution of average error with a rise in the R^2 of around 0.767 for bagging and 0.825 for boosting, respectively. The average error for MLPNN-bagging and AdaBoost boosting is 7.29 MPa and 7.05 MPa, respectively, as seen in Figure 9c–f. In addition, a major portion of testing set error is below 10 MPa, with 72.97%, 77.48%, and 74.77% of the data, respectively, for the individual, bagging, and AdaBoost MLPNN models. These ensemble-model outputs also demonstrate a rise in R^2 by exhibiting less inaccuracy than the real output. Table 4 illustrates the statistical evaluation of testing data via MLPNN ensemble modelling.

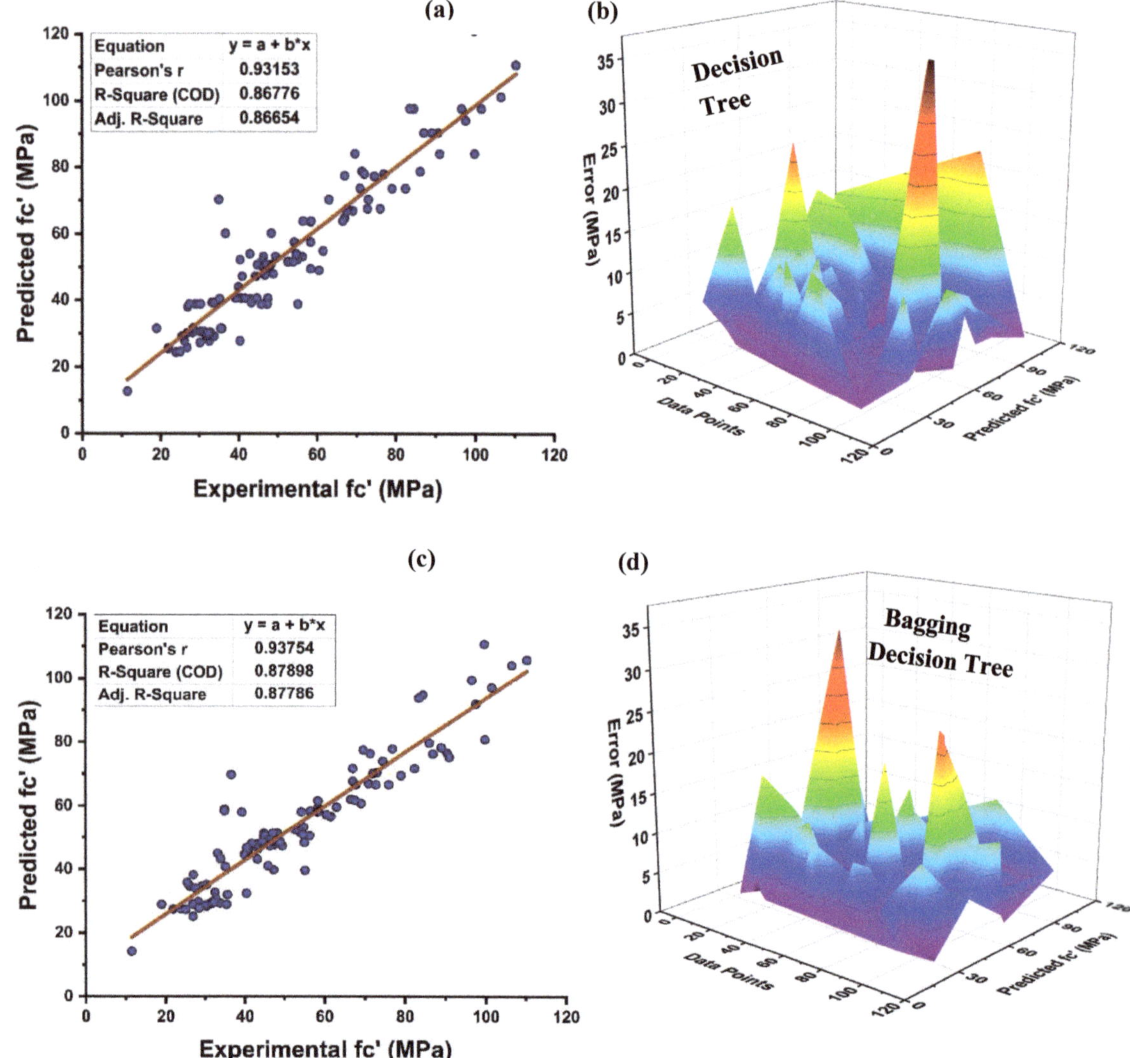

Figure 8. *Cont.*

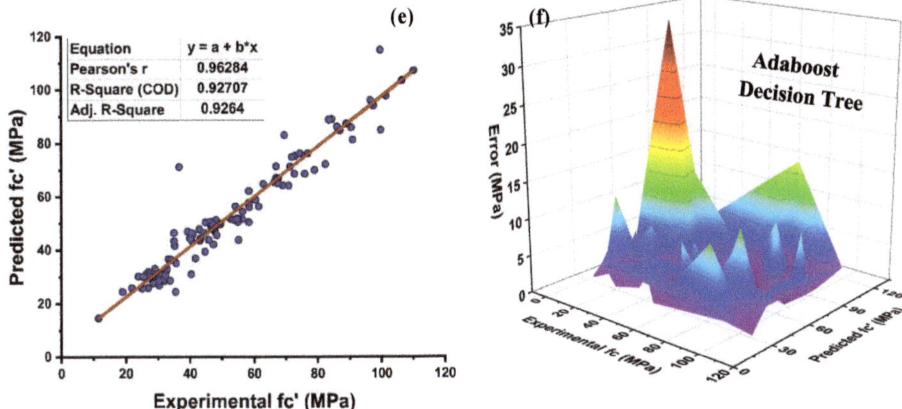

Figure 8. (**a**) DT individual base learner regression model; (**b**) DT individual base learner regression model error distribution; (**c**) DT-bagging model; (**d**) DT-bagging model error distribution; (**e**) DT-AdaBoost regression model; and (**f**) DT-AdaBoost model error distribution.

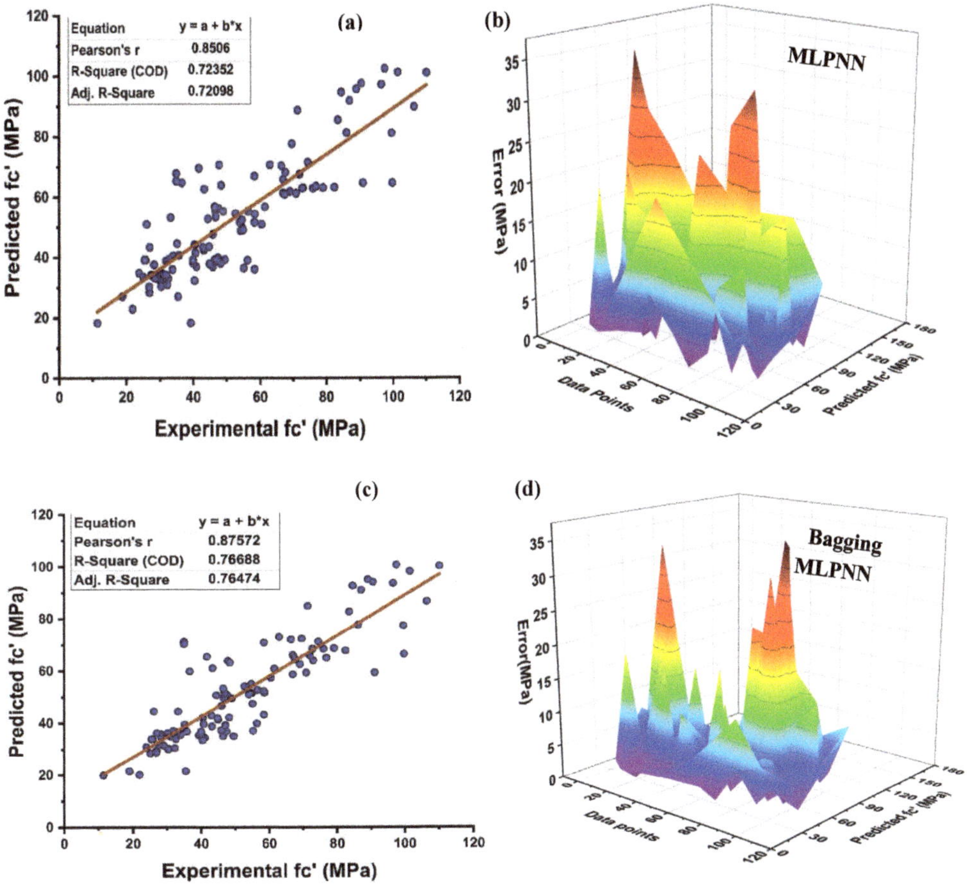

Figure 9. *Cont.*

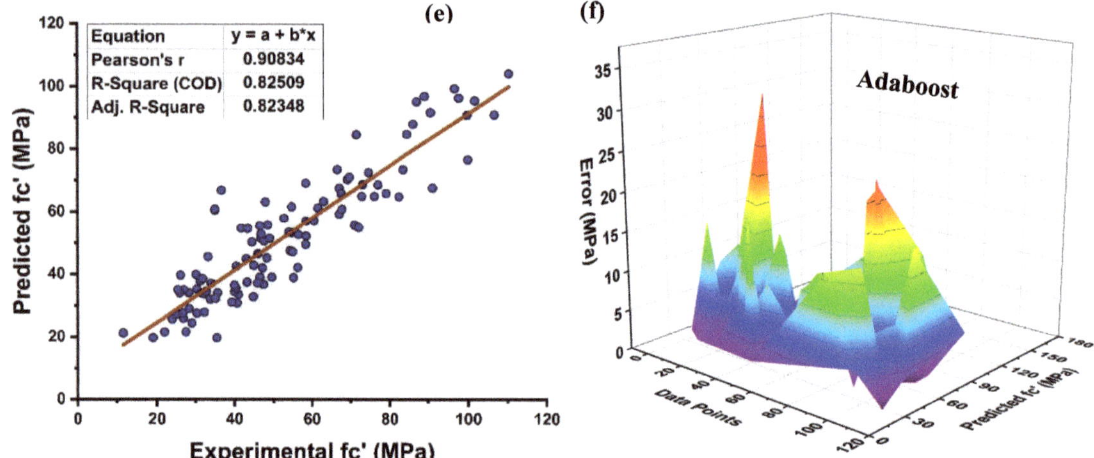

Figure 9. (**a**) MLPNN individual base learner regression model; (**b**) MLPNN individual base learner regression model error distribution; (**c**) MLPNN-bagging regression; (**d**) MLPNN-bagging regression model error distribution; (**e**) MLPNN-AdaBoost regression model; (**f**) MLPNN-AdaBoost regression model error distribution.

Table 4. MLPNN model statistical evaluation of errors.

Statistical Analysis	MLPNN	MLPNN-Bagging	MLPNN-AdaBoost
Average	8.70	7.29	7.05
Minimum	0.04	0.11	0.07
Maximum	35.15	34.74	31.31
No. of data points below 10 MPa	81	86	83
No. of data points between 10 and 20 MPa	20	18	23
No. of data points between 20 and 30 MPa	07	04	04
No. of data points between 30 and 40 MPa	03	03	01
No. of data points between 10 and 20 MPa	111	111	111
Average below 10 MPa	72.97	77.48	74.77
Average in range of 10 to 20 MPa	18.02	16.22	20.72
Average in range of 20 to 30 MPa	6.31	3.60	3.60
Average in range of 30 to 40 MPa	2.70	2.70	0.90

4.3. Random Forest Model Outcomes

Within the framework of the ensemble ML approach, RF represents a hybrid type of bagging and random feature selection, which is a technique for the production of prediction models that is both efficient and easy to use. Figure 10 depicts the prediction accuracy of the RF method for MK concrete. As it is an ensemble model, it exhibits a stubborn $R^2 = 0.929$ correlation with the target values. In addition, the RF model's prediction may also be tested using an error distribution with an average error of 3.52 MPa. In addition, 90.99 percent of the results indicate that the error falls under 10 MPa, demonstrating the precision of the non-linear estimation of the normal concrete's strength as shown in Figure 10b.

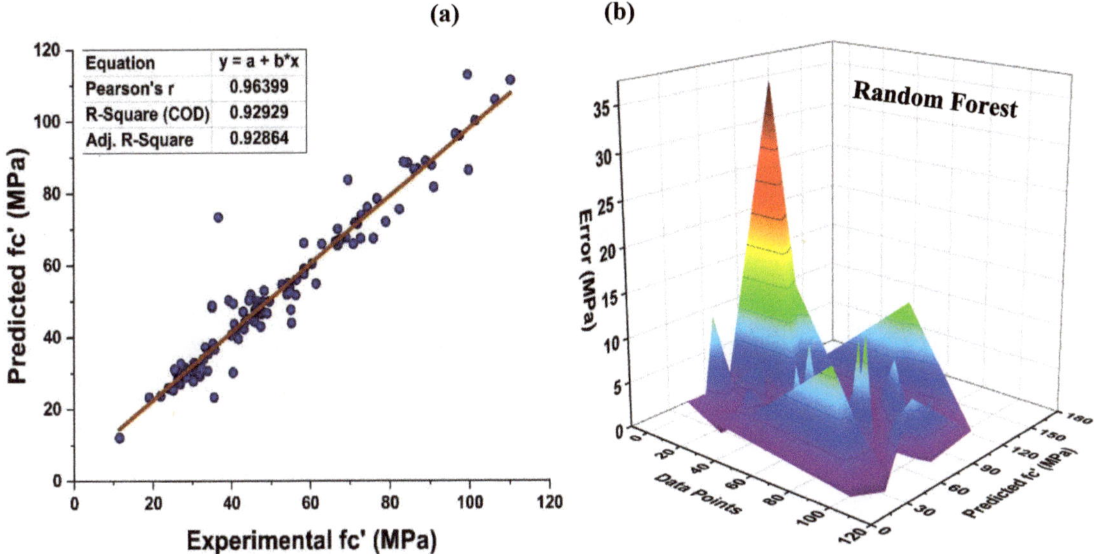

Figure 10. (**a**) RF modified learner regression model; (**b**) RF modified learner regression model error distribution.

4.4. K-Fold Results

The model's required accuracy is essential to its assessment. To verify the accuracy of prediction models, this validation is necessary. The K fold validation test is employed to validate the correctness of data using data shuffles. Randomly sampling the training data set introduces bias, hence this strategy is used to reduce it. This technique divides the samples evenly into 10 subgroups of the experimental data. One of the 10 subsets is utilised for validation, while the other nine are employed to shape up the strong learner. The procedure is done 10 times and then averaged. In general, it is commonly accepted that the 10-fold cross validation approach accurately reflects the model's generalisation and dependability [79]. The validation test of all the ensemble models is presented in Figures 11 and 12. All models exhibit a moderate to high correlational link. In addition, the outcomes of cross-validation may be evaluated based on various errors, such as R^2, MAE, RMSE, and RMSLE, as shown in Figure 11 for DT and MLPNN and Figure 12 for RF. It displays the validation representation in each 10-fold. Although variations were noticed, it retained a high degree of precision, as seen in Figures 11 and 12. For example, the lowest and highest R^2 values for all models are between 0.46 and 0.65 and 0.81 and 0.91, respectively. As demonstrated in Figure 11c–h for DT and MLPNN, MAE, RMSE, and RMSLE are also used to evaluate the accuracy of models with respect to cross-validation. Figure 11c depicts the average MAE value for DT with ensemble bagging and ensemble boosting using 10-fold validation as 11.97 MPa and 9.0 MPa, respectively. Figure 11e reveals that the RMSE offers an average error of about 14.6 MPa and 11.84 MPa for ensemble bagging and ensemble boosting using AdaBoost, respectively. Figure 11g displays RMSLE average errors of 0.106 MPa and 0.07 MPa for DT bagging and boosting, respectively. Figure 11d–f shows that the average MAE, RMSE, and RMSLE for the MLPNN bagging model are 11.06 MPa, 14.92 MPa, and 0.1 MPa, respectively. For the k fold validation of the MLPNN AdaBoost model, values of 12.03 MPa, 15.07 MPa, and 0.08 MPa were found. This demonstrates the precision of models using K-fold cross validation. Figure 12 demonstrates strong association for modified learner model with decreased error for MAE, RMSE, and RMSLE, with average errors of 8.94 MPa, 11.02 MPa, and 0.07 MPa, respectively.

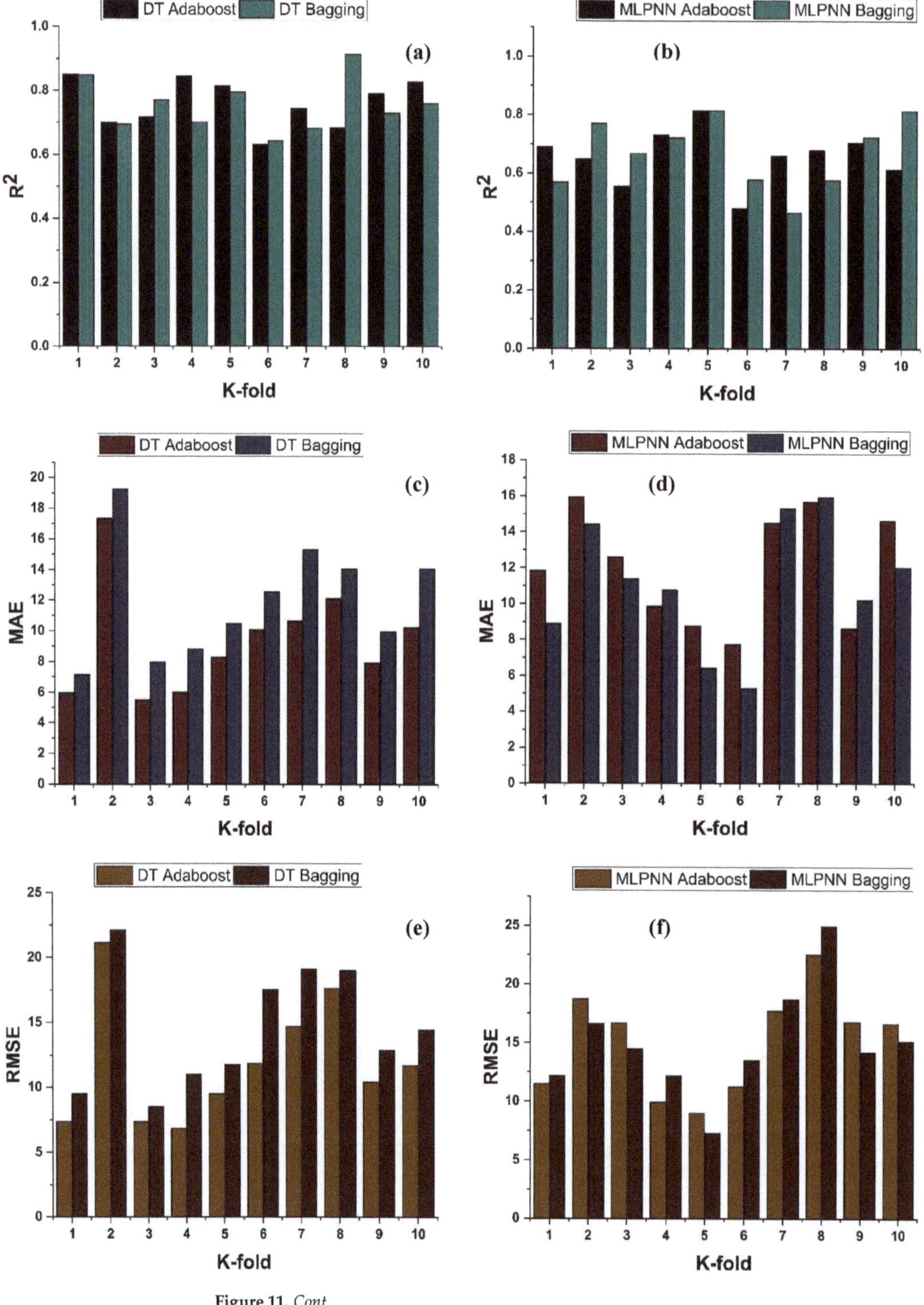

Figure 11. *Cont.*

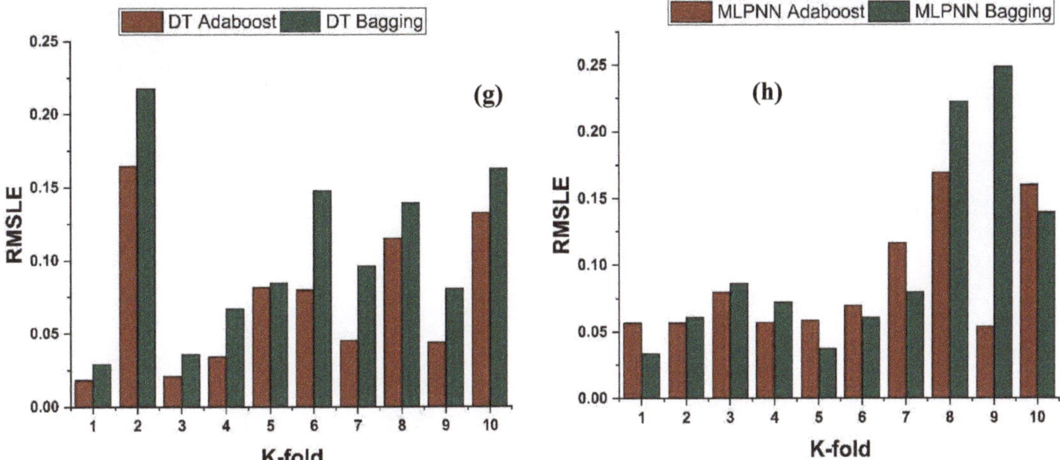

Figure 11. fc' models (**a,b**) indicate R^2 models' result validated with K fold; (**c,d**) indicate MAE models' result validated with K fold; (**e,f**) indicate RMSE models' result validated with K fold; (**g,h**) indicate RMSE models' result validated with K fold.

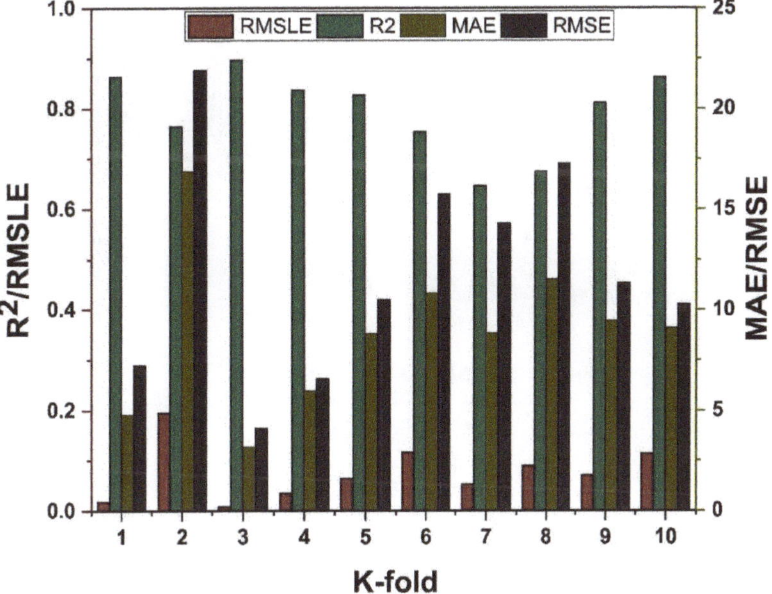

Figure 12. RF models cross validation with different statistical parameters.

4.5. Model Evaluation and Discussion Based on Statistical Metrics

Comparing the ensemble approaches to the individual ML methods helped show the ensemble algorithm's potential in comparison to them as depicted in Figure 13. This process is similar to that used for ensemble models, such as starting with a set of values and then using a grid search to find the optimal values. Table 5 shows the target and validated values for each metric. The ensemble ML models outcome have a linear trend, and their projections are more like the ones that were tested, according to this study. Using DT, and MLPNN, is a kind of individual learning, but using ML techniques such as bagging and boosting is a form of ensemble learning. High performance weak learners would gain weight, though weak learners with poor performance will lose weight, since ensemble learning is usually known to include several weak learners produced by individual ML algorithms. Because of this, it is able to provide accurate projections. MAE, RMSE, RMSLE, and R^2 are used to evaluate individual and ensemble learners. An ensemble of learners using bagging and boosting has a lower rate of error than an individual learner. A smaller error margin exists between forecasts and outcomes when using ensemble models rather than individual models alone.

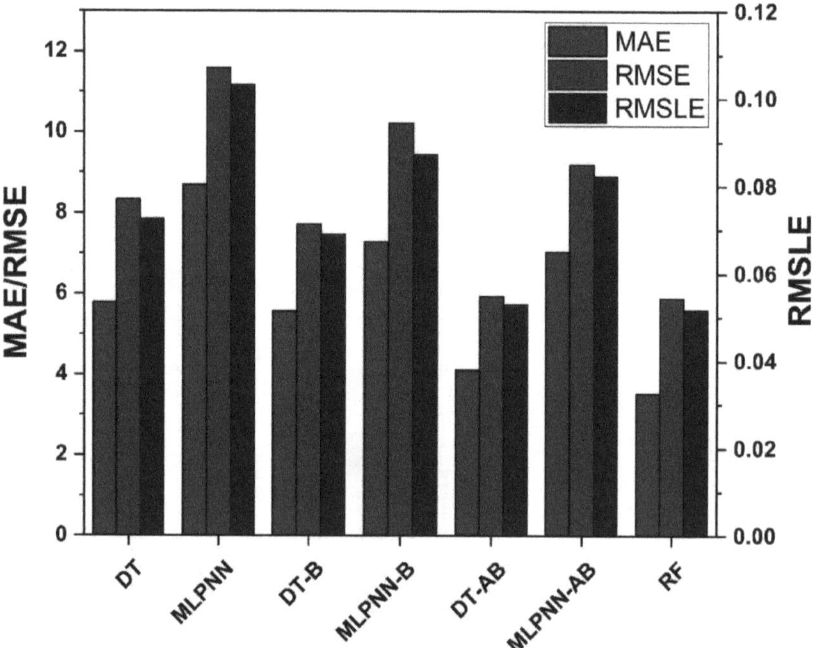

Figure 13. Statistical analysis of fc' models.

Table 5. Model's statistical errors.

Approach Employed	ML Methods	MAE	RMSE	RMSLE	R^2
Individual Learner	DT	5.79072	8.34472	0.07261	0.868
	MLPNN	8.70159	11.59452	0.10325	0.724
Ensemble Learner Bagging	DT	5.57845	7.72089	0.06911	0.879
	MLPNN	7.29168	10.21239	0.08721	0.767
Ensemble Learner Boosting	DT	4.12636	5.93813	0.05303	0.924
	MLPNN	7.04574	9.20414	0.08233	0.825
Modified Ensemble	Random Forest	3.52232	5.89161	0.05179	0.929

4.6. SHAP Analysis

The values of all of the features that were taken into consideration for the MK concrete fc' prediction are outlined in the shape of a violin, as illustrated in Figure 14. The Shapley value measures the mean marginal influence that can be attributed to each parameter value over all viable permutations of the parameters. The attributes that have substantial absolute Shapley values are regarded to have a considerable influence. In order to get the global feature effects, the absolute Shapley values for each feature throughout the data were averaged and ranked in decreasing significance as shown in Figure 14. Every single datapoint on the plot indicates a Shapley value for distinct characteristics and occurrences. The location on the x axis and the y axis is defined by the Shapley value and the feature significance, respectively. Elevated places on the y axis represent higher effect of the characteristics on the MK concrete fc' prediction and the colour scale reflects the feature relevance from low to high. Each dot in Figure 14 signifies an individual point from the dataset. The location of points along the x axis represents the effect of each parameter value on the fc' prediction. When numerous dots fall in the same location along the x-axis, the dots are stacked to illustrate the density. Age is the most influential parameter followed by coarse aggregate, superplasticizer, water, and other input parameters. Silica fume has the least impact on fc' prediction of MK concrete as illustrated in Figure 14. Higher SHAP value imply that the model forecasts higher fc' value, and vice versa. For example, high value of age (red) correlate with increased SHAP value, which suggest high fc' value. Moreover, each input parameter has positive or negative impact up to a certain limit. In Figure 14, red colour shows high impact (negative or positive) while the blue colour depicts low impact of the input feature on the predicted outcome. SHAP value at the right (greater than 0) on the x-axis shows positive impact of respective input on the fc'. For instance, in the case of input parameters like age and content of coarse aggregate, the positive effect of these factors on MK fc' can be noted from the right axis of the graph. Coarse aggregate content depicts a constructive impact till optimum content, whereas above this content, the adverse effect is shown on the left side (less than 0) on the x-axis. Super-plasticizer is also key variable for predicting the fc' of MK concrete. The effect of water on the output fc' of MK concrete is negative and increasing the water content will reduce the fc'. MK and cement show the same trend. However, SF and fine aggregate tend to have a high positive impact and a low negative impact on the fc' prediction of MK concrete. SHAP feature dependency graphs were deployed that are coloured by another interacting feature to highlight how the features interact and effect the fc' of MK concrete. This gives greater information than standard partial dependency charts. The SHAP interaction plot each considered feature is shown in Figure 15. As can be observed from Figure 15a, the dependence and interaction show that high fc' values for MK concrete can be achieved when for $50 \leq$ age ≤ 100 days when CA ≥ 700 kg/m^3. Higher fc' values for age ≤ 50 days can be achieved for $50 \leq$ CA ≤ 700 kg/m^3. Figure 15b,d,e show that 120–200 kg/m^3 of water is required for CA in for different content of CA and FA to achieve higher values of fc'. Moreover, Figure 15f,g illustrate the relation between two important constituents of MK concrete: cement and metakaolin. Higher fc' for MK concrete can be achieved for MK in the range of 20–100 kg/m^3 for concrete having density of 250–450 kg/m^3. Additionally, Figure 15h reveals that large quantity of silica fume can be used if early fc' of MK concrete is desired.

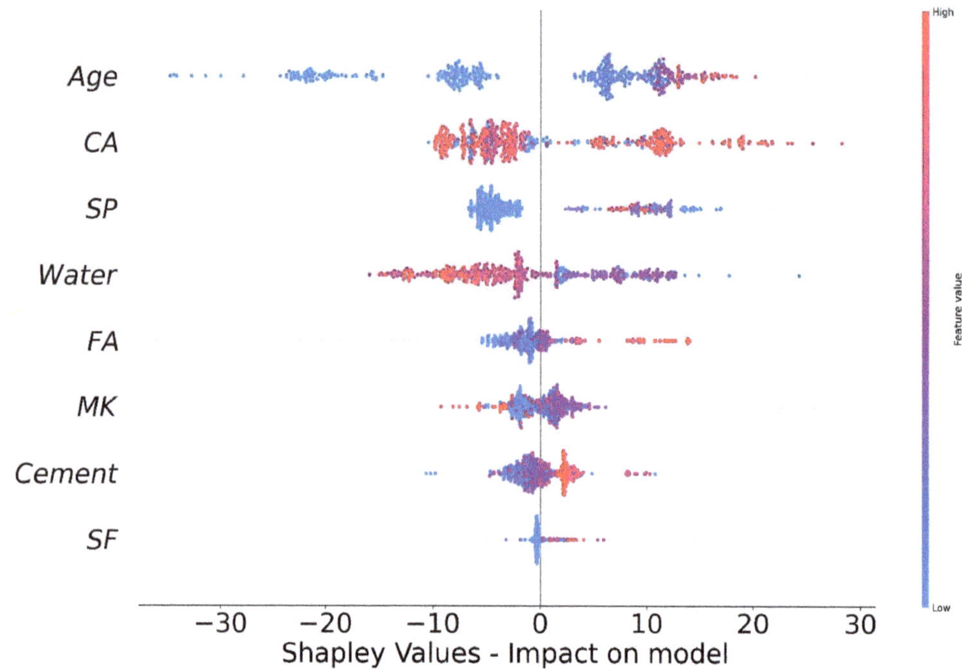

Figure 14. SHAP plot.

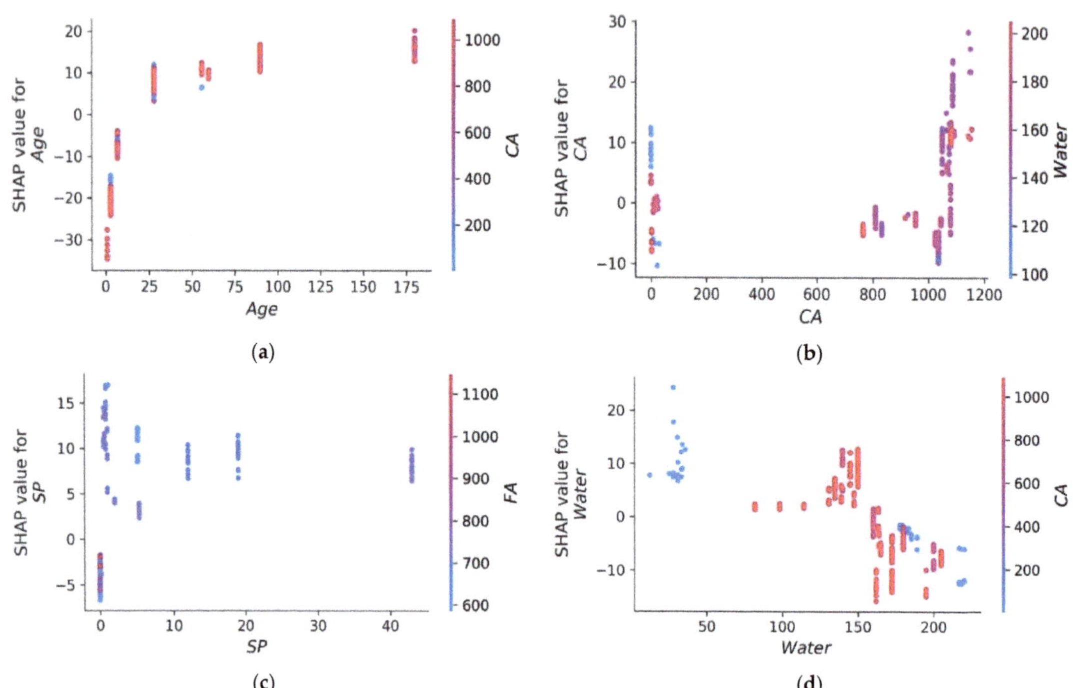

(a)

(b)

(c)

(d)

Figure 15. *Cont.*

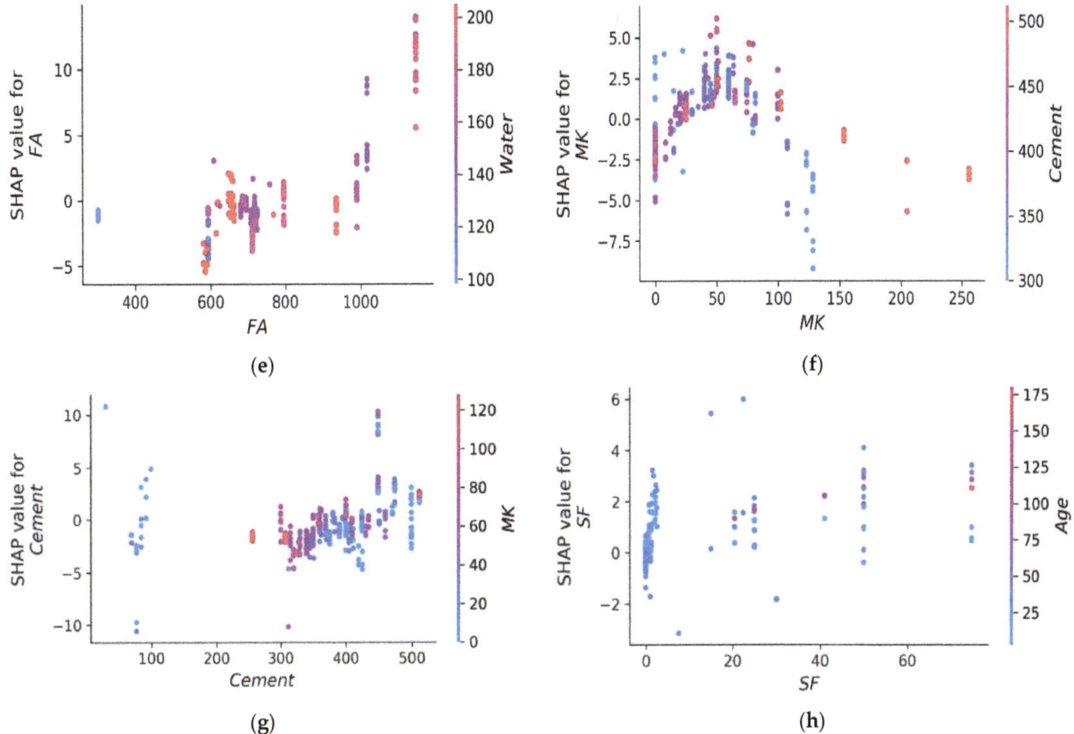

Figure 15. SHAP interaction plot of parameters: (**a**) age; (**b**) coarse aggregate; (**c**) superplasticizer; (**d**) water; (**e**) fine aggregate; (**f**) metakaolin; (**g**) cement; (**h**) silica fume.

5. Conclusions

The primary aim of this study was to assess the accuracy level achieved by various ML approaches to predict MK concrete fc′. Datasets from the literature containing 551 data points were used to train and test the models. The eight most influential constituents of MK concrete including cement, metakaolin, coarse and aggregate, water, silica fume, superplasticizer, and age were considered as input parameters. Individual and ensemble learning models for predicting the fc′ of MK concrete were investigated in this study using DT, MLPNN, and RF. Interaction of input parameters and effect of input parameters of fc′ were studied using SHAP dependency feature graphs. The results of the investigation led the authors to the following conclusions:

1. Bagging and AdaBoost models outperform the individual models. As compared to the standalone DT model, the ensemble DT model with boosting and RF demonstrates a 7% improvement. Both techniques have a significant correlation with R^2 equal to 0.92. Similarly, an improvement of 14 %, 6%, and 29% was observed in MLPNN AdaBoost, MLPNN bagging, and RF model, respectively, when compared with individual DT model;

2. Statistical measures using MAE, RMSE, RMSLE, and R^2 were also performed. Ensemble learner DT bagging and boosting depicts a smaller error of about 4%, and 29% for MAE, 8% and 29% for RMSE, 5% and 27% for RMSLE, respectively, when compared to the individual DT model. Similarly, enhancements of 16% and 19% in MAE, 12% and 21% in RMSE, and 16% and 20% in RMSLE were observed for MLPNN bagging and AdaBoost models, respectively, when compared to the individual base learner DT model;

3. RF shows improvements of 60%, 49%, and 50% in MAE, RMSE, and RMSLE when compared to the MLPNN individual model. Similarly, improvements of 39%, 29%, and 29% for the RF model, in MAE, RMSE and RMSLE, were observed in comparison to DT individual model;
4. The validity of models using R2, MAE, RMSE, and RMSLE were tested using k-fold cross-validation. Fewer inaccuracies with strong correlations were examined;
5. The DT AdaBoost model and the modified bagging model are the best techniques for forecasting MK concrete fc′ among all of the ML approaches;
6. Age has the greatest impact on calculating MK concrete fc′, followed by coarse aggregate and superplasticizer, according to the SHAP assessment. However, silica fume has the least impact on the fc′ of MK concrete. SHAP dependency feature graphs can illustrate the relationship between input parameters for various ranges;
7. Sensitivity analyses depicted that FA contributed moderately to the development of the f_c' models and f_{sts} models. Moreover, cement, SF, CA, and age played vital roles in the development of f_c' models. Tensile strength models showed to be affected least by water and CA;

These ML algorithms can accurately predict the mechanical characteristics of concrete. These models can be utilized to predict the mechanical characteristics of similar databases containing metakaolin with high accuracy. Moreover, SHAP analysis provides an insight to readers regarding the input parameters contribution towards the outcome, and inter-dependency of the input parameters. This will enable the readers to carefully select the input variables for modelling the behaviour of metakaolin concrete. Additionally, ML algorithms employed in this study may provide a sustainable way for the mix design of MK concrete. Traditionally, this process demands lengthy trials in laboratories and a significant number of raw materials in addition to a great deal of manpower.

6. Limitations and Directions for Future Work

Despite the fact that the efforts made in this research has significant limitations, it may still be regarded a data mining-based research. Completeness of dataset is essential for the efficacy of models′ prediction. The range of datasets used for this study was restricted to 551 data points. In addition, the corrosive and flexural concrete behaviour at extreme temperatures was not considered in this work. Indeed, good database management and testing are essential from a technical standpoint. To simulate high-strength concrete, this research included an extensive variety of data with eight variables. Further, it is suggested that a new dataset of concrete at increased temperatures that encompasses numerous environmental factors such as temperature, durability, and corrosion be investigated. Experimental testing data for testing of models are recommended for more accuracy. Given that concrete plays such an important role in the ecosystem, its effects under various situations should be investigated utilising various deep machine learning methods.

Author Contributions: Conceptualization, K.K.; Data curation, A.M.R.B., W.A. and A.M.A.A.; Formal analysis, A.M.R.B., A.N., M.N.A., W.A., M.U. and S.N.; Funding acquisition, K.K.; Investigation, A.N. and M.N.A.; Methodology, A.N., M.N.A. and A.M.A.A.; Project administration, K.K.; Resources, K.K.; Software, W.A.; Supervision, K.K.; Validation, M.U. and S.N.; Visualization, M.U., S.N. and A.M.A.A.; Writing—original draft, A.M.R.B., A.N. and W.A.; Writing—review & editing, K.K. All authors have read and agreed to the published version of the manuscript.

Funding: This work was supported by the Deanship of Scientific Research, Vice Presidency for Graduate Studies and Scientific Research, King Faisal University, Saudi Arabia [Grant No. 1431], through its KFU Research Summer Initiative.

Institutional Review Board Statement: Not applicable.

Informed Consent Statement: Not applicable.

Data Availability Statement: All data is available in the paper.

Acknowledgments: The authors acknowledge the Deanship of Scientific Research, Vice Presidency for Graduate Studies and Scientific Research, King Faisal University, Saudi Arabia [Grant No. 1431], through its KFU Research Summer Initiative.

Conflicts of Interest: The authors declare no conflict of interest.

References

1. Liu, T.; Nafees, A.; Khan, S.; Javed, M.F.; Aslam, F.; Alabduljabbar, H.; Xiong, J.-J.; Khan, M.I.; Malik, M. Comparative study of mechanical properties between irradiated and regular plastic waste as a replacement of cement and fine aggregate for manufacturing of green concrete. *Ain Shams Eng. J.* **2021**, *13*, 101563. [CrossRef]
2. Khan, K.; Ahmad, A.; Amin, M.N.; Ahmad, W.; Nazar, S.; Abu Arab, A.M. Comparative Study of Experimental and Modeling of Fly Ash-Based Concrete. *Materials* **2022**, *15*, 3762. [CrossRef]
3. Khan, K.; Ahmad, W.; Amin, M.N.; Ahmad, A.; Nazar, S.; Alabdullah, A.A.; Abu Arab, A.M. Exploring the Use of Waste Marble Powder in Concrete and Predicting Its Strength with Different Advanced Algorithms. *Materials* **2022**, *15*, 4108. [CrossRef] [PubMed]
4. Vu, D.; Stroeven, P.; Bui, V. Strength and durability aspects of calcined kaolin-blended Portland cement mortar and concrete. *Cem. Concr. Compos.* **2001**, *23*, 471–478. [CrossRef]
5. Batis, G.; Pantazopoulou, P.; Tsivilis, S.; Badogiannis, E. The effect of metakaolin on the corrosion behavior of cement mortars. *Cem. Concr. Compos.* **2005**, *27*, 125–130. [CrossRef]
6. Khatib, J.; Negim, E.; Gjonbalaj, E. High volume metakaolin as cement replacement in mortar. *World J. Chem.* **2012**, *7*, 7–10.
7. Ameri, F.; Shoaei, P.; Zareei, S.A.; Behforouz, B. Geopolymers vs. alkali-activated materials (AAMs): A comparative study on durability, microstructure, and resistance to elevated temperatures of lightweight mortars. *Constr. Build. Mater.* **2019**, *222*, 49–63. [CrossRef]
8. Pavlikova, M.; Brtník, T.; Keppert, M.; Černý, R. Effect of metakaolin as partial Portland-cement replacement on properties of high performance mortars. *Cem. Wapno Beton* **2009**, *3*, 115–122.
9. Kadri, E.-H.; Kenai, S.; Ezziane, K.; Siddique, R.; De Schutter, G. Influence of metakaolin and silica fume on the heat of hydration and compressive strength development of mortar. *Appl. Clay Sci.* **2011**, *53*, 704–708. [CrossRef]
10. Wianglor, K.; Sinthupinyo, S.; Piyaworapaiboon, M.; Chaipanich, A. Effect of alkali-activated metakaolin cement on compressive strength of mortars. *Appl. Clay Sci.* **2017**, *141*, 272–279. [CrossRef]
11. Khatib, J.; Wild, S. Sulphate Resistance of Metakaolin Mortar. *Cem. Concr. Res.* **1998**, *28*, 83–92. [CrossRef]
12. Van Fan, Y.; Chin, H.H.; Klemeš, J.J.; Varbanov, P.S.; Liu, X. Optimisation and process design tools for cleaner production. *J. Clean. Prod.* **2019**, *247*, 119181. [CrossRef]
13. Apostolopoulou, M.; Asteris, P.G.; Armaghani, D.J.; Douvika, M.G.; Lourenço, P.B.; Cavaleri, L.; Bakolas, A.; Moropoulou, A. Mapping and holistic design of natural hydraulic lime mortars. *Cem. Concr. Res.* **2020**, *136*, 106167. [CrossRef]
14. Asteris, P.G.; Douvika, M.G.; Karamani, C.A.; Skentou, A.D.; Chlichlia, K.; Cavaleri, L.; Daras, T.; Armaghani, D.J.; Zaoutis, T.E. A Novel Heuristic Algorithm for the Modeling and Risk Assessment of the COVID-19 Pandemic Phenomenon. *Comput. Model. Eng. Sci.* **2020**, *125*, 815–828. [CrossRef]
15. Banan, A.; Nasiri, A.; Taheri-Garavand, A. Deep learning-based appearance features extraction for automated carp species identification. *Aquac. Eng.* **2020**, *89*, 102053. [CrossRef]
16. Fan, Y.; Xu, K.; Wu, H.; Zheng, Y.; Tao, B. Spatiotemporal Modeling for Nonlinear Distributed Thermal Processes Based on KL Decomposition, MLP and LSTM Network. *IEEE Access* **2020**, *8*, 25111–25121. [CrossRef]
17. Taormina, R.; Chau, K.-W. ANN-based interval forecasting of streamflow discharges using the LUBE method and MOFIPS. *Eng. Appl. Artif. Intell.* **2015**, *45*, 429–440. [CrossRef]
18. Wu, C.; Chau, K. Prediction of rainfall time series using modular soft computingmethods. *Eng. Appl. Artif. Intell.* **2013**, *26*, 997–1007. [CrossRef]
19. Armaghani, D.; Momeni, E.; Asteris, P. Application of group method of data handling technique in assessing deformation of rock mass. *Metaheuristic Comput. Appl.* **2020**, *1*, 1–18.
20. Nafees, A.; Khan, S.; Javed, M.F.; Alrowais, R.; Mohamed, A.M.; Mohamed, A.; Vatin, N.I. Forecasting the Mechanical Properties of Plastic Concrete Employing Experimental Data Using Machine Learning Algorithms: DT, MLPNN, SVM, and RF. *Polymers* **2022**, *14*, 1583. [CrossRef]
21. Nafees, A.; Javed, M.F.; Khan, S.; Nazir, K.; Farooq, F.; Aslam, F.; Musarat, M.A.; Vatin, N.I. Predictive Modeling of Mechanical Properties of Silica Fume-Based Green Concrete Using Artificial Intelligence Approaches: MLPNN, ANFIS, and GEP. *Materials* **2021**, *14*, 7531. [CrossRef] [PubMed]
22. Nafees, A.; Amin, M.N.; Khan, K.; Nazir, K.; Ali, M.; Javed, M.F.; Aslam, F.; Musarat, M.A.; Vatin, N.I. Modeling of Mechanical Properties of Silica Fume-Based Green Concrete Using Machine Learning Techniques. *Polymers* **2021**, *14*, 30. [CrossRef] [PubMed]
23. Amin, M.N.; Ahmad, W.; Khan, K.; Ahmad, A.; Nazar, S.; Alabdullah, A.A. Use of Artificial Intelligence for Predicting Parameters of Sustainable Concrete and Raw Ingredient Effects and Interactions. *Materials* **2022**, *15*, 5207. [CrossRef] [PubMed]
24. Khan, K.; Ahmad, W.; Amin, M.N.; Ahmad, A. A Systematic Review of the Research Development on the Application of Machine Learning for Concrete. *Materials* **2022**, *15*, 4512. [CrossRef]

25. Khan, K.; Ahmad, W.; Amin, M.N.; Aslam, F.; Ahmad, A.; Al-Faiad, M.A. Comparison of Prediction Models Based on Machine Learning for the Compressive Strength Estimation of Recycled Aggregate Concrete. *Materials* **2022**, *15*, 3430. [CrossRef]
26. Khan, K.; Ahmad, W.; Amin, M.N.; Ahmad, A.; Nazar, S.; Al-Faiad, M.A. Assessment of Artificial Intelligence Strategies to Estimate the Strength of Geopolymer Composites and Influence of Input Parameters. *Polymers* **2022**, *14*, 2509. [CrossRef]
27. Khan, K.; Ahmad, W.; Amin, M.N.; Ahmad, A.; Nazar, S.; Alabdullah, A.A. Compressive Strength Estimation of Steel-Fiber-Reinforced Concrete and Raw Material Interactions Using Advanced Algorithms. *Polymers* **2022**, *14*, 3065. [CrossRef]
28. Dai, L.; Wu, X.; Zhou, M.; Ahmad, W.; Ali, M.; Sabri, M.M.S.; Salmi, A.; Ewais, D.Y.Z. Using Machine Learning Algorithms to Estimate the Compressive Property of High Strength Fiber Reinforced Concrete. *Materials* **2022**, *15*, 4450. [CrossRef]
29. Amin, M.N.; Ahmad, A.; Khan, K.; Ahmad, W.; Nazar, S.; Faraz, M.I.; Alabdullah, A.A. Split Tensile Strength Prediction of Recycled Aggregate-Based Sustainable Concrete Using Artificial Intelligence Methods. *Materials* **2022**, *15*, 4296. [CrossRef]
30. González-Taboada, I.; González-Fonteboa, B.; Martínez-Abella, F.; Pérez-Ordóñez, J.L. Prediction of the mechanical properties of structural recycled concrete using multivariable regression and genetic programming. *Constr. Build. Mater.* **2016**, *106*, 480–499. [CrossRef]
31. Asteris, P.G.; Moropoulou, A.; Skentou, A.D.; Apostolopoulou, M.; Mohebkhah, A.; Cavaleri, L.; Rodrigues, H.; Varum, H. Stochastic Vulnerability Assessment of Masonry Structures: Concepts, Modeling and Restoration Aspects. *Appl. Sci.* **2019**, *9*, 243. [CrossRef]
32. Asteris, P.G.; Apostolopoulou, M.; Skentou, A.D.; Moropoulou, A. Application of artificial neural networks for the prediction of the compressive strength of cement-based mortars. *Comput. Concr.* **2019**, *24*, 329–345.
33. Ben Chaabene, W.; Flah, M.; Nehdi, M.L. Machine learning prediction of mechanical properties of concrete: Critical review. *Constr. Build. Mater.* **2020**, *260*, 119889. [CrossRef]
34. Alkadhim, H.A.; Amin, M.N.; Ahmad, W.; Khan, K.; Nazar, S.; Faraz, M.I.; Imran, M. Evaluating the Strength and Impact of Raw Ingredients of Cement Mortar Incorporating Waste Glass Powder Using Machine Learning and SHapley Additive ExPlanations (SHAP) Methods. *Materials* **2022**, *15*, 7344. [CrossRef]
35. Ghorbani, B.; Arulrajah, A.; Narsilio, G.; Horpibulsuk, S. Experimental and ANN analysis of temperature effects on the permanent deformation properties of demolition wastes. *Transp. Geotech.* **2020**, *24*, 100365. [CrossRef]
36. Akkurt, S.; Tayfur, G.; Can, S. Fuzzy logic model for the prediction of cement compressive strength. *Cem. Concr. Res.* **2004**, *34*, 1429–1433. [CrossRef]
37. Özcan, F.; Atis, C.; Karahan, O.; Uncuoğlu, E.; Tanyildizi, H. Comparison of artificial neural network and fuzzy logic models for prediction of long-term compressive strength of silica fume concrete. *Adv. Eng. Softw.* **2009**, *40*, 856–863. [CrossRef]
38. Pérez, J.L.; Cladera, A.; Rabuñal, J.R.; Martínez-Abella, F. Optimization of existing equations using a new Genetic Programming algorithm: Application to the shear strength of reinforced concrete beams. *Adv. Eng. Softw.* **2012**, *50*, 82–96. [CrossRef]
39. Mansouri, I.; Kisi, O. Prediction of debonding strength for masonry elements retrofitted with FRP composites using neuro fuzzy and neural network approaches. *Compos. Part B Eng.* **2015**, *70*, 247–255. [CrossRef]
40. Nasrollahzadeh, K.; Nouhi, E. Fuzzy inference system to formulate compressive strength and ultimate strain of square concrete columns wrapped with fiber-reinforced polymer. *Neural Comput. Appl.* **2016**, *30*, 69–86. [CrossRef]
41. Falcone, R.; Lima, C.; Martinelli, E. Soft computing techniques in structural and earthquake engineering: A literature review. *Eng. Struct.* **2020**, *207*, 110269. [CrossRef]
42. Farooqi, M.U.; Ali, M. Effect of pre-treatment and content of wheat straw on energy absorption capability of concrete. *Constr. Build. Mater.* **2019**, *224*, 572–583. [CrossRef]
43. Farooqi, M.U.; Ali, M. Effect of Fibre Content on Compressive Strength of Wheat Straw Reinforced Concrete for Pavement Applications. *IOP Conf. Ser. Mater. Sci. Eng.* **2018**, *422*, 012014. [CrossRef]
44. Younis, K.H.; Amin, A.A.; Ahmed, H.G.; Maruf, S.M. Recycled Aggregate Concrete including Various Contents of Metakaolin: Mechanical Behavior. *Adv. Mater. Sci. Eng.* **2020**, *2020*, 8829713. [CrossRef]
45. Poon, C.; Kou, S.; Lam, L. Compressive strength, chloride diffusivity and pore structure of high performance metakaolin and silica fume concrete. *Constr. Build. Mater.* **2006**, *20*, 858–865. [CrossRef]
46. Qian, X.; Li, Z. The relationships between stress and strain for high-performance concrete with metakaolin. *Cem. Concr. Res.* **2001**, *31*, 1607–1611. [CrossRef]
47. Li, Q.; Geng, H.; Shui, Z.; Huang, Y. Effect of metakaolin addition and seawater mixing on the properties and hydration of concrete. *Appl. Clay Sci.* **2015**, *115*, 51–60. [CrossRef]
48. Ramezanianpour, A.; Jovein, H.B. Influence of metakaolin as supplementary cementing material on strength and durability of concretes. *Constr. Build. Mater.* **2012**, *30*, 470–479. [CrossRef]
49. Shehab El-Din, H.K.; Eisa, A.S.; Abdel Aziz, B.H.; Ibrahim, A. Mechanical performance of high strength concrete made from high volume of Metakaolin and hybrid fibers. *Constr. Build. Mater.* **2017**, *140*, 203–209. [CrossRef]
50. Roy, D.; Arjunan, P.; Silsbee, M. Effect of silica fume, metakaolin, and low-calcium fly ash on chemical resistance of concrete. *Cem. Concr. Res.* **2001**, *31*, 1809–1813. [CrossRef]
51. Poon, C.S.; Shui, Z.; Lam, L. Compressive behavior of fiber reinforced high-performance concrete subjected to elevated temperatures. *Cem. Concr. Res.* **2004**, *34*, 2215–2222. [CrossRef]
52. Sharaky, I.; Ghoneim, S.S.; Aziz, B.H.A.; Emara, M. Experimental and theoretical study on the compressive strength of the high strength concrete incorporating steel fiber and metakaolin. *Structures* **2021**, *31*, 57–67. [CrossRef]

53. Gilan, S.S.; Jovein, H.B.; Ramezanianpour, A.A. Hybrid support vector regression—Particle swarm optimization for prediction of compressive strength and RCPT of concretes containing metakaolin. *Constr. Build. Mater.* **2012**, *34*, 321–329. [CrossRef]

54. Silva, F.A.N.; Delgado, J.M.P.Q.; Cavalcanti, R.S.; Azevedo, A.C.; Guimarães, A.S.; Lima, A.G.B. Use of Nondestructive Testing of Ultrasound and Artificial Neural Networks to Estimate Compressive Strength of Concrete. *Buildings* **2021**, *11*, 44. [CrossRef]

55. Poon, C.S.; Azhar, S.; Anson, M.; Wong, Y.-L. Performance of metakaolin concrete at elevated temperatures. *Cem. Concr. Compos.* **2003**, *25*, 83–89. [CrossRef]

56. Nica, E.; Stehel, V. Internet of things sensing networks, artificial intelligence-based decision-making algorithms, and real-time process monitoring in sustainable industry 4. *J. Self-Gov. Manag. Econ.* **2021**, *9*, 35–47.

57. Sierra, L.A.; Yepes, V.; Pellicer, E. A review of multi-criteria assessment of the social sustainability of infrastructures. *J. Clean. Prod.* **2018**, *187*, 496–513. [CrossRef]

58. Kicinger, R.; Arciszewski, T.; De Jong, K. Evolutionary computation and structural design: A survey of the state-of-the-art. *Comput. Struct.* **2005**, *83*, 1943–1978. [CrossRef]

59. Yin, Z.; Jin, Y.; Liu, Z. Practice of artificial intelligence in geotechnical engineering. *J. Zhejiang Univ. A* **2020**, *21*, 407–411. [CrossRef]

60. Park, H.S.; Lee, E.; Choi, S.W.; Oh, B.K.; Cho, T.; Kim, Y. Genetic-algorithm-based minimum weight design of an outrigger system for high-rise buildings. *Eng. Struct.* **2016**, *117*, 496–505. [CrossRef]

61. Villalobos Arias, L. *Evaluating an Automated Procedure of Machine Learning Parameter Tuning for Software Effort Estimation*; Universidad de Costa Rica: San Pedro, Costa Rica, 2021.

62. Myles, A.J.; Feudale, R.N.; Liu, Y.; Woody, N.A.; Brown, S.D. An introduction to decision tree modeling. *J. Chemom. A J. Chemom. Soc.* **2004**, *18*, 275–285. [CrossRef]

63. Debeljak, M.; Džeroski, S. Decision Trees in Ecological Modelling. In *Modelling Complex Ecological Dynamics*; Springer: Berlin/Heidelberg, Germany, 2011; pp. 197–209.

64. Murthy, S.K. Automatic Construction of Decision Trees from Data: A Multi-Disciplinary Survey. *Data Min. Knowl. Discov.* **1998**, *2*, 345–389. [CrossRef]

65. Kheir, R.B.; Greve, M.H.; Abdallah, C.; Dalgaard, T. Spatial soil zinc content distribution from terrain parameters: A GIS-based decision-tree model in Lebanon. *Environ. Pollut.* **2010**, *158*, 520–528. [CrossRef] [PubMed]

66. Tso, G.K.F.; Yau, K.W.K. Predicting electricity energy consumption: A comparison of regression analysis, decision tree and neural networks. *Energy* **2007**, *32*, 1761–1768. [CrossRef]

67. Cho, J.H.; Kurup, P.U. Decision tree approach for classification and dimensionality reduction of electronic nose data. *Sens. Actuators B Chem.* **2011**, *160*, 542–548. [CrossRef]

68. Quinlan, J.R. *C4. 5: Programs for Machine Learning*; Elsevier: Amsterdam, The Netherlands, 2014.

69. Abidoye, L.; Mahdi, F.; Idris, M.; Alabi, O.; Wahab, A. ANN-derived equation and ITS application in the prediction of dielectric properties of pure and impure CO_2. *J. Clean. Prod.* **2018**, *175*, 123–132. [CrossRef]

70. Ozoegwu, C.G. Artificial neural network forecast of monthly mean daily global solar radiation of selected locations based on time series and month number. *J. Clean. Prod.* **2019**, *216*, 1–13. [CrossRef]

71. Hafeez, A.; Taqvi, S.A.A.; Fazal, T.; Javed, F.; Khan, Z.; Amjad, U.S.; Bokhari, A.; Shehzad, N.; Rashid, N.; Rehman, S.; et al. Optimization on cleaner intensification of ozone production using Artificial Neural Network and Response Surface Methodology: Parametric and comparative study. *J. Clean. Prod.* **2019**, *252*, 119833. [CrossRef]

72. Zhou, G.; Moayedi, H.; Bahiraei, M.; Lyu, Z. Employing artificial bee colony and particle swarm techniques for optimizing a neural network in prediction of heating and cooling loads of residential buildings. *J. Clean. Prod.* **2020**, *254*, 120082. [CrossRef]

73. Shahin, M.A.; Maier, H.R.; Jaksa, M.B. Data Division for Developing Neural Networks Applied to Geotechnical Engineering. *J. Comput. Civ. Eng.* **2004**, *18*, 105–114. [CrossRef]

74. Freund, Y.; Schapire, R.E. Experiments with a new boosting algorithm. In Proceedings of the ICML, Bari, Italy, 3–6 July 1996; pp. 148–156.

75. Guo, L.; Ge, P.; Zhang, M.-H.; Li, L.-H.; Zhao, Y.-B. Pedestrian detection for intelligent transportation systems combining AdaBoost algorithm and support vector machine. *Expert Syst. Appl.* **2012**, *39*, 4274–4286. [CrossRef]

76. Han, Q.; Gui, C.; Xu, J.; Lacidogna, G. A generalized method to predict the compressive strength of high-performance concrete by improved random forest algorithm. *Constr. Build. Mater.* **2019**, *226*, 734–742. [CrossRef]

77. Han, J.; Pei, J.; Kamber, M. *Data Mining: Concepts and Techniques*; Elsevier: Amsterdam, The Netherlands, 2011.

78. Farooq, F.; Ahmed, W.; Akbar, A.; Aslam, F.; Alyousef, R. Predictive modeling for sustainable high-performance concrete from industrial wastes: A comparison and optimization of models using ensemble learners. *J. Clean. Prod.* **2021**, *292*, 126032. [CrossRef]

79. Kohavi, R. A study of cross-validation and bootstrap for accuracy estimation and model selection. In Proceedings of the IJCAI, Montreal, QC, Canada, 20–25 August 1995; pp. 113–1145.

 *materials*

Article

Preliminary Study of the Fresh and Hard Properties of UHPC That Is Used to Produce 3D Printed Mortar

Ester Gimenez-Carbo [1,*], Raquel Torres [1], Hugo Coll [1], Marta Roig-Flores [2], Pedro Serna [1] and Lourdes Soriano [1]

1   Institute of Concrete Science and Technology (ICITECH), Universitat Politècnica de València, 46022 Valencia, Spain; ratorrem@hotmail.com (R.T.); hucolcar@cst.upv.es (H.C.); pserna@cst.upv.es (P.S.); lousomar@upvnet.upv.es (L.S.)
2   Department of Mechanical and Engineering Construction, Universitat Jaume I, 12071 Castellon de la Plana, Spain; roigma@uji.es
*   Correspondence: esgimen@cst.upv.es

Abstract: Three-dimensional printed concrete (3DPC) is a relatively recent technology that may be very important in changing the traditional construction industry. The principal advantages of its use are more rapid construction, lower production costs, and less residues, among others. The choice of raw materials to obtain adequate behavior is more critical than for traditional concrete. In the present paper a mixture of cement, silica fume, superplasticizer, setting accelerator, filler materials, and aggregates was studied to obtain a 3DPC with high resistance at short curing times. When the optimal mixture was found, metallic fibers were introduced to enhance the mechanical properties. The fresh and hard properties of the concrete were analyzed, measuring the setting time, workability, and flexural and compressive strength. The results obtained demonstrated that the incorporation of fibers (2% in volume) enhanced the flexural and compressive strength by around 163 and 142%, respectively, compared with the mixture without fibers, at 9 h of curing. At 28 days of curing, the improvement was 79.2 and 34.7% for flexural and compressive strength, respectively.

Keywords: 3D printed concrete; silica fume; setting time; workability; metallic fibers; mechanical properties

Citation: Gimenez-Carbo, E.; Torres, R.; Coll, H.; Roig-Flores, M.; Serna, P.; Soriano, L. Preliminary Study of the Fresh and Hard Properties of UHPC That Is Used to Produce 3D Printed Mortar. *Materials* **2022**, *15*, 2750. https://doi.org/10.3390/ma15082750

Academic Editor: Jorge Otero

Received: 10 March 2022
Accepted: 7 April 2022
Published: 8 April 2022

Publisher's Note: MDPI stays neutral with regard to jurisdictional claims in published maps and institutional affiliations.

Copyright: © 2022 by the authors. Licensee MDPI, Basel, Switzerland. This article is an open access article distributed under the terms and conditions of the Creative Commons Attribution (CC BY) license (https://creativecommons.org/licenses/by/4.0/).

1. Introduction

The expanded selection of additives in concrete technology has led to the development of new materials and the possibility of achieving ultra-high-performance fiber-reinforced concrete (UHPFRC). This material is the product of three technologies, self-compacting concrete, fiber-reinforced concrete, and high-strength concrete [1], and was developed with the aim of improving three important aspects, mechanical properties, durability, and workability.

UHPFRC was first developed in France in the 1990s, and, according to the Association Française de Génie Civil (AFGC) [2], this cementitious matrix material has a characteristic 28-day compressive strength of more than 150 MPa, with high flexural strength and ductile behavior. In recent years, there have been small variations in the placement of concrete, with the development of self-compacting concrete and improvements in the techniques for the use of shotcrete, which at the time represented a great advance [3]. Shotcrete can be considered as the ancestor of additive manufacturing. These techniques are the only ones that do not use formwork for the placement of concrete.

In the present work, UHPRC mixtures were developed that could be used in shotcrete as a first step until their dosages could be used to develop additive manufacturing techniques. One of these techniques was three-dimensional concrete printing (3DPC). The challenge presented by the dosages used was that ultra-high-strength concrete is manufactured with large amounts of superplasticizer additive, whose action increases the setting time. However, to use this well-projected material for 3D printing, it was necessary to reduce the setting time, which forced the need to introduce accelerating setting additives into the mixture. This could compromise the mechanical resistance achieved with the mixture [4], which will have to be studied.

3DCP is a material with numerous advantages, but it requires careful dosing. The principal advantages of its use are more rapid construction, lower production costs, and less production of residues, among others [4,5].

Additive manufacturing can be defined as a process that uses technology for automation. With this process, three-dimensional objects can be produced from digital models in a precise way and within a predetermined space. The first research on 3D printing in the construction and architecture industry dates back to 1995, when Pegna suggested incorporating cement-based materials when using these new technologies [6].

Concrete printing can be used to build complex geometric shapes. The components are designed using 3D modelling software [7]. The mixture is printed by controlled extrusion. The concrete needs to have a good degree of extrudability in order to form small concrete filaments. These filaments must bind together to form each layer. One requirement is to be able to build layers without deformation of the successive layers [8].

Lyu et al. [9] noted that the printability of 3DPC includes fluidity, extrudability, buildability, and setting time. The main important factor affecting fluidity in 3DPC is the water content. This concrete used a lower water content, and it required the use of a high-performance water reducing agent. Other factors that can modify the fluidity are the use of mineral admixtures and the effects of aggregate fineness [10].

Extrudability measures the difficulty in the extrusion process. Fresh concrete should be delivered continuously through the nozzle. According to Ma et al. [11], this property is affected by the amount and distribution of the dry mixture.

Buildability measures the degree of deformation and the stability of the printed layers. The material needs to be strong enough to retain its shape and prevent layers from collapsing under its own weight and the gravitational load [12]. Increasing the quantity of aggregates and adding mineral admixtures can improve this property [9].

The setting time for 3DCP requires a compromise between allowing sufficient time to obtain good fluidity and extrudability and sufficient time to obtain early strength.

Zhang et al. [4] noted that the characteristics required for 3DCP often conflict with one another. To obtain a material that is easily pumpable and extrudable, it needs to have low plastic viscosity and optimum yield stress. However, to obtain good buildability, the material needs to have high static yield stress. For all of these reasons, the mixtures should be carefully selected to ensure that they are thixotropic, set suitably, and are densely packed.

Supplementary cementing materials (SCMs) are used to enhance mechanical strength and durability performance, and they also have an influence in the fresh state [13–15]. Arunothayan et al. [16] studied the use of fly ash (FA) and ground blast furnace slag (BFS) as substitutes in cement for 3DCP. All of the mixtures contained 30% silica fume (SF) as a constant percentage. The authors demonstrated that the FA facilitated the flow of the mixture, in contrast with the BFS, which reduced the workability. Compressive strength was reduced when SCMs were added; however, the difference with respect to the control was reduced after 90 days of curing.

The present investigation was a preliminary study that explored the use of a mixture of cement, silica fume, superplasticizer, setting accelerator, filler quartz material, and aggregates to obtain 3DPC with high resistance at short curing times. In the first part, different setting accelerators were studied to obtain an adequate setting time, in the second part we worked with the selected accelerator and studied the influence of the setting accelerator percentage on the workability of the fresh mixture and the evolution of flexural and compressive strength. Finally, in the last part of the study, metallic fibers were added to study their influence on resistance. The incorporation of fibers can also improve the bonding and connection between the different layers of placed concrete.

2. Materials and Methods

Portland cement type CEM I 52.5R (Lafarge Holcim, Paris, France), which met the specification of the European standards [17], was used in the preparation of mortars. Elkem Microsilica 940 U (Elkem Materials, Pittsburgh, PA, USA) was used as SCM. This material is an undensified silica fume with a density between 200 and 350 kg/m^3.

To reduce the cement content and complement the granulometric curve for small sizes, quartz flour (Silbeco, Antwerp, Belgium) was also added; the main characteristics are an SiO_2 content higher than 98% and density between 200 and 300 kg/m^3.

X-ray fluorescence analysis was conducted to determine the chemical composition of silica fume and cement, and the results are shown in Table 1.

Table 1. Chemical composition of CEM I 52.5R and silica fume (% in weight).

Compound	CEM I 52.5R	Microsilica 940 U
SiO_2	19.29	95.80
Al_2O_3	5.22	0.31
Fe_2O_3	3.51	0.14
CaO	61.75	0.38
MgO	2.07	0.10
SO_3	3.55	0.02
$Na_2O + K_2O$	1.23	0.32
Others	1.42	–
Loss on ignition (950 °C)	1.96	2.93

Figure 1 shows the granulometric curves of cement (CEM), silica fume (SF), and quartz flour (QF).

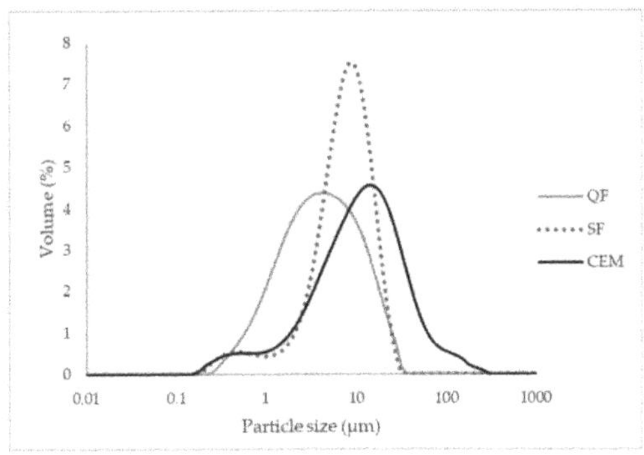

Figure 1. Granulometric curves of cement, silica fume, and quartz flour.

Two types of siliceous sand were employed, 0.8 fraction with sizes between 0.6 and 1.2 mm and 0.4 fraction with sizes between 0.2 and 0.6 mm. The distribution company was Silicam.

Sika ViscoCrete 225 P (Sika, Baar, Switzerland) was used as the superplasticizing additive. It is a superplasticizer powder water reducer and has a shorter absorption time. The typical dosage varies between 0.05 and 0.5% by weight of binder.

Four types of liquid setting accelerators were used in the first phase of testing to study the behavior of the mixture, in order to finally determine the one that would be used in the second and third phases of the study. Centrament Rapid 500 (MC Company, Scottsdale, AZ, USA) is a chloride-free additive that provides rapid hardening without affecting workability. Sikaset-3 additive (Sika, Baar, Switzerland), according to the manufacturer, is capable of doubling mechanical resistance between 5 and 10 h of curing. Master X-Seed 130 (Master Builders Solution, Mannheim, Germany) is a cement hydration activator agent composed of a suspension of C-S-H nanoparticles. Finally, AKF-63 (IQE, Cardiff, UK) is an aqueous solution of aluminum salts.

The dosage was selected following guidelines from previous studies carried out by our research group [18]. The mortar mixtures are summarized in Table 2.

Table 2. Mortar dosages.

Material	Dosage (kg/m^3)
CEM I 52.5R	800
Water	170
Sika ViscoCrete	4
Sand 0.8	562
Sand 0.4	302
Filler addition (quartz flour)	225
Active addition (microsilica)	175

All materials were poured into the mixer machine, except for additives (superplasticizer, setting accelerator, and water). The sequence and mixing time of the mixtures are shown in Table 3. The mixing machine employed was an Ibertest model that met the specifications of the European standard [19]. When fibers were incorporated, they were poured into the machine and mixed before the setting accelerator additive.

Table 3. Sequence of material incorporation.

Incorporation in Mixer Machine	Solid Materials (Without Additives)	Water	Additives		Finish Process
			Superplasticizer	Setting Accelerator	
Mixing time	30 s	30 s	150 s	600 s	Variable

To determine the setting time, specifically the start of setting of the different mixtures, which occurred in the first phase of the experimental program, two European regulations were taken as reference, UNE-EN 196-3 [20] and UNE-EN 480-2 [21]. Both regulations describe the test procedure, materials, and apparatus to be used to determine setting time. The difference between the two references is the value of the penetration of the Vicat apparatus. In the present study, the start of setting was considered to occur when the needle did not drag the material after extraction and generated a gap in the mortar (Figure 2).

Figure 2. Gap generated in paste indicating the start of setting time.

Various investigations have reported that the workability varied depending on the time when the test was carried out [22,23]. In this investigation, the workability was tested at different times after the mixing process was finished, and the European standard was taken for reference [24]. In addition, a smaller cone was used, because it is considered that a smaller sample volume is closer to the amount of concrete that would come out of the nozzle of a 3D printer. The dimensions of the standardized truncated cone-shaped mold, in accordance with European standards [24], were 7 cm inside diameter, 8.5 cm outside diameter, and 4 cm height, with a non-standardized PVC tube 3 cm in diameter and 2.5 cm in height, as shown in Figure 3.

Figure 3. Dimensions of molds used to test workability.

The mechanical strength of 40 mm × 40 mm × 160 mm prismatic mortar specimens was determined according to the normalized standard [19]. Samples were stored in molds in a humid atmosphere until the testing age in the case of short curing times and for 24 h for the rest. At the required age, the specimens were taken from storage and broken by flexure, and each half was tested for compression strength (using an Ibertest machine).

3. Results

This section is divided into three subsections; the first part describes the selection of setting accelerator, and the other two parts focus on the chosen additive.

3.1. Selection of Setting Accelerator

The selected percentages of setting accelerator were 1 and 2%, with respect to the weight of cement. Table 4 shows the mortar setting times; the control mortar without setting accelerator had a setting time of more than 90 min.

Table 4. Initial setting time of mortar with 1.5 and 2% setting accelerator.

Setting Accelerator	Initial Setting Time (Min)	
AKF-63	1.5%	2%
	10:38	8:00
Sikaset-3	1.5%	2%
	40:02	49:03
Centrament Rapid	1.5%	2%
	60:10	70:30
Master X-Seed 130	1.5%	2%
	65:13	71:00

The setting accelerator with the shortest initial setting time was AKF-63; this type of additive with aluminum salts was very effective, so it was selected as the setting accelerator for use in successive sections. Kim et al. suggested that aluminate-based activators reacting with cement minerals exhibit fast setting times; they obtained an initial setting time of around 10 min [25]. Other authors, such as Qiu et al. [26], used nano-alumina and modified alcohol amine as raw materials to prepare shotcrete with an initial setting time of only 3 min. The references consulted with regard to the use of C-S-H nanoparticles obtained results very similar to those obtained with Master X-Seed 130. Das et al. [27] reported that the reduction in setting time was a consequence of early strength development of mortar; a reduction of about 43% of the initial setting time was obtained.

3.2. Selection of Percentage of Setting Accelerator

3.2.1. Influence of Percentage of Setting Accelerator on Workability

Before measuring the workability values, we evaluated the initial setting time of the mixture using smaller quantities of setting accelerator; the values obtained were 16 and 13 min for 0.5% and 1.0% and 1.5%, respectively.

The values of the diameter obtained in the flow table were measured at different times. This time value was a previously defined variable representing the period from the end of the mixing process to the exact moment of raising the cone. After this time, the diameter of the stabilized biscuit was measured in millimeters, as shown in Figure 4.

Figure 4. Measurement of the stabilized biscuit.

The evolution of the diameter values with time for the three percentages of setting accelerator and the mixture without setting accelerator is represented in Figure 5.

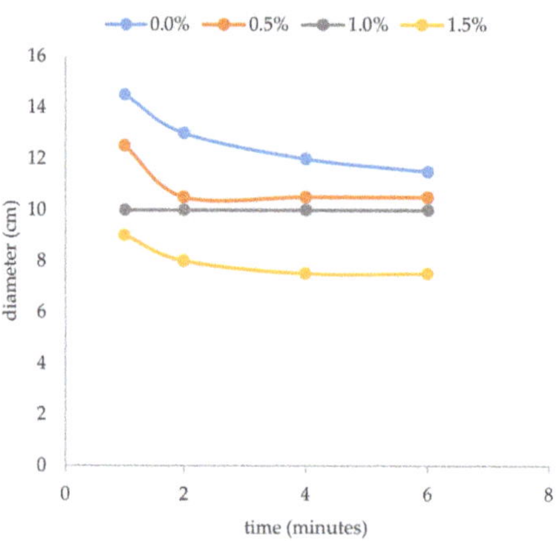

Figure 5. Evolution of workability with time and percentage of setting accelerator.

As can be seen in Figure 3, the diameter of the biscuit obtained depended on the percentage of setting accelerator used. In the mixtures with 1.5% setting accelerator, the diameter was smaller than 10 cm for all times measured, and with 1% setting accelerator, the values were always the same. The setting accelerator promoted a greater degree of hydration, therefore rigidity developed gradually, with a loss of workability [11].

The values obtained in this study were lower than those obtained by other authors as ideal values to obtain good buildability. Tay et al. [28] reported that a slump flow value between 15 and 19 cm in mixtures was optimal for a smooth surface and high buildability.

For future research, the use of other supplementary materials such as fly ash (FA) should be explored, which has demonstrated its ability to increase workability. Liu et al. [29] reported that FA has the property of lubrication and produces a reduction in cement flocculation and, therefore, greater workability.

The study of the mixture in the fresh state, through studies in rheometers, could also be performed to find better dosages for use in 3DCP mixtures. Panda et al. [30] conducted studies on 3DCP using large volumes of AF. They showed that the addition of small amounts of nanoclays improved the behavior of mixtures. The improved performance was associated with the thixotropic property of clay particles, which were responsible for better early age mechanical properties such as yield stress and stiffness.

3.2.2. Influence of Percentage of Setting Accelerator on Flexural and Compressive Strength

The influence of the setting accelerator in the first hours of curing is fundamental to obtaining stability in 3D concrete. In this subsection, the resistance of mortar was measured using 1.5 and 3% setting accelerator. The values of flexural and compressive strength at curing ages of 6, 9, 12, and 24 h and 28 days are listed in Table 5.

Table 5. Flexural strength (R_f) and compressive strength (R_c) of mortar.

% Setting Accelerator	Age of Assay	R_f (MPa)	R_c (MPa)
0.0%	6 h	1.05 ± 0.31	1.92 ± 0.23
	9 h	2.51 ± 0.27	2.20 ± 0.24
	12 h	2.76 ± 0.28	4.70 ± 0.32
	24 h	8.08 ± 0.31	51.52 ± 1.93
	7 d	13.86 ± 0.35	103.31 ± 3.2
	28 d	22.09 ± 0.35	125.77 ± 3.5
1.5%	6 h	2.45 ± 0.29	2.76 ± 0.50
	9 h	3.06 ± 0.27	7.37 ± 0.29
	12 h	4.98 ± 0.51	18.58 ± 0.96
	24 h	9.34 ± 0.78	63.04 ± 0.52
	7 d	14.80 ± 0.67	109.89 ± 1.66
	28 d	24.80 ± 1.02	137.30 ± 1.96
3.0%	6 h	3.37 ± 0.51	7.49 ± 0.97
	9 h	3.98 ± 0.26	14.20 ± 1.2
	12 h	6.58 ± 0.50	28.33 ± 0.79
	24 h	9.34 ± 0.82	60.84 ± 0.94
	7 d	14.25 ± 1.02	105.22 ± 1.21
	28 d	22.28 ± 0.68	129.14 ± 2.10

The mechanical strength of 40 mm × 40 mm × 160 mm prismatic mortar specimens was determined according to the normalized standard. The loading speed for the flexural test was 50 ± 10 N/s and the compression rate was 2400 ± 10 N/s.

As can be seen in the resistance results, the setting accelerator had positive effects during the first 24 h of curing. At 28 curing days, both types of resistance were very close to the results of mortar without setting accelerator.

Figure 6 shows the division between the results obtained by mortar with setting accelerator and mortar without additive, i.e., the division between the results of mixtures with 1.5% and 3.0% setting accelerator (R_{fi} or R_{ci}) and the mixture containing no additive ($R_{f0\%}$ or $R_{c0\%}$) for each curing age tested.

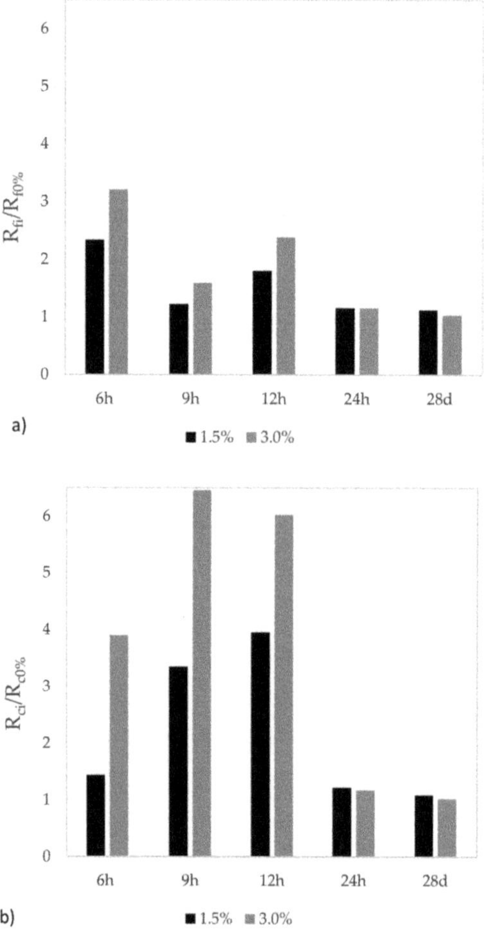

Figure 6. Ratio between mortar with and without setting accelerator: (**a**) flexural strength, (**b**) compressive strength.

The beneficial effects of adding setting accelerator were higher for compressive strength than flexural strength. For both types of resistance, the 3% setting accelerator demonstrated better results. The flexural strength values of mortar with 3% additive at 6 h and 28 days of curing were 3.20 and 1.03, respectively. At 28 curing days, the effect of the additive disappeared.

For compressive strength, the values for mortar with 3% additive at 9 h and 28 days of curing were 6.45 and 1.02, respectively. As with flexural strength, the contribution by the additive to improvements at long curing ages was nil.

The actuation of the setting accelerator in the enhancement of resistance at short curing ages was in agreement with the references consulted [31,32].

In some cases, the use of setting accelerator could reduce the strength at long curing times [33], something that did not happen in the present investigation.

Authors such as Min et al. [34] suggested that the presence of setting accelerator promoted the simultaneous hydration of C_3A and C_3S at an early age. This process was quicker as the amount of setting accelerator increased. They considered that the reaction effect was effective at an early age, until 12 h.

3.3. Incorporation of Metallic Fibers

The bibliography consulted said that the application of 1–3% vol% fibers into 3DCP is applied to obtain sufficient robustness and ductility for structural applications. This reinforcement has been studied with carbon, basalt, glass, or propylene fibers, among others [35–37].

In the last part of the present investigation, 2% of metallic fibers by volume was incorporated into the mixture containing 3% setting accelerator. The flexural and compressive strengths of the mixtures with fibers were measured at the same curing ages. Table 6 shows the ratio of mortars with and without metallic fibers (R_i/R_0).

Table 6. Ratio of resistance between mortar with and without metallic fibers.

Age of Assay	R_{fi}/R_{f0}	R_{ci}/R_{c0}
6 h	2.31	2.71
9 h	2.57	2.20
12 h	2.07	1.22
24 h	2.46	1.25
7 d	2.20	1.26
28 d	1.54	1.35

The incorporation of metallic fibers enhanced both flexural and compressive strength, and flexural strength improved by a greater extent. The incorporation of fibers in 3D concrete printing is a good solution, because in this system the use of classical steel reinforcement is difficult.

Hambach and Volkmer studied the use of three types of fiber in 3D fiber-reinforced Portland cement paste [38]. They determined that the most effective fiber was carbon fiber at 1 vol%, which achieved flexural strength of 30 MPa. Other fibers (glass and basalt) did not significantly increase flexural strength. Zhu et al. [35] used polyethylene fibers and obtained flexural strength of around 19 MPa.

4. Discussion

The results obtained in this study were promising, but this was a preliminary study in which tests were carried out in molds and not in samples obtained by 3D printing. The next step would be to test this dosage using 3D printing equipment and measure the properties of a structure made in this way. Some authors, such as Rehman and Kim [39], pointed out that the compression strength of printed samples could be up to 22% lower than that of samples placed in molds. If the dosage was not suitable for 3D printing, it could be used for the manufacture of shotcrete.

We also believe that rheology studies are essential in future investigations to know how the mixture will behave when used in 3D printing equipment. The behaviour of the fibers in the use of the equipment could also be a critical point; their use could lead to obstruction, segregation, and lack of uniformity in the distribution of them. We consider it essential to expand the study of how to improve the workability of the use of FA without excessively compromising the mechanical resistances. Since FA has a slower pozzolanic reaction than SF and this could mean a decrease in resistances in the early ages of curing.

5. Conclusions

The main conclusions of the presented investigation are as follows:

- Mortars with fast setting start were achieved, specifically in a time of 10 min.
- The loss of workability over time was pronounced due to the high reactivity of the setting accelerator used.
- The percentage of setting accelerator had an influence on the flexural and compressive strength, mainly at short curing ages, but no negative effects at longer curing ages.
- With the inclusion of metal fibers, flexural strength of more than 23 MPa and compressive strength of around 76 MPa at 24 h of curing were obtained.

Author Contributions: Conceptualization, E.G.-C. and P.S.; methodology, R.T., H.C. and M.R.-F.; validation, H.C., E.G.-C. and P.S.; formal analysis, R.T.; investigation, R.T. and M.R.-F.; resources, H.C.; data curation, R.T. and L.S.; writing—original draft preparation, L.S. and E.G.-C.; writing—review and editing, M.R.-F. and E.G.-C.; visualization, H.C. and R.T.; supervision, P.S.; project administration, P.S.; funding acquisition, P.S. All authors have read and agreed to the published version of the manuscript.

Funding: This research received no external funding.

Institutional Review Board Statement: Not applicable.

Informed Consent Statement: Not applicable.

Data Availability Statement: Data are contained within the article.

Acknowledgments: The authors thank Research & Development Concretes S.L. for supplying some of the additives used to manufacture the samples.

Conflicts of Interest: The authors declare no conflict of interest.

References

1. Serna, P.; López, J.A.; Camacho, E. UHPFRC: De los componentes a la estructura. In *I Simposio Latino Americano sobre Concreto Autoadensavel, 54° Congreso Brasileiro do Concreto*; IBRACON: Maceio, Brasil, 2012; pp. 59–81, ISSN 2175-8182.
2. Resplendino, J. Ultra high performance concrete: New AFGC recommendations. In *Designing and Building with UHPFRC*; Wiley Library: Hoboken, NJ, USA, 2011; Chapter 47; pp. 713–721, ISBN 9781848212718.
3. Yoggi, G.D. *History of Shotcrete*; American Concrete Institute (ACI): Indianapolis, IN, USA, 2002.
4. Zhang, C.; Naidu, V.; Krishma, A.; Wang, S.; Zhang, Y.; Mechtcherine, V.; Banthia, N. Mix design concepts for 3D printable concrete: A review. *Cem. Concr. Compos.* **2021**, *122*, 104155. [CrossRef]
5. Zhang, J.; Wang, J.; Dong, S.; Yu, X.; Han, B. A review of the current progress and application of 3D printed concrete. *Compos. Part A* **2019**, *125*, 105533. [CrossRef]
6. Pegna, J. Exploratory investigation of solid freeform construction. *Autom. Constr.* **1997**, *5*, 427–437. [CrossRef]
7. Buswell, R.A.; Soar, R.C.; Gibb, A.G.F.; Thorpe, A. Freeform construction: Mega-scale rapid manufacturing for construction. *Autom. Constr.* **2007**, *16*, 224–231. [CrossRef]
8. Le, T.T.; Austin, S.A.; Lim, S.; Buswell, R.A.; Gibb, A.G.F.; Thorpe, T. Mix design and fresh properties for high-performance printing concrete. *Mater. Stuct.* **2012**, *45*, 1221–1232. [CrossRef]
9. Lyu, F.; Zhao, D.; Hou, X.; Sun, L.; Zhang, Q. Overview of the development of 3D-Printing concrete: A review. *Appl. Sci.* **2021**, *11*, 9822. [CrossRef]
10. Lin, J.C.; Wu, X.; Yang, W.; Zhao, R.X.; Qiao, L.G. The influence of fine aggregates on the 3D printing performance. *IOP Conf. Ser. Mater. Sci. Eng.* **2018**, *292*, 012079. [CrossRef]
11. Ma, G.; Li, Z.; Wang, L. Printable properties of cementitious material containing copper tailings for extrusion based 3D printing. *Constr. Build. Mater.* **2018**, *162*, 613–627. [CrossRef]
12. Mechtcherine, V.; Bos, F.P.; Perrot, A.; Leal da Silva, W.R.; Nerella, V.N.; Fataei, S.; Wolfs, R.J.M.; Sonebi, M.; Roussel, N. Extrusion-based additive manufacturing with cement-based materials. Production steps, processes, and their underlying physics: A review. *Cem. Concr. Res.* **2020**, *132*, 613–627. [CrossRef]
13. Bhattacherje, S.; Basavaraj, A.S.; Rahul, A.V.; Santhanam, M.; Gettu, R.; Panda, B.; Schalangen, E.; Chen, Y.; Copuroglu, O.; Ma, G.; et al. Sustainable materials for 3D concrete printing. *Cem. Concr. Compos.* **2021**, *122*, 104156. [CrossRef]
14. Papachristoforou, M.; Mitsopoulos, V.; Stefanidou, M. Use of by-products for partial replacement of 3D printed concrete constituents; rheology, strength and shrinkage performance. *Frat. Integrità Strutt.* **2019**, *50*, 526–536. [CrossRef]
15. Manikandan, K.; Wi, K.; Zhang, X.; Wang, K.; Qin, H. Characterizing cement mixtures for concrete 3D printing. *Manuf. Lett.* **2020**, *24*, 33–37. [CrossRef]

16. Arunothayan, A.R.; Nematollahi, B.; Ranade, R.; Khayat, K.H.; Sanjayan, J.G. Digital fabrication of eco ultra-high performance fiber-reinforced concrete. *Cem. Concr. Compos.* **2022**, *125*, 104281. [CrossRef]

17. *UNE-EN 197-1*; Cement. Part I: Composition, Specifications and Conformity Criteria for Common Cements. AENOR: Madrid, Spain, 2011.

18. Camacho, E. Dosage Optimization and Bolted Connections for UHPFRC Ties. Ph.D. Thesis, Universitat Politècnica de València, Valencia, Spain, 2013.

19. *UNE-EN 196-1*; Methods of Testing Cement. Part 1: Determination of Strength. AENOR: Madrid, Spain, 2018.

20. *UNE-EN 196-3*; Methods of Testing Cement. Part 3: Determination of Setting Times and Soundness. AENOR: Madrid, Spain, 2017.

21. *UNE-EN 480-2*; Admixtures for concrete, mortar and grout. Test Methods. Part 2: Determination of Setting Time. AENOR: Madrid, Spain, 2007.

22. Kim, B.; Jiang, S.; Aïtcin, P. Slump improvement mechanism of alkalies in PNS superplasticized cement pastes. *Mater. Struct.* **2000**, *33*, 363–369. [CrossRef]

23. Lee, H.; Seo, E.A.; Kim, W.W.; Moon, J.H. Experimental study on time dependent changes in rheological properties and flow rate of 3D concrete printing. *Materials* **2021**, *14*, 6278. [CrossRef] [PubMed]

24. *UNE-EN 12350-5*; Testing Fresh Concrete. Part 5. Flow Table Test. AENOR: Madrid, Spain, 2020.

25. Kim, J.; Ryu, J.; Hooton, R.D. Evaluation of strength and set behaviour of mortar containing shotcrete set accelerators. *Can. J. Civ. Eng.* **2008**, *35*, 400–407. [CrossRef]

26. Qiu, Y.; Ding, B.; Gan, J.; Guo, Z.; Zheng, C.; Jiang, H. Mechanism and preparation of liquid alkali-free liquid setting accelerator for shotcrete. *IOP Conf. Ser. Mater. Sci. Eng.* **2017**, *182*, 012034. [CrossRef]

27. Das, S.; Ray, S.; Sarkar, S. Early strength development in concrete using preformed CSH nano crystals. *Constr. Build. Mater.* **2020**, *233*, 117214. [CrossRef]

28. Tay, Y.W.D.; Qian, Y.; Tan, M.J. Printability region for 3D concrete printing using slump and slump flow test. *Compos. B. Eng.* **2019**, *174*, 106968. [CrossRef]

29. Liu, J.; An, R.; Jiang, Z.; Jin, H.; Zhu, J.; Liu, W.; Huang, Z.; Xing, F.; Liu, J.; Fan, X.; et al. Effects of w/b ratio, fly ash, limestone calcined clay, seawater and sea-sand on workability, mechanical properties, drying shrinkage behavior and micro-structural characteristics of concrete. *Constr. Build. Mater.* **2022**, *321*, 126333. [CrossRef]

30. Panda, B.; Mohamed, N.; Paul, S.C.; Bhagath, G.V.P.; Tan, M.; Savija, B. The effect of material fresh properties and process parameters on buildability and interlayer adhesion of 3D printed concrete. *Materials* **2019**, *12*, 2149. [CrossRef]

31. Pan, Z.; Wang, X.; Liu, W. Properties and acceleration mechanism of cement mortar added with low alkaline liquid state setting accelerator. *J. Wuhan Univ. Technol. Mater.* **2014**, *29*, 1196–1200. [CrossRef]

32. Luo, B.; Luo, Z.; Wang, D.; Shen, C.; Xia, M. Influence of alkaline and alkali-free accelerators on strength, hydration and microstructure characteristics of ultra-high performance concrete. *J. Mater. Res. Technol.* **2021**, *15*, 3283–3295. [CrossRef]

33. Park, H.G.; Sung, S.K.; Park, C.G.; Won, J.P. Influence of a $C_{12}A_7$ mineral-based accelerator on the strength and durability of shotcrete. *Cem. Concr. Res.* **2008**, *38*, 379–385. [CrossRef]

34. Min, T.-B.; Cho, I.-S.; Park, W.-J.; Choi, H.-K.; Lee, H.-S. Experimental study on the development of compressive strength of early concrete age using calcium-based hardening accelerator and high early strength cement. *Constr. Build. Mater.* **2014**, *64*, 208–214. [CrossRef]

35. Zhu, B.; Pan, J.; Nematollahi, B.; Zhou, Z.; Zhang, Y.; Sanjayan, J. Development of 3D printable engineered cementitious composites with ultra-high tensile ductility for digital construction. *Mater. Des.* **2019**, *181*, 1080088. [CrossRef]

36. Bos, F.; Wolfs, R.; Ahmed, Z.; Salet, T. Additive manufacturing of concrete in construction: Potentials and challenges of 3D concrete printing. *Virtual Phys. Prototyp.* **2016**, *11*, 209–225. [CrossRef]

37. Ye, J.; Cui, C.; Yu, J.; Yu, K.; Dong, F. Effect of polyethylene fiber content on workability and mechanical-anisotropic properties of 3D printed ultra-high ductile concrete. *Constr. Build. Mater.* **2021**, *281*, 122586. [CrossRef]

38. Hambach, M.; Volkmer, D. Properties of 3D-printed fiber-reinforced Portland cement paste. *Cem. Concr. Compos.* **2017**, *79*, 62–72. [CrossRef]

39. Rehman, A.U.; Kim, J.H. 3D Concrete printing: A systematic review of rheology, mix designs, mechanical, microstructural and durability characteristics. *Materials* **2021**, *14*, 3800. [CrossRef]

Review

A Scientometric Review on Mapping Research Knowledge for 3D Printing Concrete

Chuan He [1,*], Shiyu Zhang [1], Youwang Liang [1], Waqas Ahmad [2,*], Fadi Althoey [3], Saleh H. Alyami [3], Muhammad Faisal Javed [2,*] and Ahmed Farouk Deifalla [4]

1   School of Architecture, Changsha University of Science and Technology, Changsha 410015, China; zsy627377809@sina.com (S.Z.); liangyouwang66@sina.com (Y.L.)
2   Department of Civil Engineering, COMSATS University Islamabad, Abbottabad 22060, Pakistan
3   Department of Civil Engineering, Najran University, Najran, Saudi Arabia; fmalthoey@nu.edu.sa (F.A.); shalsalem@nu.edu.sa (S.H.A.)
4   Structural Engineering and Construction Management Department, Faculty of Engineering and Technology, Future University in Egypt, Cairo 11835, Egypt; ahmed.deifalla@fue.edu.eg
*   Correspondence: hechuan@csust.edu.cn (C.H.); waqasahmad@cuiatd.edu.pk (W.A.); arbab_faisal@yahoo.com (M.F.J.)

Abstract: The scientometric analysis is statistical scrutiny of books, papers, and other publications to assess the "output" of individuals/research teams, organizations, and nations, to identify national and worldwide networks, and to map the creation of new (multi-disciplinary) scientific and technological fields that would be beneficial for the new researchers in the particular field. A scientometric review of 3D printing concrete is carried out in this study to explore the different literature aspects. There are limitations in conventional and typical review studies regarding the capacity of such studies to link various elements of the literature accurately and comprehensively. Some major problematic phases in advanced level research are: co-occurrence, science mapping, and co-citation. The sources with maximum articles, the highly creative researchers/authors known for citations and publications, keywords co-occurrences, and actively involved domains in 3D printing concrete research are explored during the analysis. VOS viewer application analyses bibliometric datasets with 953 research publications were extracted from the Scopus database. The current study would benefit academics for joint venture development and sharing new strategies and ideas due to the graphical and statistical depiction of contributing regions/countries and researchers.

Keywords: 3D printing; concrete; scientometric analysis; cementitious composites

Citation: He, C.; Zhang, S.; Liang, Y.; Ahmad, W.; Althoey, F.; Alyami, S.H.; Javed, M.F.; Deifalla, A.F. A Scientometric Review on Mapping Research Knowledge for 3D Printing Concrete. *Materials* **2022**, *15*, 4796. https://doi.org/10.3390/ma15144796

Academic Editor: Jorge Otero

Received: 25 May 2022
Accepted: 6 July 2022
Published: 8 July 2022

Publisher's Note: MDPI stays neutral with regard to jurisdictional claims in published maps and institutional affiliations.

Copyright: © 2022 by the authors. Licensee MDPI, Basel, Switzerland. This article is an open access article distributed under the terms and conditions of the Creative Commons Attribution (CC BY) license (https://creativecommons.org/licenses/by/4.0/).

1. Introduction

Charles Hull, in 1986, initially introduced the 3D printing or additive manufacturing (AM) technology in stereolithography (SLA). Afterwards, it gained the attention of everyone, either from industry or an individual hobbyist [1]. The enhanced popularity of 3D printing is primarily because of its potentially freeform design, minimizing waste materials, mass customization, complex geometries manufacturing, and accelerating the fabrication procedure [2]. In the current era, the application of 3D printing technology in construction is becoming very prevalent [3,4]. Kim et al. [5] used the 3D printing technology to determine reinforced concrete beams' shear strength having multiple interfaces before initial setting. Three-dimensional printing technology can offer new prospects in the construction sector, such as geometrical flexibility, labor cost reduction, safety and efficiency improvement, and hard/harsh area/environment construction [6,7]. The primary distinguishing component of 3D printing technology is the flexibility in geometry that enables the improved architectural appearance. Three-dimensional printing technology also offers the independency of shape on cost, ultimately providing design freedom [3]. Further, the additive/3D printing technology enables the creation of multi-functional components of a building and links the digital building and designing process [4,8]. The cost reduction, coupled with human resources, is also an essential component of said technology. It is linked with enhancement in safety and efficiency. Three-dimensional technology offers higher cost-effectiveness and accuracy with respect to traditional technology [4,9]. The need for formwork, a significant component in conventional construction, is also eliminated using 3D printing [4,10]. The enhancement in safety levels by reducing the injury rates can also be achieved by eliminating the formwork stage [10,11]. Furthermore, it also helps to reduce the on-site construction time [10]. The last and most important advantage of 3D printing is sustainability. The construction waste, specifically generated from formwork, is also significantly reduced by using this technology [3,10]. Initially, the lesser material would be consumed for casting and molding, followed by the possible optimization of construction provided by this technology and the reduction in materials consumed by this process itself. A further benefit of this technology also includes the reduction in transportation costs. In addition to that, this technology also comes up with reduced CO_2 emissions by declining the inadequacies throughout the process of building. Multiple studies are going on for achieving sustainable development by using recycled/waste materials such as natural fibers, supplementary cementitious materials, construction and demolition waste, marble and ceramic waste powders, functionally graded materials etc., to conserve the natural resources [12–22].

Hence, it can be concluded that the rising agreement of using the 3D printing technology over conventional methods is due to multiple benefits such as highly accurate complex geometry fabrication, design flexibility, personal customization, and maximum material conservation. A wide variety of materials are applied in 3D printing such as, concrete, polymers, metals, and ceramics. Acrylonitrile butadiene styrene (ABS) and polylactic acid (PLA) are among the significant polymers that are utilized for composites 3D printing. Advanced alloys and metals are usually used in aerospace to reduce the time and cost consumption involved in conventional methods. Three-dimensional printing of scaffolds mainly consumes ceramics, whereas concrete is the primary material for building additive manufacturing. However, large-scale printing is still quite limited due to poorer anisotropic behavior and mechanical characteristics of the 3D printed parts. Accordingly, there is a need to have an optimized 3D priming pattern to restrict anisotropic behavior and error sensitivity [23].

The finished products quality is also dependent on the printing environment [24]. The multiple sizes, i.e., micro to macro scale, of parts fabrication can be performed by using additive manufacturing (AM). Whereas the printed parts accuracy is mainly dependent on the precision of the applied printing scale and method. For example, 3D printing at the micro level offers challenges with the layer bonding, surface finish and resolution that usually need sintering like post-processing treatments [25]. The limited 3D printing materials provide challenges in employing 3D printing technology in different industrial sectors. Therefore, appropriate materials are needed to be utilized in 3D printing. In addition, the improvement techniques for 3D printed parts' mechanical characteristics are also to be developed [26].

The enhancement in additive manufacturing leads to the development of research on 3D printing concrete. The obstacles in scholarly collaboration and creative investigation are created due to researchers' information restraints. Accordingly, the creation and application of a process for the researchers/scientists to obtain important information from dependable sources are vital. Applying a scientometric method via the software may support overcoming this loophole and research gap. The main aim of the current study is to provide a detailed review of 3D printing methods with a focus on utilized materials, primary techniques applied, their applications, and the current state in different industry sectors. The research challenges and gaps in accepting 3D technology are also provided in this paper. The scientometric analysis of research published in 3D printing concrete up to 2022 is intended in this study. The quantitative evaluation of the bulk research dataset may be undertaken with the help of scientometric analysis using appropriate software [27,28].

The traditional review natured studies are somehow weak in their respective capability of connecting the various literature segments thoroughly and accurately. Co-occurrence, science mapping, and co-citation are key factors of exploration in the current era [29–31]. Identifying sources with co-occurrence of keywords, most research articles, the main credible researchers in terms of citations and papers, and actively engaged research areas in 3D printing concrete can also be performed with scientometric analysis. The bibliometric data of 953 related research articles is extracted using the Scopus dataset, which is determined afterwards using VOSviewer. The current study would assist academics in the engineering field belonging to various geographical locations in exchanging ground-breaking novel methods/ideas, creating joint ventures, and forming research alliances due to the graphical and statistical depiction of countries and authors.

2. Methodology

In this study, the scientometric analysis is carried out for the research dataset to evaluate the different aspects of bibliographic data [32–36]. Multiple studies have been conducted and reported on this matter depicting the questionable application of a reputable search engine. The two highly precise search engines, i.e., Web of Science and Scopus, are specifically explored for said aim [37,38]. The research data to conduct the current study on 3D printing concrete were collected using the academically highly recommended search engine, i.e., Scopus [39,40]. As of May 2022, the Scopus search for "3D printing concrete" found 1837 articles from 1998 to 2022. Multiple filters, depending upon preferences, are applied to avoid unnecessary data. The "journal research article", "journal review", "conference review", and "conference paper" are opted as the document type. The "source type" selected is "conference proceeding" and "Journal". The chosen period restriction for "publication year" is set to "2022", and "English" is set as "language" constraint. For more scrutiny, the "engineering", "environmental science", and "material science", are selected as "subject areas". Following the employment of said desirables, a total of 953 records are kept. Similarly, multiple studies have been conducted by using same method [41–45].

In academics, scientific mapping is developed to analyze bibliometric data, which is usually employed to analyze scientometric inquiries [46–48]. Comma separated values (CSV) files are used to save Scopus records for further determination with the help of a suitable software tool. The quantitative evaluation of the recovered records' literature and scientific visualization are generated using VOSviewer (version: 1.6.17). In academics, the VOSviewer is a majorly suggested and mainly used tool over a broader range of areas, and this mapping tool (open-source) has easy availability [49–52]. Therefore, the application of VOSviewer in the current study satisfies its goals. Loading of attained CSV files in VOSviewer is performed, and further evaluation is conducted to retain the consistency and integrity of data. At the same time, assessment of bibliographic data, countries' participation, the publication sources, the researchers having more citations and publications, the frequently appearing keywords, and the country's involvement are assessed. The various aspects and their co-occurrence and relationships are graphically represented, whereas the figures' statistics are listed in tables. Figure 1 presents the strategical flowchart for scientometric analysis.

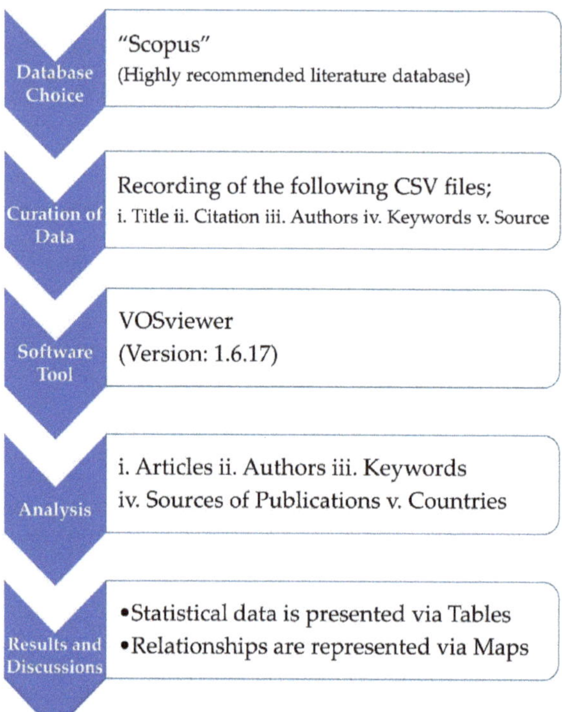

Figure 1. Sequential research methodology.

3. Analysis of Results

3.1. Annual Publications and Related Subject Areas

The analysis for discovering the most appropriate research areas is carried out by applying the Scopus analyzer. The three leading articles producing sections are engineering, materials science, and computer science, having almost 41%, 27%, and 8% articles, respectively, bearing the overall contribution of 76% depending on document count, as presented in Figure 2. Furthermore, Figure 3 shows the evaluation of paper type in the Scopus database for searched terms. According to the current study, conference review papers, journal review articles, and journal articles bear around 3%, 9%, 27%, and 61% of documents. The annual publication trend in the current research field from 1998 to 2022 is shown in Figure 4, as the occurrence of the first respective article was revealed in 1998. A mild rise in publication trend for the said research area, i.e., 3D printing concrete, is observed, with an approximate average of three annual articles until 2014. Afterwards, a gradual rise in annual publications is observed, with an approximate average of twenty articles per annum from 2015 to 2019. However, a significant enhancement in annual publications has been observed during the last three years, (i.e., 2020–2021). Recently, a drastic increase in 3D printing concrete for building and concrete research has been observed, depicting the initiative for all-rounded and comprehensive research work in the said field [53]. Scientific research globalization might be the reason behind increasing trend development in 3D printing concrete.

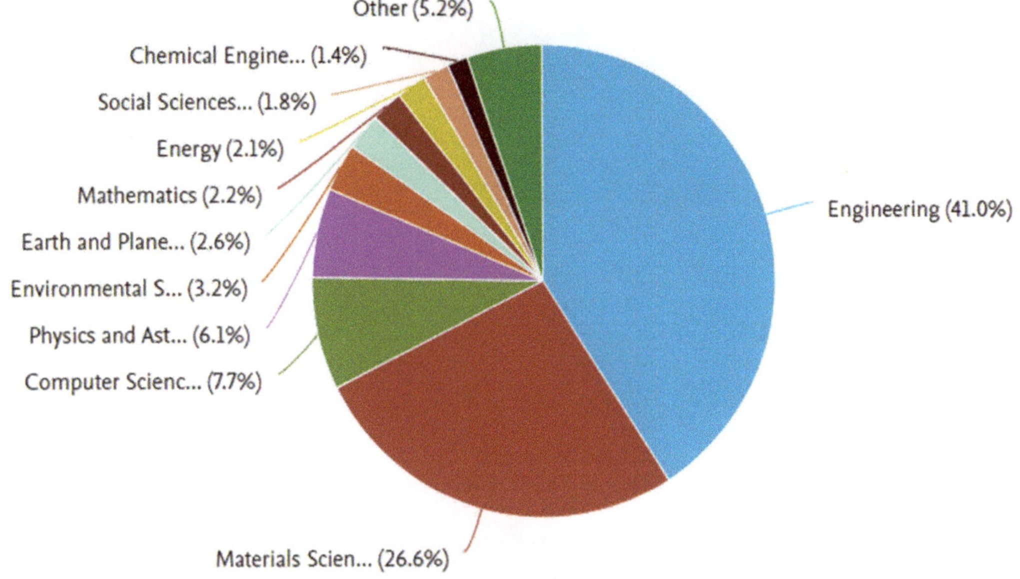

Figure 2. Articles subject area.

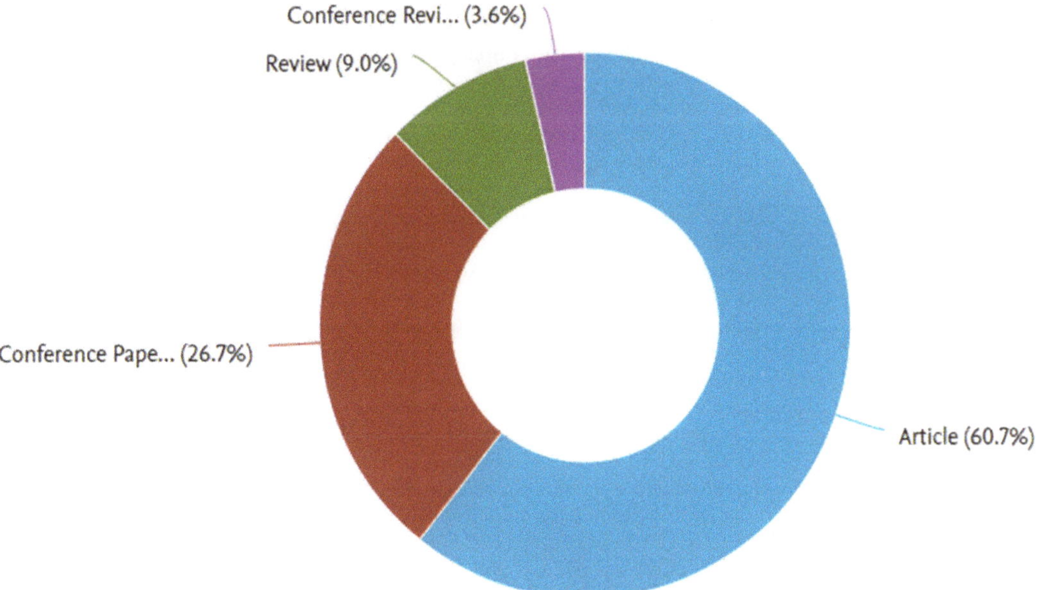

Figure 3. Published document types in the relevant field of study.

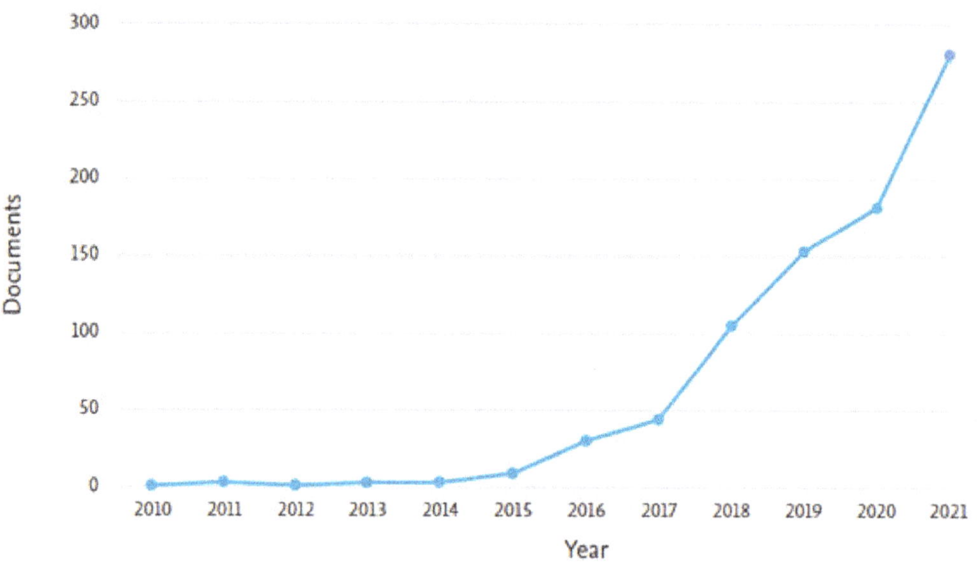

Figure 4. Annual publication trend of articles.

3.2. Publication Sources

The VOSviewer was utilized on the gathered bibliographic database to evaluate the published sources. While performing analysis, the sources are taken as "unit of analysis", whereas the "bibliographic coupling" is opted as a "kind of analysis". The minimum quantum of articles per source is set to ten. The sources of publication that met the said requirement are 14 out of 265. The publication sources are listed in Table 1, with at least ten published articles presenting data on 3D printing concrete until 2022 and the citation's quantum acquired in the said period. The three main journals/sources, depending upon the paper count, are i. "Construction and Building Materials" having 60 papers, ii. "Additive Manufacturing" has 39 documents, and iii. "Automation in Construction" with 35 articles. Furthermore, the primary three sources having overall maximum citations are "Automation in Construction", "Additive Manufacturing", and "Buildings", born 1580, 871, and 798 citations, respectively. Automation in construction also covers the aspect of multiple software applications such as building information modelling (BIM) for 3D printing concrete [54]. The coupling of 3D printing concrete with BIM to monitor and track novel variables was also performed by Azhar [55] and Bryde, et al. [56]. Combining 3D printing and BIM may make the creation of customized building components easier and facilitate sophisticated and complex design [53]. Davtalab, Kazemian and Khoshnevis [54] also declared that robotic construction is a construction industry revolution by using 3D printing concrete. This significant research exploration in the area of 3D printing concrete is come out to be the reason for intended scientometric analysis in the said research area. Further, conventional review studies are not enough to develop scientific visualization maps. The mapping journals with a minimum of ten articles in understudied research areas is presented in Figure 5. The quantum of research in 3D printing concrete in the form of articles is directly proportional to the size of the box showing the impact of the journal. The bigger the dimension of the box, the effect is more superior. For example, the biggest box in terms of sizes is for "Construction and Building Materials" showing the significant importance of this source in the considered field. Based on the type, five groups are developed, and all of them are offered in different colors, i.e., purple, red, green, blue, and yellow. The formation of groups is based on the similar article co-citations frequency [57]. The patterns of published articles' co-citation are the basis of group creation in VOSviewer. For example, the red group comprises three sources having frequent co-citations in similar works. In addition, the space among frames/journals in a group shows significant relationships compared to the other far-spaced frames. For example, "Additive Manufacturing" is more firmly correlated with "Rapid Prototyping Journal" than with "Materials Today: Proceedings".

Table 1. Sources of publication having minimum 10 publications in the considered research area till 2022.

S/N	Publication Source	Number of Publications	Total Number of Citations
1	*Automation in Construction*	35	1580
2	*Additive Manufacturing*	39	871
3	*Buildings*	13	798
4	*Lecture Notes in Civil Engineering*	22	727
5	*Materials and Design*	12	716
6	*Advanced Functional Materials*	12	381
7	*Construction and Building Materials*	60	309
8	*Cement and Concrete Research*	20	307
9	*Cement and Concrete Composites*	18	165
10	*Journal of Building Engineering*	12	131
11	*Rapid Prototyping Journal*	22	115
12	*Polymers*	10	74
13	*3D Printing and Additive Manufacturing*	11	71
14	*Materials*	45	56
15	*Applied Sciences (Switzerland)*	12	56

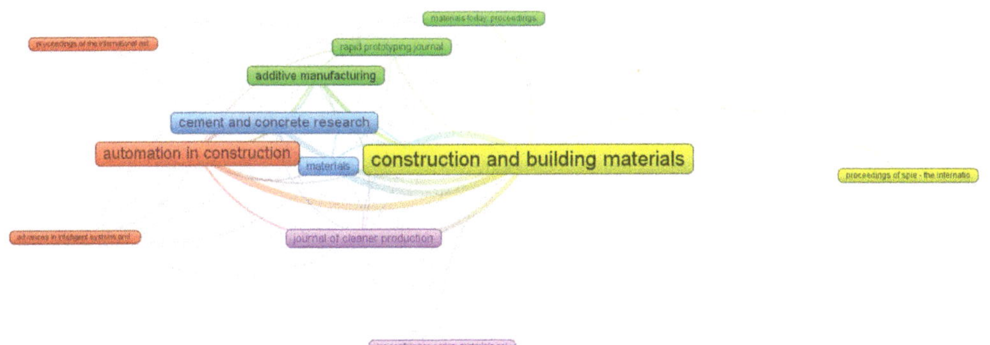

Figure 5. Scientific visualization of publication sources with at least 10 publications in the related research area.

3.3. Keywords

The fundamental subject of a study domain is highlighted and defined with the help of keywords in the research [58]. For the evaluation, the "analysis type" is selected as "co-occurrence", whereas the "analysis unit" is opted to "all keywords". The minimum repetition restriction is set at 20 for a keyword. Accordingly, 96 keywords out of 4185 are taken. The leading and most frequently used 20 keywords in published papers on relevant research areas are provided in Table 2. The terms 3D printers, concretes, 3-D printing, 3D printing, and concrete printings are among the most frequent five keywords in the considered area of research. As per the analysis of keywords, 3D printing concrete has been mainly studied for concrete mixtures, its rheology, and mechanical properties. Furthermore, it has also been explored for multiple types of building systems. Duballet et al. [59] classified the 3D printing concrete building systems based on five parameters: extrusion scale, object scale, printing environment, assembly parameter, and printing support. This classification was mainly featured for reinstating techniques apart from a single extrusion phase for concrete 3D printing at a larger scale. The keywords visualization map in terms of linkages, co-occurrences, and the occurrence frequency-related density is shown in Figure 6. The frequency of keywords is depicted by the size of the circle for the respective keyword, while the co-occurrence in papers is shown by its position (Figure 6a). It is evident from the graph that the comparatively bigger circles are for leading keywords depicting their significance for research on 3D printing. The formation of groups is also made for keywords to reflect the keywords' co-occurrence over several research publications. The multiple keywords' co-occurrence in published articles is the basis of color-coded grouping. Four different colors, i.e., green, red, yellow and blue, indicate the group's existence (Figure 6a). The concentrations for density of keywords are indicated by different colors (Figure 6b). The colors are aligned with respect to respective density concentrations. The red color shows the highest, whereas the blue color shows the lowest density concentration. Three-dimensional printers and concretes show red symbols depicting significant density concentration. This finding may aid ambitious researchers in selecting keywords that would enable the published data identification in a specific area.

Table 2. Fifty leading frequently used keywords in 3D printing concrete research.

S/N	Keyword	Occurrences
1	3DPrinters	558
2	Concretes	272
3	3-D Printing	209
4	3D Printing	188
5	Concrete Printings	184
6	3D Concrete Printing	154
7	Additive Manufacturing	139
8	Compressive Strength	104
9	Construction Industry	89
10	Additives	85
11	Rheology	81
12	Concrete	75
13	3D-Printing	70
14	Concrete Mixtures	70
15	Extrusion	66
16	Yield Stress	66
17	Reinforcement	61
18	Cementitious Materials	60
19	Cements	58
20	Reinforced Concrete	57
21	Mechanical Properties	55
22	Printing	51
23	Mixtures	50
24	Mortar	50
25	Tensile Strength	50
26	Concrete Products	48
27	Concrete Industry	47
28	Construction	45
29	Concrete Construction	39
30	Fly Ash	37
31	Rheological Property	37
32	Portland Cement	36
33	Geopolymers	35
34	Inorganic Polymers	35
35	Digital Fabrication	34
36	Structural Design	34
37	Buildability	32
38	Digital Construction	30
39	Sustainable Development	30
40	Mechanical Performance	28

Table 2. *Cont.*

S/N	Keyword	Occurrences
41	Geopolymer	27
42	Hardening	27
43	Shrinkage	26
44	Fabrication	25
45	Cement Based Material	24
46	Concrete Buildings	24
47	Fibers	24
48	Anisotropy	23
49	Binders	23
50	Concrete Additives	22

3.4. Authors

A researcher's influence in a specific study area is depicted from the citations [60]. Accordingly, the "co-authorship" is selected as a "kind of analysis", whereas; "authors" is chosen as the "unit of analysis" for the authors' assessment. The efficacy of a researcher is hard to determine while considering all parameters, such as total citations, the number of publications, and average citations. Contrary to this, a researcher's evaluation is performed by considering each factor independently, i.e., total citations, total publications, and average citations. The leading researcher is Tan, M.J., having 34 publications, followed by 29 publications each by Panda, B. and Mechtcherine, V. Afterward, Sanjayan, J. and Ma, G. are prominent, with 28 publications each. However, in terms of total citations, Tan, M.J leads the field with 2453 citations, followed by Panda, B. having 2362 citations in the 3D printing concrete research area. In addition, upon comparing the citations average, Paul, S.C. stands out with an average of 113, followed by Panda, B. having an average of 81 and Tan, M.J with a 72 average. The correlation between most eminent researchers and authors with a minimum of 10 publications is illustrated in Figure 7. The noticed largest network of interconnected researchers is seven. It is revealed from this analysis that a few researchers are inter-connected in terms of citations in the 3D printing concrete research area.

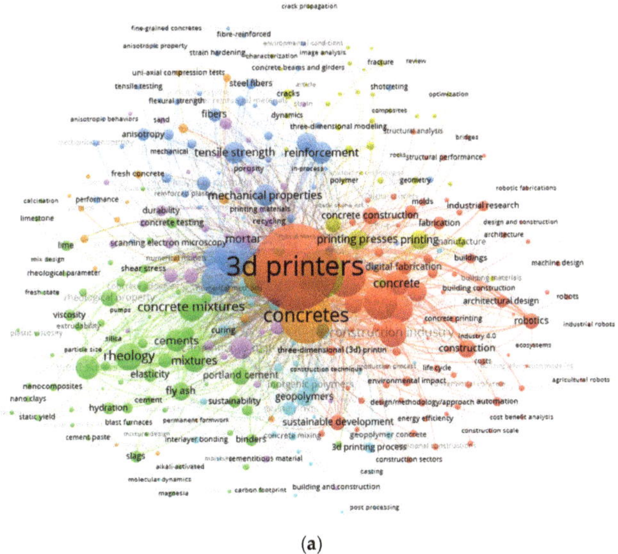

(**a**)

Figure 6. *Cont.*

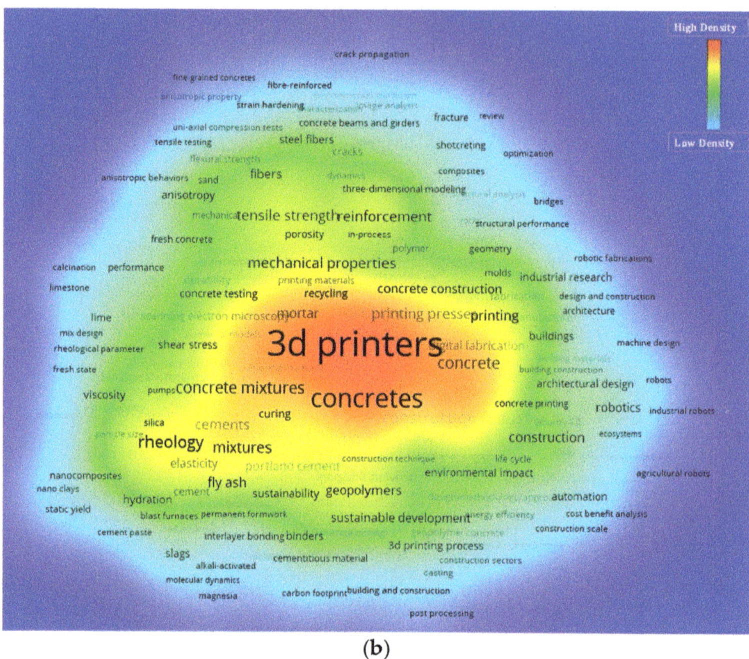

(b)

Figure 6. Analysis of keywords; (**a**) scientific visualization and (**b**) density visualization.

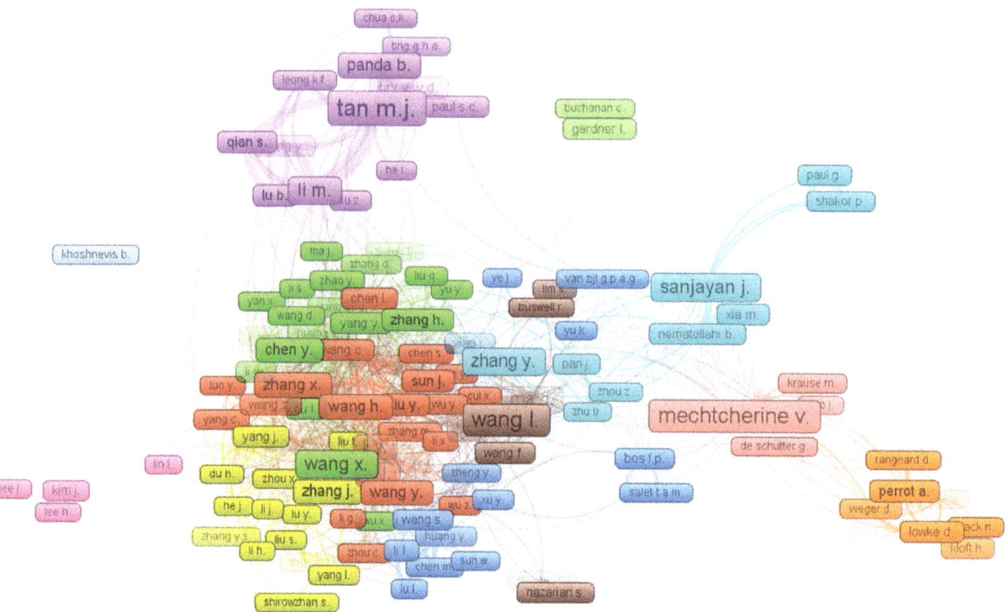

Figure 7. Scientific visualization of researchers who published articles in relevant research area.

3.5. Articles

The number of article citations influences a specific research area. Articles with a higher citation count are known as pioneers in relevant research areas. To evaluate articles, the "bibliographic coupling" is set for "kind of analysis", and "documents" is designated as "unit of analysis". The set limitation of most minor citations for an article is 50. In the 3D printing concrete research area, the top 10 articles, as per citations, are presented in Table 3 with respective citations and authors' detailing. Ngo, Kashani, Imbalzano, Nguyen and Hui [26] have 2520 citations for the research article titled; "Additive manufacturing (3D printing): A review of materials, methods, applications and challenges". For their relevant publications, Stansbury and Idacavage [61] and Buswell et al. [62] have 793 and 466 citations, respectively, and are placed in the first three positions. Furthermore, the linked articles mapping, based on citations and their density in the considered area, is shown in Figure 8. The inter-connected articles citation mapping is presented in Figure 8a, whereas, in Figure 8b, the enhancement of density concentration by top articles is revealed from the density mapping.

Table 3. Top 10 highly cited published articles up to 2021 in the research of RHA concrete.

S/N	Article	Title	Total Number of Citations Received
1	Ngo, Kashani, Imbalzano, Nguyen and Hui [26]	Additive manufacturing (3D printing): A review of materials, methods, applications and challenges	2520
2	Stansbury and Idacavage [61]	3D printing with polymers: Challenges among expanding options and opportunities	793
3	Buswell, De Silva, Jones and Dirrenberger [62]	3D printing using concrete extrusion: A roadmap for research	466
4	Bos, et al. [63]	Additive manufacturing of concrete in construction: potentials and challenges of 3D concrete printing	453
5	Gosselin, et al. [64]	Large-scale 3D printing of ultra-high performance concrete—a new processing route for architects and builders	424
6	Perrot, et al. [65]	Structural built-up of cement-based materials used for 3D-printing extrusion techniques	384
7	Tay, Panda, Paul, Noor Mohamed, Tan and Leong [53]	3D printing trends in building and construction industry: a review	310
8	De Schutter, et al. [66]	Vision of 3D printing with concrete—Technical, economic and environmental potentials	305
9	Kazemian, et al. [67]	Cementitious materials for construction-scale 3D printing: Laboratory testing of fresh printing mixture	285
10	Wolfs, et al. [68]	Early age mechanical behaviour of 3D printed concrete: Numerical modelling and experimental testing	284

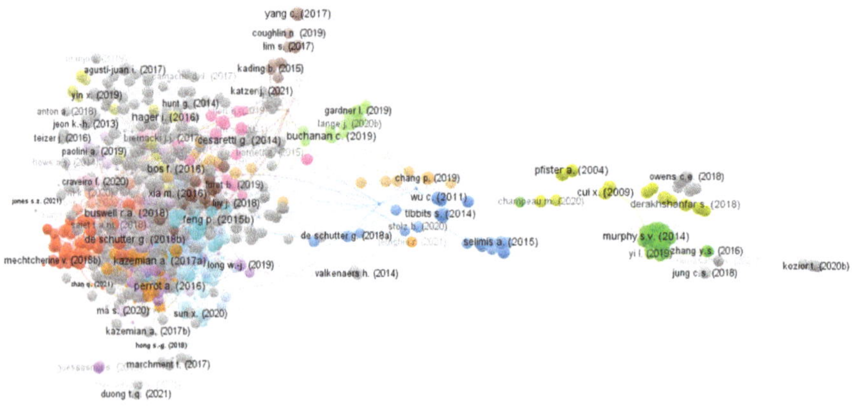

(**a**)

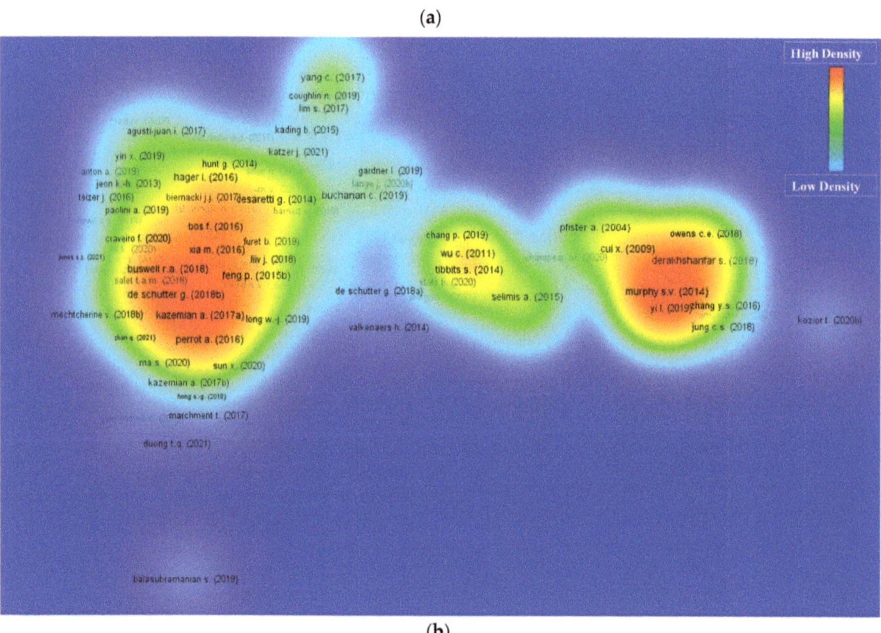

(**b**)

Figure 8. Published articles scientific mapping in relevant area of research until 2022; (**a**) connected documents based on citations (**b**) connected articles density.

3.6. Countries

The contribution of multiple countries is comparatively more towards 3D printing concrete research than others, and different expectations are there for enhancement in contribution. A network map is developed to help researchers access the areas related to 3D printing concrete research. Again, "Bibliographic coupling" is taken as a "kind of analysis", whereas, "countries" are opted for as a "unit of analysis". The limitation of the minor article for a nation is set at 10, and the countries met the desired limitation are 38 (Table 4). China, the United States, and Germany have the most articles with 377, 348, and 148 documents. Furthermore, the top three countries with the most considered research area citations of 10,514, 6179, and 3435 are the United States, China and Australia. The science mapping visualization and nation density inter-connected with citations is illustrated in Figure 9. The box size is directly proportional to a country's effect on the considered area of research (Figure 9a). The most engaging countries have more density, as illustrated in the map of density visualization (Figure 9b). It may be noted that the publication trend in developed countries such as the USA, China, Australia, Germany and UK is significantly more than that in developing countries such as India, Pakistan, etc. [69]. As in developed countries, there are diverse applications of 3D printing; however, in recent years, this technology is also gaining attention in countries. There is a huge potential for 3D printing in developing countries [70,71]. The graphical and statistical analysis of the contributing countries may help concerned scientists form joint ventures, develop scientific alliances, and exchange novel ideas and methods. Scientists from different countries contributing for enhancing research on 3D printing concrete may collaborate with specialists in the said research area and yield from their expertise.

Table 4. Leading countries in published articles 3D printing concrete research area until 2022.

S/N	Country	Number of Publications	Total Number of Citations
1	China	377	6179
2	United States	348	10,514
3	Germany	148	2813
4	United Kingdom	114	2540
5	Australia	113	3435
6	Singapore	72	2725
7	India	70	433
8	Russian Federation	69	132
9	South Korea	67	1268
10	France	58	1807
11	Italy	57	1022
12	Netherlands	56	1958
13	Poland	56	466
14	Spain	41	362
15	Canada	37	978
16	Belgium	30	698
17	Brazil	28	338
18	Japan	28	502
19	Switzerland	28	1557
20	Hong Kong	26	220
21	Czech Republic	25	113
22	Portugal	23	251
23	South Africa	21	382
24	Norway	20	540
25	United Arab Emirates	20	279
26	Romania	18	48
27	Taiwan	17	111
28	Greece	16	400
29	Iran	16	228
30	Turkey	15	50
31	Austria	14	70
32	Malaysia	13	110
33	Denmark	12	518
34	Sweden	12	60
35	Finland	11	56
36	Chile	10	35
37	Lebanon	10	119
38	Saudi Arabia	10	229

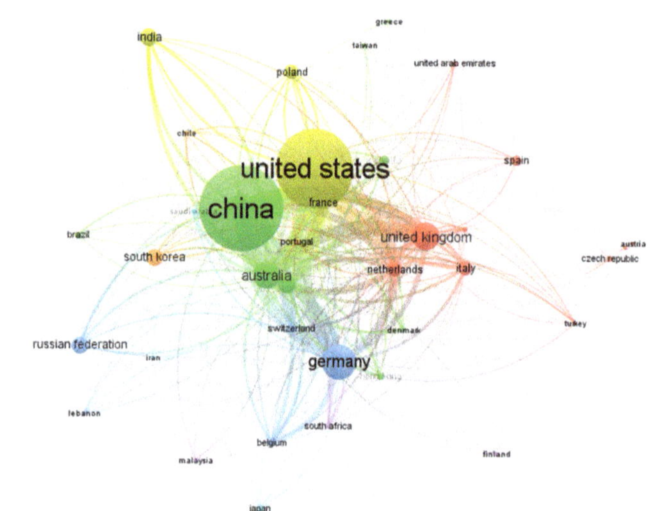

(a)

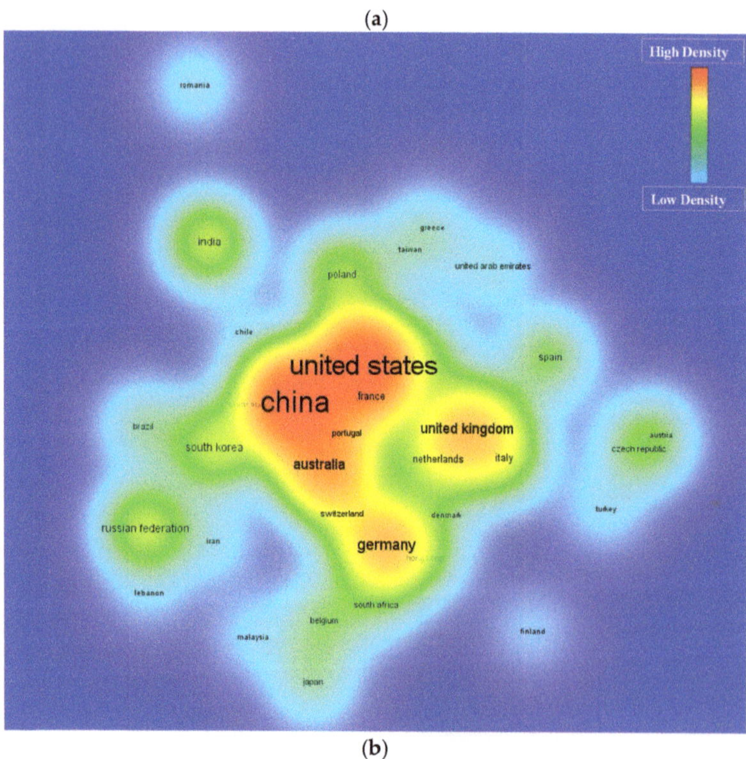

(b)

Figure 9. Scientific visualization regions having minimum ten publications in relevant research area until 2022 (**a**) network visualization and (**b**) density visualization.

4. Discussions and Future Perspectives

The mapping and statistical overview of different aspects of the 3D printing concrete literature are presented in the current study. The conventional and manually conducted review studies have limited capability in terms of comprehensiveness and precise interconnectivity among the various literature segments. The identification of most articles publishing journals, the frequently applied/used keywords in articles, countries having significant contributions, and authors and articles with most citations in the research field of 3D printing concrete is made in the current study. It is revealed from the keyword analysis that 3D printing concrete has been mainly explored in terms of its mechanical and rheological properties [72–76]. Furthermore, 3D printing is also researched for manufacturing geopolymer concrete [76–78]. Three-dimensional printing has various benefits upon utilization as concrete. The new prospects that can be utilized by 3D printing construction such as labor cost reduction, geometrical flexibility, efficiency improvement, safety and hard area construction [6,7]. In addition, physical construction consumes a bulk quantity of energy that comes out with higher CO_2 emissions [79]. As a result, there are rising concerns regarding natural resource depletion. Thus, 3D printing concrete reduces the cement requirement, resulting in sustainable construction with reduced CO_2 emissions [80–82]. The application of 3D printing concrete may also have resolved difficulties in waste management, specifically in the formwork [3,10]. The above-mentioned 3D printing concrete applications are yet in the phase of development. Detailed analyses are still required before their application broadens. Presently, the available research on 3D printing concrete is mainly based on their insight for extracting the optimal dosage of mix ingredients for desirable properties. Additionally, due to inferior properties and anisotropic behavior, the applicability of 3D printing at a larger scale is restricted. Therefore, it can be said that the information in said field is developing yet and needs to pass specific transition stages to accomplish optimum commercial applications and replace conventional manufacturing techniques. Therefore, the following research horizons in the field of 3D printing concrete may further be explored:

- Three-dimensional printed components' structural integrity, especially in regions vulnerable to natural disasters, seismic activity, military attacks and extreme climatic conditions, needs to be ensured by performing structural testing and developing specified standards and codes.
- Due to the provision of controlled environmental conditions for performing the experiment on 3D printed components, its behavior may not depict the true performance. As in real site conditions, the components are exposed to variable climatic factors such as humidity, temperature and precipitation, debris and dust, and varied lighting, etc. [83]. Therefore, the performance of 3D printing concrete components may be evaluated under real environmental conditions to ensure the global efficiency of this method.
- The effective implementation of the 3D printing approach is dependent on reliable, strong and printer-compatible materials [84,85]. Hence, the in situ materials that are available locally should be used for printing to have compatible and effective printing.
- Further research should also be conducted for large-scale building construction and experimentation to ensure the real capacity of 3D printing technology and to depict the application of this technology in the industry.
- Furthermore, nowadays, calcium carbonate ($CaCO_3$) whisker is used as micro-fiber in cementitious composites to improve the micromechanical properties of concrete [86–94]. Hence, the exploration of $CaCO_3$ whisker for 3D printing concrete would be an interesting horizon to explore.
- Three-dimensional printing is still a new technology; therefore, the information on its life cycle cost, including the maintenance and upfront costs, is limited yet. Further, there may be variations in the costs of design, planning, machinery, labor, and materials from country to country [95]. Therefore, a thorough life cycle cost analysis should be conducted for 3D printing technology to have detailed insight into its cost–benefit ratio with respect to conventional construction [96].

- Furthermore, the information regarding the life cycle assessment (LCA) of 3D printing concrete is also limited and demands thorough exploration in terms of its sustainability impact, preparation of material, construction, utilization, and ultimately the structures' demolition. This information is necessary to explore to have a clear picture of 3D printing concrete environmental impacts [97,98].

5. Conclusions

The abundance of scientific information produced in recent years, along with new communication channels, prompted the research community to propose the metric that gave origin to the new field of bibliometrics. This utilizes mathematical and statistical analysis techniques that permit getting dependable quality indicators. Thus, it is feasible to determine the number of documents published by an institution, nation, research group, or individual with the highest scientific output. A bibliometric study is an appropriate tool for identifying the volume and growth trend of literature focusing on concrete for the further 3D printing-related investigation that would be helpful for early-stage researchers.

The main aim of the current study is to perform a scientometric analysis of the literature available on 3D printing concrete to assess different measures. The Scopus database is enquired for 953 related articles, and the outcomes are evaluated by applying the VOSviewer program. It is revealed from the conducted analysis that the top three journals are "Construction and Building Materials", "Additive Manufacturing", and "Automation in Construction", having 60, 39, and 35 articles, respectively. Further, the top three journals having the most citations of 1580, 871, and 798 are "Automation in Construction", "Additive Manufacturing", and "Buildings", respectively. The analysis of keywords regarding the considered research area depicts that 3D printers, concretes, 3D printing, 3-D printing and concrete printing are the five most frequently appearing keywords. The keyword analysis revealed that 3D printing is mainly explored as concrete in the construction industry.

The top researchers are also classified based on the number of citations, publications, and average citations. Tan, M.J, with 34, Panda B., and Mechtcherine, V., with 29 each, and Sanjayan, J., and Ma, G., with 28 articles each, are among the leading three researchers with the most publications. With 2453 citations, Tan, M.J. leads the field, followed by 2362 citations of Panda, B. and 1441 citations o Bos, F.P. untill 2022. Furthermore, comparing average citations, the stand-out authors are Paul, S.C., who has almost 113, Tay, Y.W.D., who has around 95, and Panda, B., who has 85 average citations. In the analysis of articles related to 3D printing concrete, Ngo, Kashani, Imbalzano, Nguyen and Hui [26] have 2520 citations for the article "Additive manufacturing (3D printing): A review of materials, methods, applications and challenges". Stansbury and Idacavage [61] and Buswell, De Silva, Jones and Dirrenberger [62] have 793 and 466 citations for the respective publications and are among the best three.

The leading countries are also determined by their contribution to the 3D printing concrete research area. China, the United States, and Germany have contributed 377, 348, and 148 articles. Further, the countries, i.e., the United States, China, and Australia, have received citations of 10,514, 6179, and 3435, respectively. The 3D printing concrete application in the construction industry would develop sustainable construction by having reduced demand for cement, waste, and formwork requirements, ultimately saving natural sources and declining CO_2 emissions. The applicability of 3D printing concrete at a larger scale is still quite limited, and most of its applications are under exploration. Further analysis is also vital for broadening the effective applications of 3D printing concrete.

Author Contributions: C.H.: conceptualization, data curation, software, methodology, investigation, validation, writing—original draft. S.Z.: conceptualization, funding acquisition, resources, project administration, supervision, writing, reviewing, and editing. Y.L.: conceptualization, data curation, software, methodology, investigation, validation, writing—original draft. W.A.: methodology, investigation, supervision, writing, reviewing, and editing. F.A.: resources, funding acquisition, visualization, writing, reviewing, and editing. S.H.A.: resources, project administration, funding acquisition, validation, writing, reviewing, and editing. M.F.J.: formal analysis, resources, project administration, supervision, writing, reviewing, and editing. A.F.D.: validation, resources, visualization, writing, reviewing, and editing. All authors have read and agreed to the published version of the manuscript.

Funding: The authors are thankful to the Deanshipof Scientific Research at Najran University for funding this work under the Research Collaboration Funding program grant code (NU/RC/SERC/11/2).

Institutional Review Board Statement: Not applicable.

Informed Consent Statement: Not applicable.

Data Availability Statement: The data used in this research has been properly cited and reported in the main text.

Acknowledgments: This work was sponsored in part by The Key Scientific Research Project of Education Department of Hunan Province, China (No. 21A0189).The authors are thankful to the Deanship of Scientific Research at Najran University for funding this work under the Research Collaboration Funding program grant code (NU/RC/SERC/11/2).

Conflicts of Interest: The authors declare no conflict of interest.

References

1. Tibbits, S.; McKnelly, C.; Olguin, C.; Dikovsky, D.; Hirsch, S. 4D Printing and Universal Transformation. In Proceedings of the 34th Annual Conference of the Association for Computer Aided Design in Architecture (ACADIA), Los Angeles, CA, USA, 23–25 October 2014. [CrossRef]
2. Panda, B.; Tan, M.J.; Gibson, I.; Chua, C.K. The disruptive evolution of 3D printing. In Proceedings of the 2nd International Conference on Progress in Additive Manufacturing, Singapore, 17–19 May 2016; pp. 16–19.
3. Sanjayan, J.G.; Nematollahi, B.; Xia, M.; Marchment, T. Effect of surface moisture on inter-layer strength of 3D printed concrete. *Constr. Build. Mater.* **2018**, *172*, 468–475. [CrossRef]
4. Labonnote, N.; Rønnquist, A.; Manum, B.; Rüther, P. Additive construction: State-of-the-art, challenges and opportunities. *Autom. Constr.* **2016**, *72*, 347–366. [CrossRef]
5. Kim, K.; Park, S.; Kim, W.; Jeong, Y.; Lee, J. Evaluation of Shear Strength of RC Beams with Multiple Interfaces Formed before Initial Setting Using 3D Printing Technology. *Materials* **2017**, *10*, 1349. [CrossRef]
6. Panda, B.; Tan, M.-J. Rheological behavior of high volume fly ash mixtures containing micro silica for digital construction application. *Mater. Lett.* **2018**, *237*, 348–351. [CrossRef]
7. Soltan, D.G.; Li, V.C. A self-reinforced cementitious composite for building-scale 3D printing. *Cem. Concr. Compos.* **2018**, *90*, 1–13. [CrossRef]
8. Katzer, J.; Szatkiewicz, T. Properties of concrete elements with 3-D printed formworks which substitute steel reinforcement. *Constr. Build. Mater.* **2019**, *210*, 157–161. [CrossRef]
9. Hager, I.; Golonka, A.; Putanowicz, R. 3D printing of buildings and building components as the future of sustainable construction? *Procedia Eng.* **2016**, *151*, 292–299. [CrossRef]
10. Nematollahi, B.; Xia, M.; Sanjayan, J. Current progress of 3D concrete printing technologies, ISARC. In Proceedings of the International Symposium on Automation and Robotics in Construction, Taipei, Taiwan, 28 June–1 July 2017.
11. Yao, Y.; Hu, M.; Di Maio, F.; Cucurachi, S. Life cycle assessment of 3D printing geo-polymer concrete: An ex-ante study. *J. Ind. Ecol.* **2020**, *24*, 116–127. [CrossRef]
12. Farooqi, M.U.; Ali, M. Effect of pre-treatment and content of wheat straw on energy absorption capability of concrete. *Constr. Build. Mater.* **2019**, *224*, 572–583. [CrossRef]
13. Farooqi, M.U.; Ali, M. Effect of Fibre Content on Splitting-Tensile Strength of Wheat Straw Reinforced Concrete for Pavement Applications. *Key Eng. Mater.* **2018**, *765*, 349–354. [CrossRef]
14. Farooqi, M.U.; Ali, M. Effect of Fibre Content on Compressive Strength of Wheat Straw Reinforced Concrete for Pavement Applications. *IOP Conf. Ser. Mater. Sci. Eng.* **2018**, *422*, 012014. [CrossRef]
15. Awoyera, P.O.; Ndambuki, J.M.; Akinmusuru, J.O.; Omole, D. Characterization of ceramic waste aggregate concrete. *HBRC J.* **2018**, *14*, 282–287. [CrossRef]
16. Ahmad, W.; Farooq, S.H.; Usman, M.; Khan, M.; Ahmad, A.; Aslam, F.; Al Yousef, R.; Al Abduljabbar, H.; Sufian, M. Effect of Coconut Fiber Length and Content on Properties of High Strength Concrete. *Materials* **2020**, *13*, 1075. [CrossRef] [PubMed]
17. Li, X.; Qin, D.; Hu, Y.; Ahmad, W.; Ahmad, A.; Aslam, F.; Joyklad, P. A systematic review of waste materials in cement-based composites for construction applications. *J. Build. Eng.* **2021**, *45*, 103447. [CrossRef]
18. Alyousef, R.; Ahmad, W.; Ahmad, A.; Aslam, F.; Joyklad, P.; Alabduljabbar, H. Potential use of recycled plastic and rubber aggregate in cementitious materials for sustainable construction: A review. *J. Clean. Prod.* **2021**, *329*, 129736. [CrossRef]
19. Althoey, F.; Hosen, M.A. Physical and mechanical characteristics of sustainable concrete comprising industrial waste materials as a replacement of conventional aggregate. *Sustainability* **2021**, *13*, 4306. [CrossRef]

20. Althoey, F.; Farnam, Y. Performance of Calcium Aluminate Cementitious Materials in the Presence of Sodium Chloride. *J. Mater. Civ. Eng.* **2020**, *32*, 1–10. [CrossRef]

21. Ali, E.; Althoey, F. A Simplified Stress Analysis of Functionally Graded Beams and Influence of Material Function on Deflection. *Appl. Sci.* **2021**, *11*, 11747.

22. Althoey, F.; Balapour, M.; Farnam, Y. Reducing detrimental sulfate-based phase formation in concrete exposed to sodium chloride using supplementary cementitious materials. *J. Build. Eng.* **2022**, *45*, 103639. [CrossRef]

23. Korniejenko, K.; Łach, M. Geopolymers reinforced by short and long fibres–innovative materials for additive manufacturing. *Curr. Opin. Chem. Eng.* **2020**, *28*, 167–172. [CrossRef]

24. Ivanova, O.; Williams, C.; Campbell, T.A. Additive manufacturing (AM) and nanotechnology: Promises and challenges. *Rapid Prototyp. J.* **2013**, *19*, 353–364. [CrossRef]

25. Vaezi, M.; Seitz, H.; Yang, S. A review on 3D micro-additive manufacturing technologies. *Int. J. Adv. Manuf. Technol.* **2013**, *67*, 1721–1754. [CrossRef]

26. Ngo, T.D.; Kashani, A.; Imbalzano, G.; Nguyen, K.T.Q.; Hui, D. Additive Manufacturing (3D Printing): A Review of Materials, Methods, Applications and Challenges. *Compos. Part B Eng.* **2018**, *143*, 172–196. [CrossRef]

27. Amin, M.N.; Ahmad, W.; Khan, K.; Sayed, M.M. Mapping Research Knowledge on Rice Husk Ash Application in Concrete: A Scientometric Review. *Materials* **2022**, *15*, 3431. [CrossRef] [PubMed]

28. Huang, S.; Wang, H.; Ahmad, W.; Ahmad, A.; Vatin, N.I.; Mohamed, A.M.; Deifalla, A.F.; Mehmood, I. Plastic Waste Management Strategies and Their Environmental Aspects: A Scientometric Analysis and Comprehensive Review. *Int. J. Environ. Res. Public Health* **2022**, *19*, 4556. [CrossRef]

29. Zakka, W.P.; Lim, N.H.A.S.; Khun, M.C. A scientometric review of geopolymer concrete. *J. Clean. Prod.* **2020**, *280*, 124353. [CrossRef]

30. Udomsap, A.D.; Hallinger, P. A bibliometric review of research on sustainable construction, 1994–2018. *J. Clean. Prod.* **2020**, *254*, 120073. [CrossRef]

31. Yang, H.; Liu, L.; Yang, W.; Liu, H.; Ahmad, W.; Ahmad, A.; Aslam, F.; Joyklad, P. A comprehensive overview of geopolymer composites: A bibliometric analysis and literature review. *Case Stud. Constr. Mater.* **2021**, *16*, e00830. [CrossRef]

32. Xu, Y.; Zeng, J.; Chen, W.; Jin, R.; Li, B.; Pan, Z. A holistic review of cement composites reinforced with graphene oxide. *Constr. Build. Mater.* **2018**, *171*, 291–302. [CrossRef]

33. Xiao, X.; Skitmore, M.; Li, H.; Xia, B. Mapping Knowledge in the Economic Areas of Green Building Using Scientometric Analysis. *Energies* **2019**, *12*, 3011. [CrossRef]

34. Darko, A.; Chan, A.P.; Huo, X.; Owusu-Manu, D.-G. A scientometric analysis and visualization of global green building research. *Build. Environ.* **2018**, *149*, 501–511. [CrossRef]

35. Zhang, B.; Ahmad, W.; Ahmad, A.; Aslam, F.; Joyklad, P. A scientometric analysis approach to analyze the present research on recycled aggregate concrete. *J. Build. Eng.* **2021**, *46*, 103679. [CrossRef]

36. Ahmad, W.; Ahmad, A.; Ostrowski, K.A.; Aslam, F.; Joyklad, P. A scientometric review of waste material utilization in concrete for sustainable construction. *Case Stud. Constr. Mater.* **2021**, *15*, e00683. [CrossRef]

37. Aghaei Chadegani, A.; Salehi, H.; Yunus, M.M.; Farhadi, H.; Fooladi, M.; Farhadi, M.; Ale Ebrahim, N. A comparison between two main academic literature collections: Web of science and scopus databases. *Asian Soc. Sci.* **2013**, *9*, 18–26. [CrossRef]

38. Afgan, S.; Bing, C. Scientometric review of international research trends on thermal energy storage cement based composites via integration of phase change materials from 1993 to 2020. *Constr. Build. Mater.* **2021**, *278*, 122344. [CrossRef]

39. Bergman, E.M.L. Finding Citations to Social Work Literature: The Relative Benefits of Using Web of Science, Scopus, or Google Scholar. *J. Acad. Libr.* **2012**, *38*, 370–379. [CrossRef]

40. Meho, L.I. Using Scopus's CiteScore for assessing the quality of computer science conferences. *J. Inf.* **2019**, *13*, 419–433. [CrossRef]

41. Zuo, J.; Zhao, Z.-Y. Green building research–current status and future agenda: A review. *Renew. Sustain. Energy Rev.* **2014**, *30*, 271–281. [CrossRef]

42. Darko, A.; Zhang, C.; Chan, A.P. Drivers for green building: A review of empirical studies. *Habitat Int.* **2017**, *60*, 34–49. [CrossRef]

43. Ahmad, W.; Khan, M.; Smarzewski, P. Effect of Short Fiber Reinforcements on Fracture Performance of Cement-Based Materials: A Systematic Review Approach. *Materials* **2021**, *14*, 1745. [CrossRef]

44. Luhar, I.; Luhar, S.; Abdullah, M.M.A.B.; Razak, R.A.; Vizureanu, P.; Sandu, A.V.; Matasaru, P.-D. A State-of-the-Art Review on Innovative Geopolymer Composites Designed for Water and Wastewater Treatment. *Materials* **2021**, *14*, 7456. [CrossRef] [PubMed]

45. Pal, K.; Sarkar, P.; Anis, A.; Wiszumirska, K.; Jarzębski, M. Polysaccharide-Based Nanocomposites for Food Packaging Applications. *Materials* **2021**, *14*, 5549. [CrossRef] [PubMed]

46. Markoulli, M.P.; Lee, C.I.; Byington, E.; Felps, W.A. Mapping Human Resource Management: Reviewing the field and charting future directions. *Hum. Resour. Manag. Rev.* **2017**, *27*, 367–396. [CrossRef]

47. Amin, M.N.; Ahmad, W.; Khan, K.; Ahmad, A. A Comprehensive Review of Types, Properties, Treatment Methods and Application of Plant Fibers in Construction and Building Materials. *Materials* **2022**, *15*, 4362. [CrossRef] [PubMed]

48. Maier, D. Building Materials Made of Wood Waste a Solution to Achieve the Sustainable Development Goals. *Materials* **2021**, *14*, 7638. [CrossRef]

49. Jin, R.; Gao, S.; Cheshmehzangi, A.; Aboagye-Nimo, E. A holistic review of off-site construction literature published between 2008 and 2018. *J. Clean. Prod.* **2018**, *202*, 1202–1219. [CrossRef]

50. Park, J.Y.; Nagy, Z. Comprehensive analysis of the relationship between thermal comfort and building control research—A data-driven literature review. *Renew. Sustain. Energy Rev.* **2017**, *82*, 2664–2679. [CrossRef]

51. Oraee, M.; Hosseini, M.R.; Papadonikolaki, E.; Palliyaguru, R.; Arashpour, M. Collaboration in BIM-based construction networks: A bibliometric-qualitative literature review. *Int. J. Proj. Manag.* **2017**, *35*, 1288–1301. [CrossRef]

52. Van Eck, N.J.; Waltman, L. Software survey: VOSviewer, a computer program for bibliometric mapping. *Scientometrics* **2009**, *84*, 523–538. [CrossRef]

53. Tay, Y.W.D.; Panda, B.; Paul, S.C.; Mohamed, N.A.N.; Tan, M.J.; Leong, K.F. 3D printing trends in building and construction industry: A review. *Virtual Phys. Prototyp.* **2017**, *12*, 261–276. [CrossRef]

54. Davtalab, O.; Kazemian, A.; Khoshnevis, B. Perspectives on a BIM-integrated software platform for robotic construction through Contour Crafting. *Autom. Constr.* **2018**, *89*, 13–23. [CrossRef]

55. Azhar, S. Building information modeling (BIM): Trends, benefits, risks, and challenges for the AEC industry. *Leadersh. Manag. Eng.* **2011**, *11*. [CrossRef]

56. Bryde, D.; Broquetas, M.; Volm, J.M. The project benefits of Building Information Modelling (BIM). *Int. J. Proj. Manag.* **2013**, *31*, 971–980. [CrossRef]

57. Wuni, I.Y.; Shen, G.Q.; Osei-Kyei, R. Scientometric review of global research trends on green buildings in construction journals from 1992 to 2018. *Energy Build.* **2019**, *190*, 69–85. [CrossRef]

58. Su, H.-N.; Lee, P.-C. Mapping knowledge structure by keyword co-occurrence: A first look at journal papers in Technology Foresight. *Scientometrics* **2010**, *85*, 65–79. [CrossRef]

59. Duballet, R.; Baverel, O.; Dirrenberger, J. Classification of building systems for concrete 3D printing. *Autom. Constr.* **2017**, *83*, 247–258. [CrossRef]

60. Yu, F.; Hayes, B.E. Applying data analytics and visualization to assessing the research impact of the Cancer Cell Biology (CCB) Program at the University of North Carolina at Chapel Hill. *J. E Sci. Librariansh.* **2018**, *7*, 4. [CrossRef]

61. Stansbury, J.W.; Idacavage, M.J. 3D Printing with Polymers: Challenges among Expanding Options and Opportunities. *Dent. Mater.* **2016**, *32*, 54–64. [CrossRef]

62. Buswell, R.A.; De Silva, W.R.L.; Jones, S.Z.; Dirrenberger, J. 3D printing using concrete extrusion: A roadmap for research. *Cem. Concr. Res.* **2018**, *112*, 37–49. [CrossRef]

63. Bos, F.; Wolfs, R.; Ahmed, Z.; Salet, T. Additive manufacturing of concrete in construction: Potentials and challenges of 3D concrete printing. *Virtual Phys. Prototyp.* **2016**, *11*, 209–225. [CrossRef]

64. Gosselin, C.; Duballet, R.; Roux, P.; Gaudillière, N.; Dirrenberger, J.; Morel, P. Large-scale 3D printing of ultra-high performance concrete—A new processing route for architects and builders. *Mater. Des.* **2016**, *100*, 102–109. [CrossRef]

65. Perrot, A.; Rangeard, D.; Pierre, A. Structural built-up of cement-based materials used for 3D-printing extrusion techniques. *Mater. Struct.* **2015**, *49*, 1213–1220. [CrossRef]

66. De Schutter, G.; Lesage, K.; Mechtcherine, V.; Nerella, V.N.; Habert, G.; Agusti-Juan, I. Vision of 3D printing with concrete—Technical, economic and environmental potentials. *Cem. Concr. Res.* **2018**, *112*, 25–36. [CrossRef]

67. Kazemian, A.; Yuan, X.; Cochran, E.; Khoshnevis, B. Cementitious materials for construction-scale 3D printing: Laboratory testing of fresh printing mixture. *Constr. Build. Mater.* **2017**, *145*, 639–647. [CrossRef]

68. Wolfs, R.J.M.; Bos, F.P.; Salet, T.A.M. Early age mechanical behaviour of 3D printed concrete: Numerical modelling and experimental testing. *Cem. Concr. Res.* **2018**, *106*, 103–116. [CrossRef]

69. Schuldt, S.J.; Jagoda, J.A.; Hoisington, A.J.; Delorit, J.D. A systematic review and analysis of the viability of 3D-printed construction in remote environments. *Autom. Constr.* **2021**, *125*, 103642. [CrossRef]

70. Geneidy, O.; Ismaeel, W.S.; Abbas, A. A critical review for applying three-dimensional concrete wall printing technology in Egypt. *Arch. Sci. Rev.* **2019**, *62*, 438–452. [CrossRef]

71. Ahmed, A.; Azam, A.; Bhutta, M.M.A.; Khan, F.A.; Aslam, R.; Tahir, Z. Discovering the technology evolution pathways for 3D printing (3DP) using bibliometric investigation and emerging applications of 3DP during COVID-19. *Clean. Environ. Syst.* **2021**, *3*, 100042. [CrossRef]

72. Zahabizadeh, B.; Pereira, J.; Gonçalves, C.; Pereira, E.N.B.; Cunha, V.M.C.F. Influence of the printing direction and age on the mechanical properties of 3D printed concrete. *Mater. Struct.* **2021**, *54*, 73. [CrossRef]

73. Alchaar, A.S.; Al-Tamimi, A.K. Mechanical properties of 3D printed concrete in hot temperatures. *Constr. Build. Mater.* **2020**, *266*, 120991. [CrossRef]

74. Paul, S.C.; Tay, Y.W.D.; Panda, B.; Tan, M.J. Fresh and hardened properties of 3D printable cementitious materials for building and construction. *Arch. Civ. Mech. Eng.* **2018**, *18*, 311–319. [CrossRef]

75. Bohuchval, M.; Sonebi, M.; Amziane, S.; Perrot, A. Rheological properties of 3D printing concrete containing sisal fibres. *Acad. J. Civ. Eng.* **2019**, *37*, 249–255.

76. Li, L.; Wei, Y.-J.; Li, Z.; Farooqi, M.U. Rheological and viscoelastic characterizations of fly ash/slag/silica fume-based geopolymer. *J. Clean. Prod.* **2022**, *354*, 131629. [CrossRef]

77. Ziejewska, C.; Marczyk, J.; Korniejenko, K.; Bednarz, S.; Sroczyk, P.; Łach, M.; Mikuła, J.; Figiela, B.; Szechyńska-Hebda, M.; Hebda, M. 3D Printing of Concrete-Geopolymer Hybrids. *Materials* **2022**, *15*, 2819. [CrossRef] [PubMed]

78. Ilcan, H.; Sahin, O.; Kul, A.; Yildirim, G.; Sahmaran, M. Rheological properties and compressive strength of construction and demolition waste-based geopolymer mortars for 3D-Printing. *Constr. Build. Mater.* **2022**, *328*, 127114. [CrossRef]

79. Gustavsson, L.; Sathre, R. Variability in energy and carbon dioxide balances of wood and concrete building materials. *Build. Environ.* **2006**, *41*, 940–951. [CrossRef]

80. Gislason, S.; Bruhn, S.; Breseghello, L.; Sen, B.; Liu, G.; Naboni, R. *Lightweight 3D Printed Concrete Beams Show an Environmental Promise: A Cradle-to-Grave Comparative Life Cycle Assessment*; Springer: Berlin/Heidelberg, Germany, 2022.

81. Bedarf, P.; Szabo, A.; Zanini, M.; Heusi, A.; Dillenburger, B. Robotic 3D printing of mineral foam for a lightweight composite concrete slab. In Proceedings of the 27th CAADRIA Conference, Sydney, NSW, Australia, 9–15 April 2022; pp. 61–70.

82. Mohan, M.K.; Rahul, A.; van Dam, B.; Zeidan, T.; De Schutter, G.; Van Tittelboom, K. Performance criteria, environmental impact and cost assessment for 3D printable concrete mixtures. *Resour. Conserv. Recycl.* **2022**, *181*, 106255. [CrossRef]

83. Kazemian, A.; Yuan, X.; Davtalab, O.; Khoshnevis, B. Computer vision for real-time extrusion quality monitoring and control in robotic construction. *Autom. Constr.* **2019**, *101*, 92–98. [CrossRef]

84. Panda, B.; Tay, Y.W.D.; Paul, S.C.; Tan, M.J. Current challenges and future potential of 3D concrete printing: Aktuelle Herausforderungen und Zukunftspotenziale des 3D-Druckens bei Beton. *Mater. Werkst.* **2018**, *49*, 666–673. [CrossRef]

85. Tay, Y.W.; Panda, B.; Paul, S.C.; Tan, M.J.; Qian, S.Z.; Leong, K.F.; Chua, C.K. Processing and Properties of Construction Materials for 3D Printing. *Mater. Sci. Forum* **2016**, *861*, 177–181. [CrossRef]

86. Khan, M.; Cao, M.; Chu, S.; Ali, M. Properties of hybrid steel-basalt fiber reinforced concrete exposed to different surrounding conditions. *Constr. Build. Mater.* **2022**, *322*, 126340. [CrossRef]

87. Cao, M.; Khan, M.; Ahmed, S. Effectiveness of Calcium Carbonate Whisker in Cementitious Composites. *Period. Polytech. Civ. Eng.* **2020**, *64*, 265–275. [CrossRef]

88. Khan, M.; Cao, M.; Ai, H.; Hussain, A. Basalt Fibers in Modified Whisker Reinforced Cementitious Composites. *Period. Polytech. Civ. Eng.* **2022**, *66*, 344–354. [CrossRef]

89. Xie, C.; Cao, M.; Guan, J.; Liu, Z.; Khan, M. Improvement of boundary effect model in multi-scale hybrid fibers reinforced cementitious composite and prediction of its structural failure behavior. *Compos. Part B Eng.* **2021**, *224*, 109219. [CrossRef]

90. Cao, M.; Khan, M. Effectiveness of multiscale hybrid fiber reinforced cementitious composites under single degree of freedom hydraulic shaking table. *Struct. Concr.* **2020**, *22*, 535–549. [CrossRef]

91. Khan, M.; Cao, M.; Xie, C.; Ali, M. Effectiveness of hybrid steel-basalt fiber reinforced concrete under compression. *Case Stud. Constr. Mater.* **2022**, *16*, e00941. [CrossRef]

92. Khan, M.; Cao, M.; Xie, C.; Ali, M. Hybrid fiber concrete with different basalt fiber length and content. *Struct. Concr.* **2021**, *23*, 346–364. [CrossRef]

93. Cao, M.; Xie, C.; Li, L.; Khan, M. Effect of different PVA and steel fiber length and content on mechanical properties of $CaCO_3$ whisker reinforced cementitious composites. *Mater. Construcción* **2019**, *69*, e200. [CrossRef]

94. Khan, M.; Cao, M.; Hussain, A.; Chu, S. Effect of silica-fume content on performance of $CaCO_3$ whisker and basalt fiber at matrix interface in cement-based composites. *Constr. Build. Mater.* **2021**, *300*, 124046. [CrossRef]

95. Rouhana, C.M.; Aoun, M.S.; Faek, F.S.; Eljazzar, M.S.; Hamzeh, F.R. The reduction of construction duration by implementing contourontour crafting (3D printing). In Proceedings of the 22nd Annual Conference of the International Group for Lean Construction: Understanding and Improving Project Based Production, IGLC, Oslo, Norway, 25–27 June 2014; pp. 1031–1042.

96. Wu, P.; Zhao, X.; Baller, J.H.; Wang, X. Developing a conceptual framework to improve the implementation of 3D printing technology in the construction industry. *Arch. Sci. Rev.* **2018**, *61*, 133–142. [CrossRef]

97. Ford, S.; Despeisse, M. Additive manufacturing and sustainability: An exploratory study of the advantages and challenges. *J. Clean. Prod.* **2016**, *137*, 1573–1587. [CrossRef]

98. Verhoef, L.A.; Budde, B.W.; Chockalingam, C.; Nodar, B.G.; van Wijk, A.J. The effect of additive manufacturing on global energy demand: An assessment using a bottom-up approach. *Energy Policy* **2018**, *112*, 349–360. [CrossRef]

 *materials*

Article

Investigating the Ultrasonic Pulse Velocity of Concrete Containing Waste Marble Dust and Its Estimation Using Artificial Intelligence

Dawei Yang [1,*], Jiahui Zhao [1], Salman Ali Suhail [2], Waqas Ahmad [3,*], Paweł Kamiński [4], Artur Dyczko [5], Abdelatif Salmi [6] and Abdullah Mohamed [7]

[1] Civil & Architecture Engineering, Xi'an Technological University, Xi'an 710021, China; xingqitian@sina.com
[2] Department of Civil Engineering, University of Lahore (UOL), 1-Km Defence Road, near Bhuptian Chowk, Lahore 54000, Pakistan; salmanalisuhail@gmail.com
[3] Department of Civil Engineering, COMSATS University Islamabad, Abbottabad 22060, Pakistan
[4] Faculty of Civil Engineering and Resource Management, AGH University of Science and Technology, 30-059 Krakow, Poland; pkamin@agh.edu.pl
[5] Mineral and Energy Economy Research Institute of the Polish Academy of Sciences, J. Wybickiego 7a, 31-261 Krakow, Poland; arturdyczko@min-pan.krakow.pl
[6] Department of Civil Engineering, College of Engineering, Prince Sattam bin Abdulaziz University, Al-Kharj 16273, Saudi Arabia; a.salmi@psau.edu.sa
[7] Research Centre, Future University in Egypt, New Cairo 11845, Egypt; mohamed.a@fue.edu.eg
* Correspondence: gaoxinlu001@sina.com (D.Y.); waqasahmad@cuiatd.edu.pk (W.A.)

Citation: Yang, D.; Zhao, J.; Suhail, S.A.; Ahmad, W.; Kamiński, P.; Dyczko, A.; Salmi, A.; Mohamed, A. Investigating the Ultrasonic Pulse Velocity of Concrete Containing Waste Marble Dust and Its Estimation Using Artificial Intelligence. *Materials* **2022**, *15*, 4311. https://doi.org/10.3390/ma15124311

Academic Editors: Rui Vasco Silva and Jorge Otero

Received: 3 April 2022
Accepted: 20 May 2022
Published: 17 June 2022

Publisher's Note: MDPI stays neutral with regard to jurisdictional claims in published maps and institutional affiliations.

Copyright: © 2022 by the authors. Licensee MDPI, Basel, Switzerland. This article is an open access article distributed under the terms and conditions of the Creative Commons Attribution (CC BY) license (https://creativecommons.org/licenses/by/4.0/).

Abstract: Researchers and engineers are presently focusing on efficient waste material utilization in the construction sector to reduce waste. Waste marble dust has been added to concrete to minimize pollution and landfills problems. Therefore, marble dust was utilized in concrete, and its prediction was made via an artificial intelligence approach to give an easier way to scholars for sustainable construction. Various blends of concrete having 40 mixes were made as partial substitutes for waste marble dust. The ultrasonic pulse velocity of waste marble dust concrete (WMDC) was compared to a control mix without marble dust. Additionally, this research used standalone (multiple-layer perceptron neural network) and supervised machine learning methods (Bagging, AdaBoost, and Random Forest) to predict the ultrasonic pulse velocity of waste marble dust concrete. The models' performances were assessed using R^2, RMSE, and MAE. Then, the models' performances were validated using k-fold cross-validation. Furthermore, the effect of raw ingredients and their interactions using SHAP analysis was evaluated. The Random Forest model, with an R^2 of 0.98, outperforms the MLPNN, Bagging, and AdaBoost models. Compared to all the other models (individual and ensemble), the Random Forest model with greater R^2 and lower error (RMSE, MAE) has a superior performance. SHAP analysis revealed that marble dust content has a positive and direct influence on and relationship to the ultrasonic pulse velocity of concrete. Using machine learning to forecast concrete properties saves time, resources, and effort for scholars in the engineering sector.

Keywords: waste; marble dust; building materials; mortar; concrete

1. Introduction

Keeping in mind sustainable development, the need is to curtail excessive industrial processes, along with the enhancement of cost efficiency in parallel with a reduction in environmental pollution [1]. Industrial waste, when incorporated in concrete, can contribute towards sustainable development in terms of environmentally friendly and economical construction materials [2,3]. The partial replacement of cement and other constituents of concrete has already been made extensively by industrial byproducts in various studies [4–8]. Several types of waste materials have been studied for their potential use in building materials, such as marble [9–12], super-absorbent polymer [13,14], glass [15–17], slag [18], bagasse ash [19], rubber [20,21], plastic [22], ceramic [23,24], natural fiber [25–28], and recycled aggregate [29–32]. Among these, marble dust, which is produced during cutting processes in mines, has also been used in the production of concrete. The use of marble dust, either as a natural aggregate [33–35] or as a replacement for Portland cement (PC) [36–38], has been studied in various research. The major focus of existing studies has been replacing cement with alternative sustainable materials to reduce emissions caused by PC. Marble waste has been used as a cement replacement in concrete by various researchers [33,35,39–42]. However, Li, et al. [43] reported the reduced emissions with 10% marble dust replacement in concrete. Li, et al. [44] and Li, et al. [43] also proposed a paste replacement method for reducing significant (i.e., 33%) cement content and enhancing the utilization of marble dust waste with enhanced durability and strength. Marvila, et al. [10] conducted research on cement and lime mortars using marble waste as a complementary binder. The authors observed that the results were satisfactory, with an increase in mechanical strength with the use of marble waste. However, as a result of technology advancements, laboratory testing is increasingly inadequate and uneconomical due to the time and expense involved.

The mechanical characteristics of concrete can now be predicted using machine learning (ML) methods, owing to advances in artificial intelligence (AI) [45]. Classification, clustering, and regression are examples of machine learning approaches that can be used to estimate a variety of parameters with varying degrees of effectiveness and predict the precise ultrasonic pulse velocity of concrete. As a result of recently evolved artificial intelligence, the mechanical properties of different material types can be forecasted with the help of supervised machine learning (ML) algorithms [46]. ML approaches, e.g., classification, regression, and clustering, are deployed for statistical processes and for the prediction of compressive strength with high accuracy [47]. The accuracy of the prediction can be enhanced by the integration of standalone models, which yields an ensemble machine learning (EML) model, as depicted by other fields of study [48,49]. The employment of ensemble learning for the prediction of concrete parameters has been studied with a limited scope. Random Forest and adaptive boosting (AdaBoost) are EML techniques that can enhance prediction accuracy through the combination of voting and various regression tree forecasting on the ultimate result [50]. Song, et al. [51] determined the compressive strength of ceramic-waste-modified concrete both experimentally and with standalone techniques. The marginal variation in the experimental results and the prediction model outcomes were reported. Accordingly, the current study aims at investigating the usage of advanced techniques for forecasting concrete properties. Ahmad, et al. [50] performed both EML and standalone techniques for the prediction of concrete's compressive strength and accuracy comparison. It was reported that the outcome predicted by the EML techniques had more accuracy than that of the standalone technique. However, the range of the standalone technique results was also acceptable.

Taking into account the above-mentioned issues, NDT techniques are becoming an emerging alternative solution nowadays. Rebound hammer and ultrasonic pulse velocity (UPV) are the most commonly employed techniques [52,53], both in situ and in the laboratory, as per European standards [54,55]. The quality and homogeneity of different materials such as rocks, wood, and concrete can be evaluated using a nondestructive test named ultrasonic pulse velocity (UPV). In the said test, computation of the velocity using an ultrasonic wave pulse that travels through the considered concrete structure is considered to determine the quality and strength of concrete. The time required for the said pulse to dissipate through the test specimen is measured. The ratio of the test specimen's width to the time consumed by the wave pulse for dissipation is called pulse velocity. The ultrasonic wave speed relies on Young's modulus and the density of the testing element. Great care must be given while performing the test, although it is easy to conduct a UPV test. The applicability of a UPV test is in the field, as well as in the laboratory. Both deterioration analysis and quality control can be conducted using UPV. However, higher accuracy can be achieved by considering both values to predict the strength of concrete. Even so, it has been revealed from experimental outcomes that the developed individual machine learning models can achieve predictions with more accuracy. However, ensemble machine learning models are gaining popularity these days; therefore, a performance comparison between these models is necessary. In addition, in the designing phase of projects, it may be an effective alternative for assisting civil engineers.

Only data regarding concrete composite mix proportions are usually accounted for in various studies as input variables, instead of performing other additional measurements. However, knowledge about the combined application of prediction models with NDT techniques is still missing, pointing towards a research gap. Accordingly, the main aim of the current study is to explore a reliable yet simple method for predictions of UPV for waste marble concrete composites. Waste marble dust in concrete is explored in terms of ultrasonic pulse velocity prediction through the application of artificial intelligence, as presented in the current study. Nondestructive testing data are used for this prediction, and its performance with existing artificial intelligence models, considering the effect of raw ingredients and their interactions using SHAP analysis, is claimed to be the novelty of the current research. To tackle challenges such as the excessive consumption of time and money, novel machine learning algorithms are presented for anticipating the behavior of waste concrete in terms of NDT. The focus of this research is to examine the UPV of marble waste concrete and its estimation using an artificial intelligence approach. The current work is unique in that it conducts experiments on waste marble concrete and uses computational models for the prediction of UPV. This study is important for understanding the significance of input parameters and their correctness in ML algorithm results. The findings of the experimental work are also compared to the results of individual ML and ensemble techniques in this study. Each model's performance is additionally assessed using k-fold cross-validation and statistical tests. Furthermore, a technique [56] is also employed for the attainment of the implemented ML models' enhanced explanation with the help of global feature influence classification and the respective feature dependencies and interactions. This technique discovers a novel area of knowledge in the form of marble dust concrete ingredients' influences on UPV, which is beneficial to researchers for classifying suitable design mixes for marble dust concrete and for rapidly forecasting the UPV of marble dust concrete without performing trial and error experimentation. The above-mentioned knowledge area is also helpful for conducting studies in the future for the strategic establishment of marble dust concrete with advanced functional and mechanical features depending upon numerous limitations, such as time, cost, materials, and UPV requirements, for various projects in the construction industry.

2. Materials and Methods

The raw materials included cement and marble dust, as well as fine and coarse aggregates. For Type I OPC, the Blaine fineness value was 2196 m^2/kg, and the relative density was 2.43 g/cm^3. The marble powder had a large specific surface area, which suggests that adding it to concrete would improve its cohesiveness. An XRF technique was performed in order to check the chemical composition. The physical properties were determined using ASTM standards, i.e., ASTM C136, ASTM C29, ASTM C566, and ASTM C128/C127. Table 1 lists the chemical content of the used marble dust, and Figure 1 shows the physical appearance of the marble dust. Silicon dioxide in an amount of 73% was found in the sand sample using an XRF technique. Locally accessible coarse aggregates up to 25.4 mm in nominal size were employed. The Type I cement's surface area was 385 m^2/kg. The specific gravities of the sand and aggregate were 2670 and 2650 kg/m^3, respectively. Detailed information about the properties of the raw materials is available in a previous study [51]. Figure 2 depicts the frequency distribution of each component used in the mixes. It is related to distribution probability, which represents the number of observations linked with a set of values or a single value. Table 2 also shows the physical parameters of the fine and coarse aggregates. This research compares two mix designs, i.e., 20 different mixes for controlled concrete and 20 different mixes for marble-replaced concrete. A marble content of 10% has been suggested in the literature for optimized properties. Therefore, 10% marble waste was used in all the mixes for prediction using the artificial intelligence approach. The study was designed to estimate the UPV using machine learning techniques, and this was the main reason for selecting different types of mixes. Three cube specimens of 150 mm^3 were prepared for each mix. After demolding, the specimens were water-cured for 28 days. The ASTM C192/C192M was followed for the making and curing of the test specimens of the concrete. Then, ASTM C597 was followed to determine the ultrasonic pulse velocity of the concrete, as shown in Figure 3.

The test results showed that an increase in UPV was observed with the addition of marble dust in the concrete. The UPV results of the controlled and waste marble dust mixes are presented in Figure 4a,b, respectively. The UPV of the waste marble dust concrete was higher than that of the controlled concrete. Calcium carbo-aluminate, which is formed in concrete due to a reaction with the $CaCO_3$ in marble dust, accelerates both the hydration rate and strength development [37]. A greater pulse velocity indicated homogeneity and excellent quality, whereas a lower pulse velocity indicated nonhomogeneity. The methodology of the current research with the application of machine learning is shown in Figure 5.

The dataset comprised 6 inputs: cement, marble dust, w/c ratio, coarse aggregates, sand, and days. Table 3 describes the statistical analysis of the input parameters. Except for age, which was evaluated in days, all the characteristics were weighted in kg/m^3. The findings of the descriptive analysis were dependent on many input factors. The table provides the lowest and maximum values and ranges for each variable utilized in the model. Other analytic parameters used to show the relevant values include standard deviation, mean, mode, and a total of all the data points for each variable.

Figure 1. Marble dust.

Table 1. Chemical composition.

Components	Marble Dust	Cement
SiO_2	14.08	18.93
Al_2O_3	2.69	9.89
MgO	2.77	1.67
CaO	42.14	59.6
K_2O	0.63	1.13
Na_2O	0.61	0.90
Fe_2O_3	1.94	3.59

Table 2. Physical properties of raw materials.

Parameters	Maximum Size	Fineness Modulus	Moisture Content	Density
	mm	-	%	kg/m^3
Cement	-	-	-	1432
Marble dust	-	1.86	-	1118
Sand	-	2.72	1.57	1790
Coarse aggregate	25.4	-	1.49	1591

Table 3. Details of input data.

	Input Data						
	Cement (kg/m^3)	Marble Dust (kg/m^3)	Sand (kg/m^3)	Aggregate (kg/m^3)	Water (kg/m^3)	Days	UPV (m/s)
Standard Error	9.73	2.95	19.55	33.62	4.55	1.18	42.34
Median	472.84	17.24	615.26	1116.36	220.83	17.50	3334
Minimum	310.15	0.00	129.47	659.33	130.97	7.00	3110
Maximum	708.80	70.89	1020.65	1750.97	303.96	28.00	4502
Mode	486.95	0.00	620.06	1201.29	185.03	7.00	3357
Mean	484.40	25.49	618.70	1202.28	217.13	17.50	3518
Standard Deviation	86.98	26.39	174.90	300.70	40.67	10.57	378.68
Range	398.65	70.89	891.17	1091.64	172.99	21.00	1392

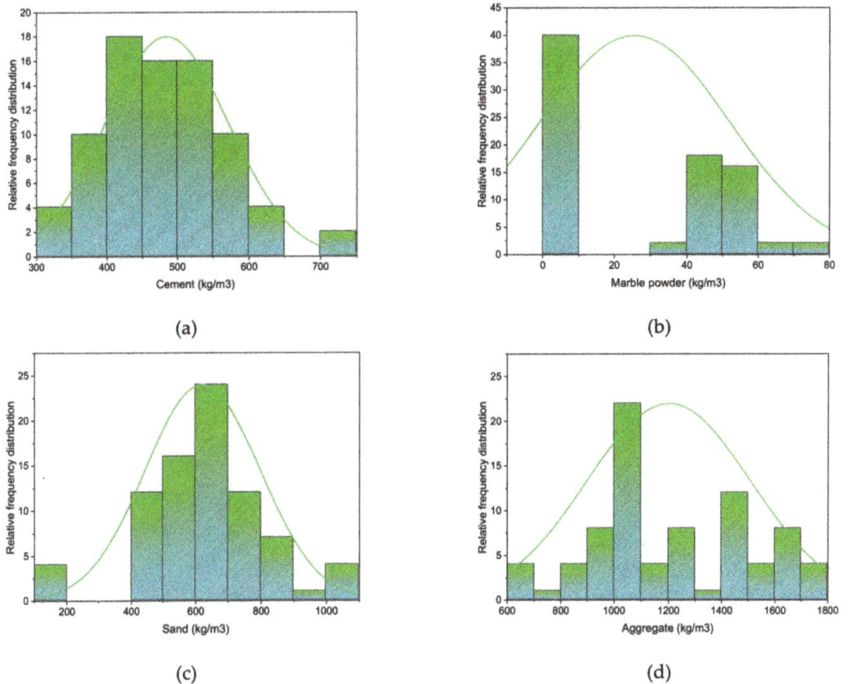

Figure 2. Relative frequency distribution of input parameters: (**a**) cement; (**b**) marble dust; (**c**) sand; and (**d**) coarse aggregate.

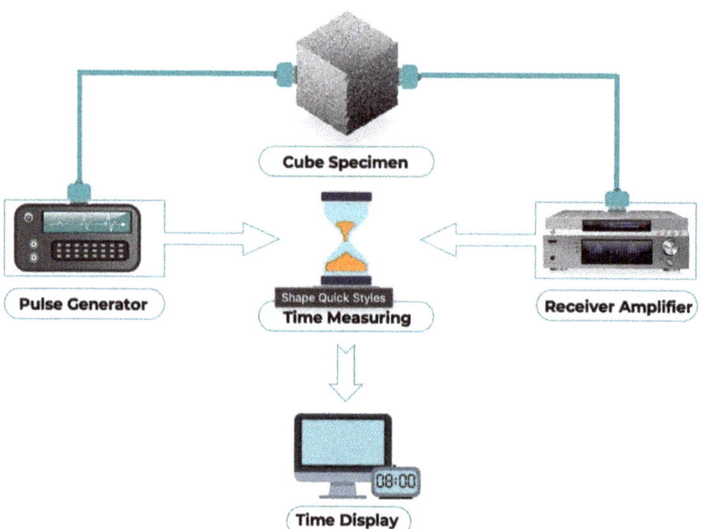

Figure 3. UPV testing procedure.

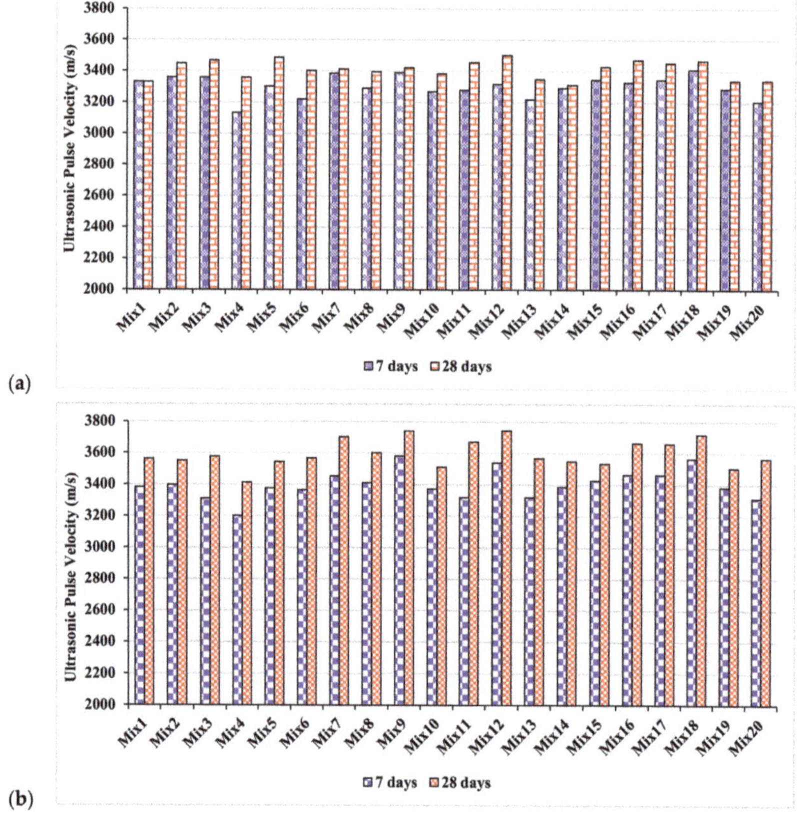

Figure 4. Experimental ultrasonic pulse velocity of mixes: (**a**) control; and (**b**) waste marble dust.

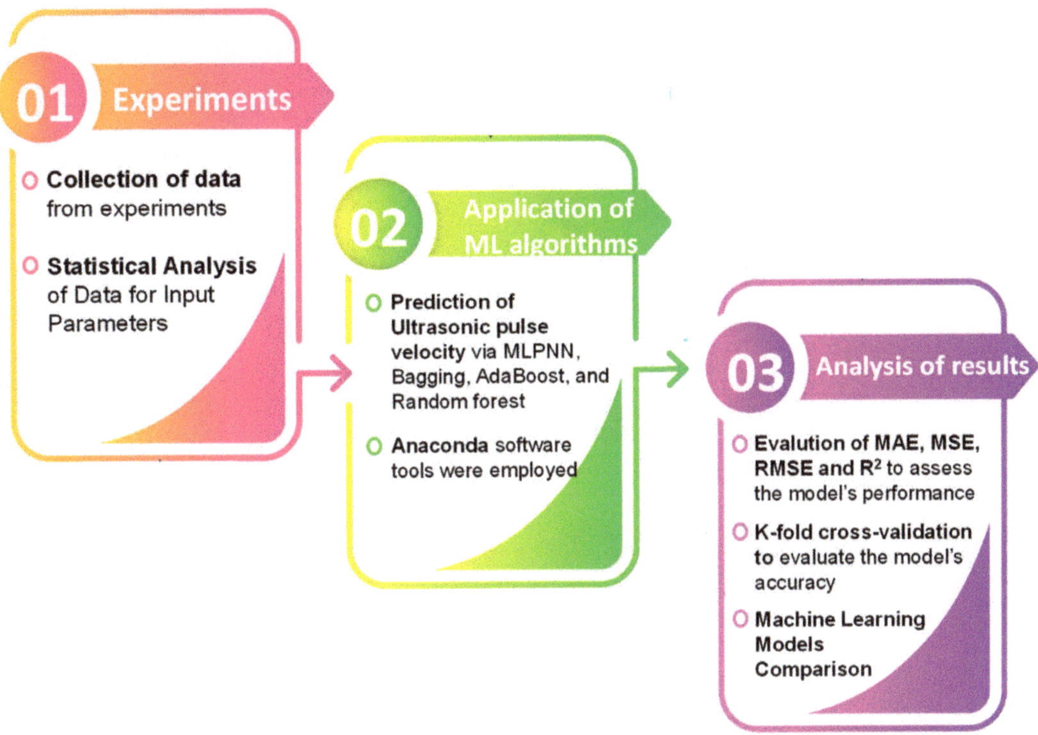

Figure 5. Research methodology with application of machine learning for this study.

3. Results and Discussion

This section addresses the ultrasonic pulse velocity prediction algorithms. A single-layer perceptron neural network (MLPNN) was used as an individual algorithm, while Bagging, AdaBoost, and Random Forest models were implemented as ensemble ML approaches using Python code with Anaconda software. These algorithms are generally used to anticipate outcomes based on input factors. All the techniques used six input parameters and one output parameter (ultrasonic pulse velocity) during the modeling phase. All the ensemble models were shown to be accurate and valid, as discussed below.

3.1. Multiple-Layer Perceptron Neural Network (MLPNN) Algorithm

Figure 6 depicts the statistical analysis of the predicted and actual results regarding the UPV of WMDC for MLPNN modeling. A reasonably précised output and very low variation between the anticipated and actual values was obtained with the MLPNN technique. The accuracy of predicting results was assessed as having a 0.88 R^2 value. The dispersions for the predicted and experimental values (targets) with the MLPNN model errors are shown in Figure 7. The average, highest, and lowest values of the training set were 6.20, 20.7, and 0.07 MPa, respectively. A total of 45% of the error values were less than 500 m/s, 45% were from 500 to 1000 m/s, and 10% were higher than 1000 m/s.

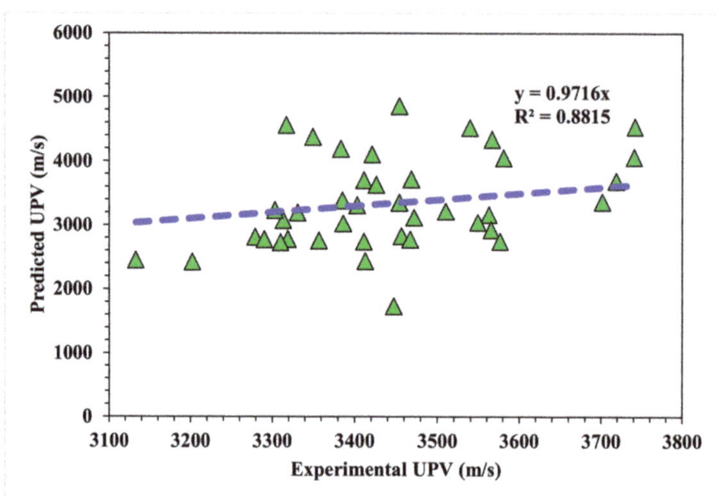

Figure 6. MLPNN model experimental and predicted results.

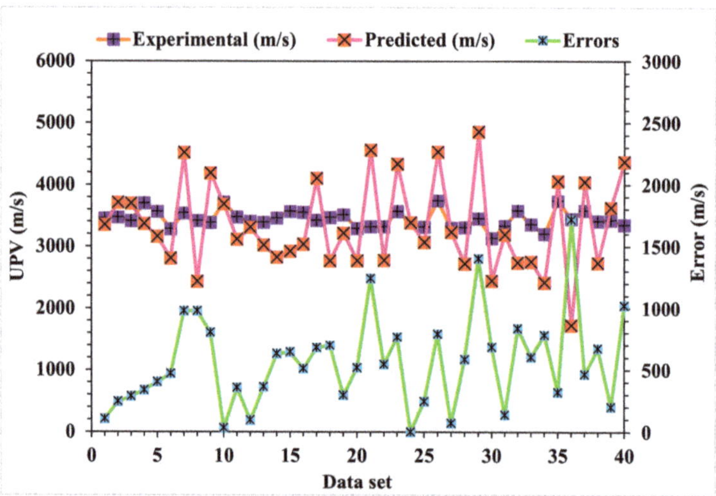

Figure 7. MLPNN model experimental and predicted values with the errors.

3.2. Bagging Algorithm

The correlation between the projected and actual results of the Bagging model is shown in Figure 8. The R^2 value for the Bagging model was 0.94, which represents the highly precise and more accurate Bagging model with respect to the MLPNN model. Furthermore, the dispersion of the projected values, the actual targeted values, and the errors for the Bagging model are shown in Figure 9. It was noted that 45% of the error data was below 500 m/s, 47.5% was from 500 to 1000 m/s, and only 7.5% was higher than 1000 m/s. The higher accuracy of the Bagging model with respect to the MLPNN model was revealed from this analysis. It was also depicted by lower error and greater R^2 values. In addition, twenty submodels were employed using EML methods (MLPNN, AdaBoost, and Random Forest) to obtain an optimized value that produced a firm output.

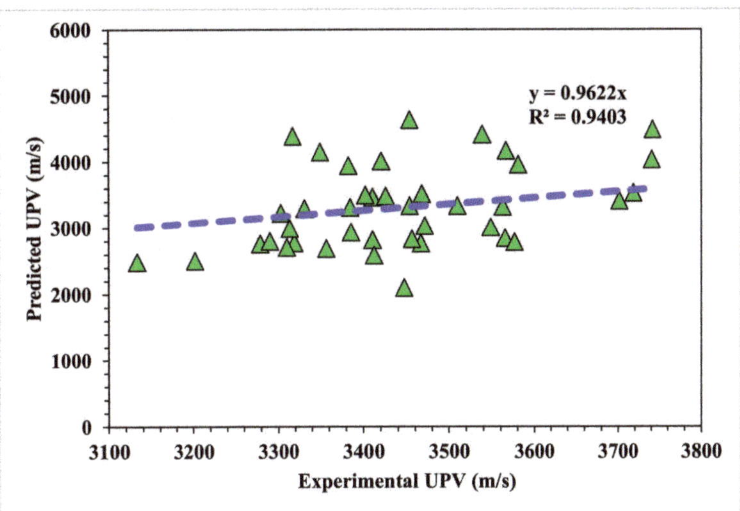

Figure 8. Bagging model experimental and predicted results.

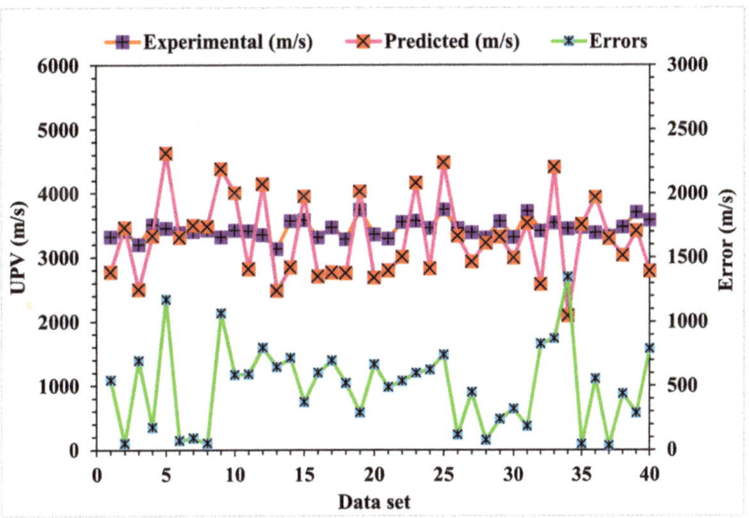

Figure 9. Bagging model experimental and predicted values with the errors.

3.3. AdaBoost Algorithm

A comparison of the projected and actual outputs for the AdaBoost model is shown in Figure 10. The R^2 value was 0.91, which showed a better outcome when compared to the MLPNN model. The dispersions of the actual and predicted values with the errors for the AdaBoost model are illustrated in Figure 11. However, 47.5% of the error values were below 500 m/s, 45% ranged from 500 to 1000 m/s, and only 7.5% were higher than 1000 m/s. The higher accuracy of the AdaBoost model in comparison with the MLPNN model was also depicted by lower error values.

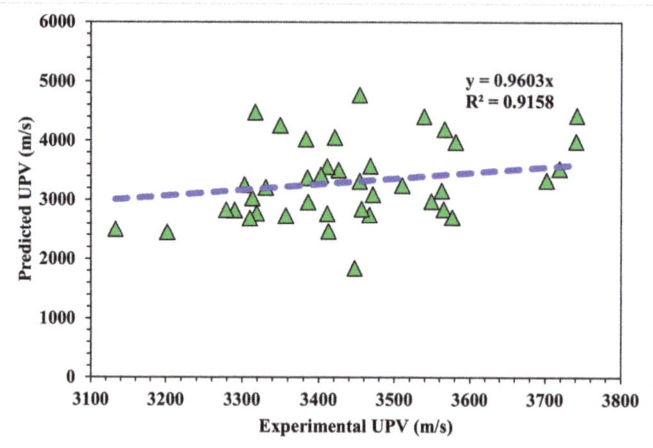

Figure 10. AdaBoost model experimental and predicted results.

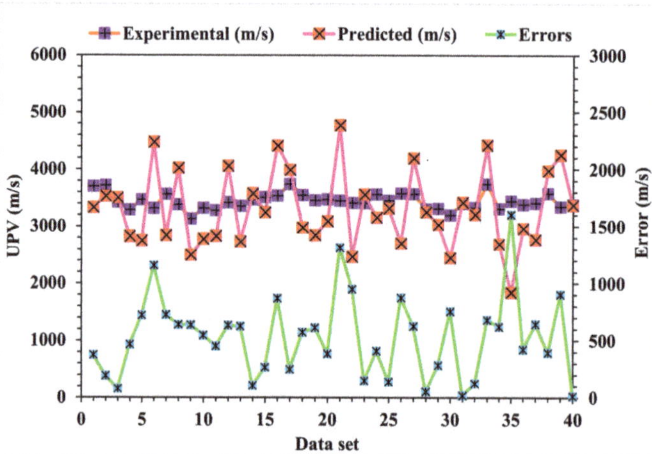

Figure 11. AdaBoost model experimental and predicted values with the errors.

3.4. Random Forest Algorithm

The correlation between the predicted and actual output values for the Random Forest model is provided in Figure 12. The R^2 value for this model came out to be 0.98, showing considerable accuracy compared to the MLPNN, Bagging, and AdaBoost models. The dispersions of the actual and predicted values with the errors for the Random Forest model are shown in Figure 13. Only 57.5% of the error values were below 500 m/s, 42.5% of the values ranged from 500 to 900 m/s, and no values were found above 900 m/s. The error distribution and R^2 values were more accurate than the MLPNN, Bagging, and AdaBoost models for the UPV prediction of WMDC. The R^2 values, along with the error values, obtained from all the considered ensemble ML models were in an acceptable range, depicting better prediction outcomes. Hence, it was observed in this study that EML techniques (Random Forest, followed by Bagging and Adaboost) predicted high-accuracy outcomes when compared to a standalone MLPNN technique.

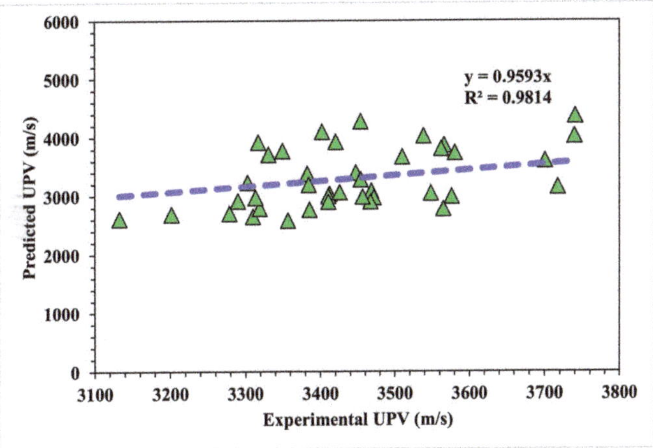

Figure 12. Random Forest model experimental and predicted results.

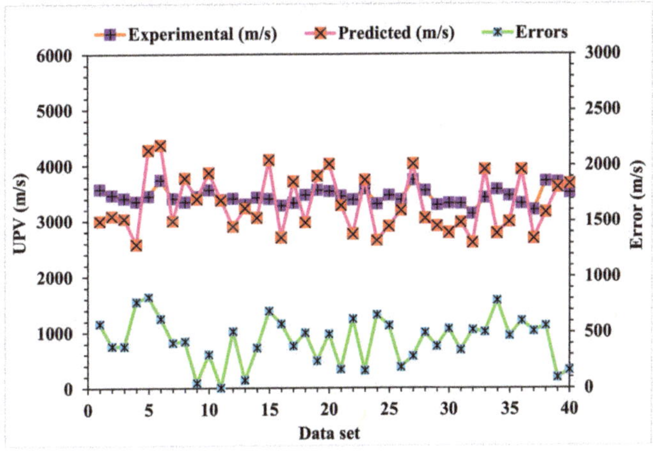

Figure 13. Random Forest model experimental and predicted values with the errors.

4. Model Performance Assessment

4.1. K-Fold Cross-Validation Checks

Statistical analyses with Equations (1) and (2) were utilized to predict the responses of the models. The legitimacy of the models was evaluated by utilizing a k-fold cross-validation approach during execution. Usually, the validity of a model is performed with a k-fold cross-validation process [57] in which random dispersion is perfomed by splitting the model into 10 groups. The greater the R^2 value and the fewer the errors (RMSE and MAE), the higher the accuracy of the model. Furthermore, this process should be repeated multiple (i.e., 10) times for a satisfactory result. The exceptional precision of a model can be achieved by using this comprehensive approach. In addition, statistical analyses (i.e., RMSE and MSE) were also performed for all the models (Table 4). The Random Forest model accuracy (inversely related to error values) compared to the AdaBoost, Bagging, and MLPNN models was also supported by these checks. Statistical analysis as reported in the literature [47,58] is used to assess the response of a model to prediction. The k-fold cross-validation is assessed by utilizing R^2, RMSE, and MAE. Respective dispersions for the DT, Random Forest, AdaBoost, and Bagging models are presented in Figure 14. The average and maximum values of R^2 for the MLPNN were 0.55 and 0.88, respectively (refer to Figure 14a). The maximum and average values of R^2 for the Bagging model were 0.94 and 0.66, respectively, as shown in Figure 14b. Contrary to this, the maximum and average R^2 values of the AdaBoost model were 0.91 and 0.62, respectively, as portrayed in Figure 14c. In comparison, the maximum and average values of R^2 for Random Forest were 0.98 and 0.76, respectively (see Figure 14d). To compare the error values (RMSE and MAE), the RMSE and MAE values for all the models are shown in Table 4. The Random Forest model, with the lowest error and a higher R^2 value, performed better in results prediction.

$$\text{MAE} = \frac{1}{n}\sum_{i=1}^{n}|x_i - x| \tag{1}$$

$$\text{RMSE} = \sqrt{\sum \frac{\left(y_{pred} - y_{ref}\right)^2}{N}} \tag{2}$$

where n is the number of total data samples, x and y_{ref} are the data sample reference values, and x_i and y_{pred} are the model prediction values.

Table 4. Statistical descriptions of MLPNN, Bagging, AdaBoost, and Random Forest models.

Models	MAE (m/s)	RMSE (m/s)	R^2
MLPNN	564.4	676.7	0.88
Bagging	500.8	594.7	0.94
AdaBoost	531.4	637.6	0.91
Random Forest	429.3	475.7	0.98

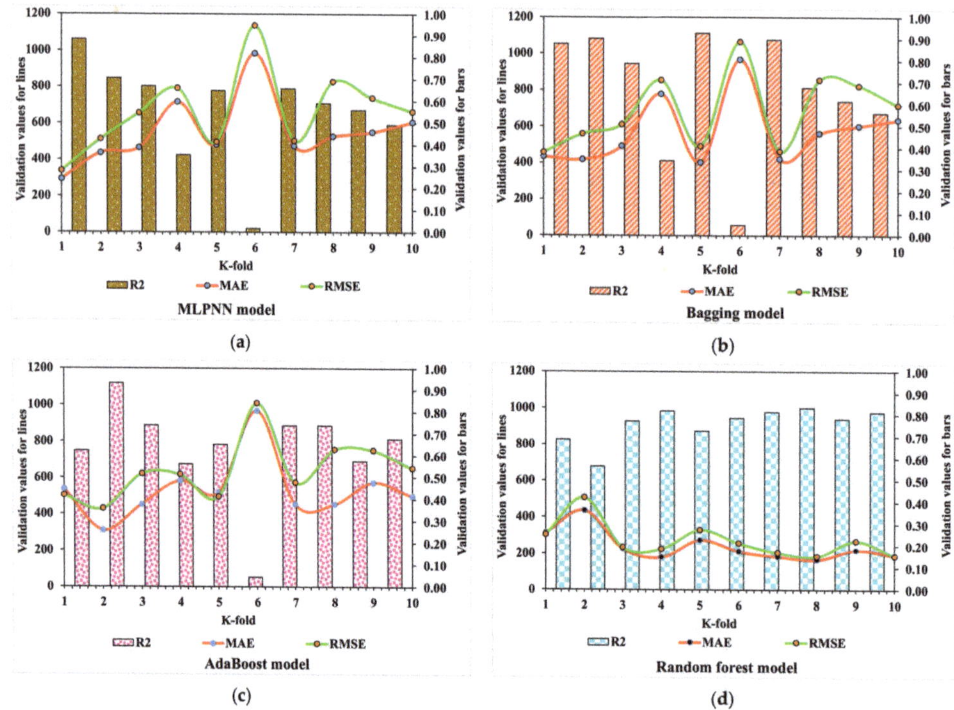

Figure 14. K-fold cross-validation: (**a**) MLPNN model; (**b**) Bagging model; (**c**) AdaBoost model; and (**d**) Random Forest model.

4.2. Comparison of Machine Learning Models

Both ensemble ML and individual approaches were explored in this study for the estimation of WMDC with the aim of sustainable development in terms of environment-friendly construction materials. Random Forest, Bagging, AdaBoost, and MLPNN machine learning techniques were used in this study to predict the compressive strength of WMDC. The goal of the MLPNN algorithm was the development of a model that could predict the target variable accurately. On the other hand, for the Bagging technique, a random sample was selected from the data of the training set, i.e., the selection of individual data points could be made multiple times. The individual training of the said weak models was conducted in the pursuance of numerous data sample generation and based on task type, such as classification or regression or average or majority of these predictions to give an estimate with high accuracy. For the establishment of an algorithm's prediction superiority, the employed algorithms were compared for targeted performance. MLPNN and Random Forest are two alternative learning techniques that can be utilized in similar applications. The main rationale for using a Random Forest rather than an individual decision tree or MLPNN was that it allowed the aggregation of predictions of multiple decision trees in a single model. The theory was that a single model comprised of numerous poor models is still preferable to a single good model. Given the widespread performance of Random Forests, this s true. As a result, Random Forests are less prone to overfitting. Random Forest's major benefit is that it relies on a collection of different decision trees to arrive at any solution. It is an ensemble method that takes into account the findings of multiple classifying algorithms of the same or different types. It is capable of both regression and classification. A Random Forest generates accurate predictions that are simple to comprehend. It is capable of effectively handling huge datasets. In comparison to the individual MLPNN method, the Random Forest algorithm is more accurate at predicting outcomes. The sklearn (Scikit-learn) library was used, and 50% of the data were taken for training purposes and 50% for testing. The output of the Random Forest model was more accurate, having a 0.98 R^2 value, in comparison to Bagging with 0.94 R^2, AdaBoost with 0.91 R^2, and MLPNN with 0.88 R^2. Furthermore, the performances of the MLPNN, Bagging, AdaBoost, and Random Forest models were also evaluated by utilizing a k-fold cross-validation technique and statistical analysis. The performance of the model was higher with low error levels. However, it was difficult to assess optimized machine learning regressors to forecast results from a wide range of topics because the performance of the model was very much dependable on the datapoints and the model's input parameters. On the other hand, for ensemble ML techniques, submodels were generated to leverage the weak learner that could be optimized and trained with data for achieving a higher value of R^2. Other researchers have also observed that AdaBoost, Bagging, and RF models are more accurate in predicting outcomes than individual machine learning techniques [45,50,59–61]. Feng, et al. [45] observed that an AdaBoost model outperformed individual models, including an artificial neural network (ANN) and a support vector machine (SVM), in terms of R^2 and error values. In addition, Ahmad, et al. [50] compared the performances of Bagging, AdaBoost, gene expression programming (GEP), and DT and concluded the best predictor was the Bagging algorithm, with an R^2 of 0.92. Similarly, Farooq, et al. [60] compared the performance of Random Forest with those of ANN, GEP, and DT approaches and found that the Random Forest model had greater precision than the others, with an R^2 of 0.96. A higher accuracy for Random Forest was also reported in the literature, having an R^2 of 0.98 to calibrate a low-cost particle monitor. The dispersion of values for the determinant coefficient of the Bagging, AdaBoost, and Random Forest submodels is shown in Figure 15. The values of R^2 for all the submodels of Random Forest were greater than 0.76, as shown in Figure 15, while most values of R^2 in the cases of the submodels for AdaBoost and Bagging were less than 0.63 and 0.51 (Figure 15), respectively. It depicts the higher accuracy of the Random Forest technique for results prediction, showing a maximum value of R^2, i.e., 0.98. Therefore, the Random Forest model was suggested to predict the ultrasonic pulse velocity of waste marble dust concrete.

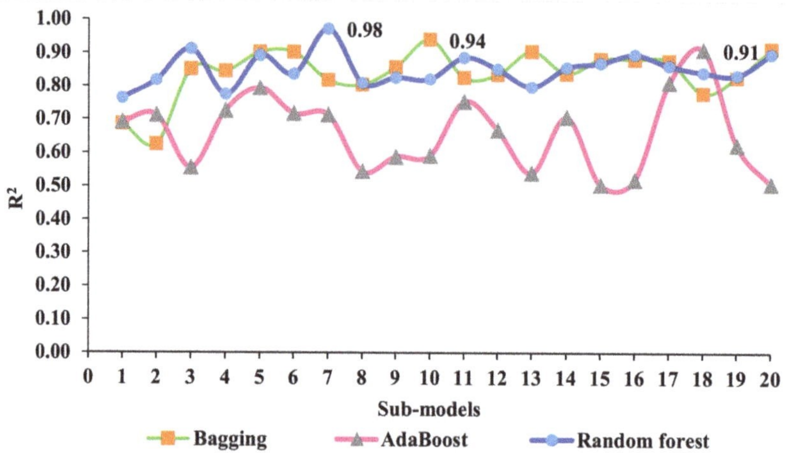

Figure 15. R^2 values of submodels.

4.3. Effect of Raw Ingredients and Their Interactions Using SHAP Analysis

An in-depth ML model explanation was made in the current research. In addition to this, the respective feature dependencies and interactions were also discovered. Initially, the implementation of a SHAP tree explainer for the entire dataset was performed for the provision of an enhanced global feature impact description by the mergence of SHAP descriptions. A tree explainer, i.e., a tree-like SHAP approximation technique, was employed [62]. In this technique, the tree-based model's internal structure, i.e., the sum of the calculation set linked with a leaf node of the tree model that leads to low-order complexity, is assessed [62]. The highest-precision prediction model was obtained by the Random Forest algorithm for the UPV of marble dust concrete. Accordingly, the model interpretation was made for the UPV of marble dust concrete with the help of SHAP analysis.

Figure 16 depicts the violin SHAP-plot values of the considered features for the prediction of UPV for marble dust concrete. A unique color is used to show the feature values in this plot, and the x-axis-corresponding SHAP value represents the output contribution. For example, for marble dust, the content input feature had a higher impact and positive influence, showing the direct relation of this feature with the UPV of marble dust concrete. This means that an increasing content of marble would result in a higher UPV value. A SHAP value of more than 100 in the form of red points (high-value color) at the rightmost side depicts that higher marble dust content enhanced the marble dust concrete UPV. In the case of the curing age feature, a positive influence was seen here as well. At 7 days of age, it is depicted in blue, showing a lower value. Whereas, at 28 days, it increased, as depicted from the higher, i.e., red, values on the right side of the axis. However, in the case of the water content feature, both positive and negative influences are depicted. The water content up to the optimum content was influenced positively; beyond that, there was a negative influence on the UPV of marble dust concrete. In the case of considerably decreased water content, it was also negatively influenced due to affected compaction, resulting in enhanced porosity and, ultimately, a decreased UPV of marble dust concrete. Similarly, sand, aggregate, and cement had more or less the same influence and were on the border of having both positive and negative influences. This evaluation relied on the dataset employed in this study, and high-precision outcomes may also be achieved with more datapoints.

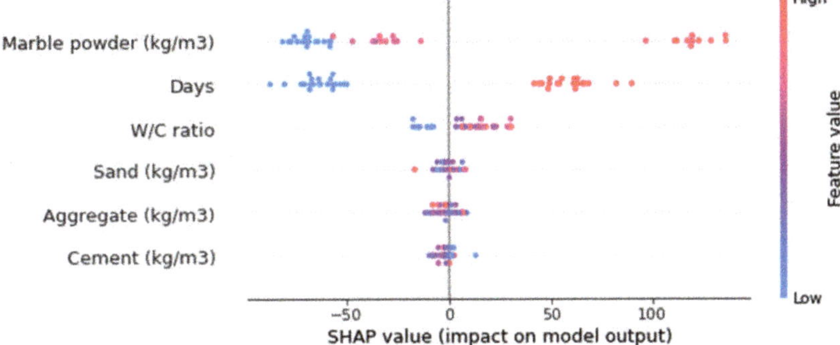

Figure 16. SHAP plot.

The feature interactions with the UPV of marble dust concrete are presented in Figure 17. The marble dust feature interaction is shown in Figure 17a. It can be observed from the plot that marble dust positively interacted with the UPV of marble dust concrete and was in a positive–direct relationship. It may also be noted that, among all the features, marble dust majorly interacted with cement, as it was used as a cement replacement. In Figure 17b, the positive influence of curing days on the UPV of marble dust concrete is observed because more interaction of days with the cement hydration process ultimately increased the strength and UPV of the concrete. The w/c feature interaction is plotted in Figure 17c. The w/c indicated both negative and positive impacts, depending upon its content. The major interaction of w/c was with the cement content, as both water and cement have a link to the hydration process, which is mainly dependent on curing age (days). Then, the cement content feature interaction with sand did not show any particular trend (Figure 17d) and showed almost the same pattern.

Although SHAP was used for the interpretations in this study, there are numerous other post hoc explanatory models that can be used for the same purpose. As a result, we recommend comparing the interpretations obtained using various explanation methodologies. The SHAP-plot values estimated using SHAP, for example, may differ from those obtained using other explanation approaches. Furthermore, the research focused on concrete's UPV. The study, however, can be applied to other strength parameters as well, such as compressive strength, etc. Other strength features need to be predicted using ML in conjunction with post hoc explainable approaches, and the underlying rationales are required to be explained. As a result, the influencing parameters that are required for the design stage can be discovered using this approach, but they still need to be investigated in the future.

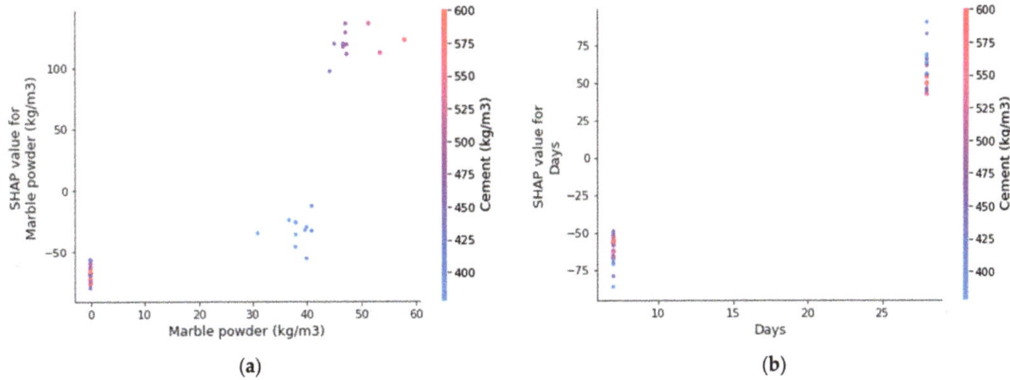

(a)

(b)

Figure 17. *Cont.*

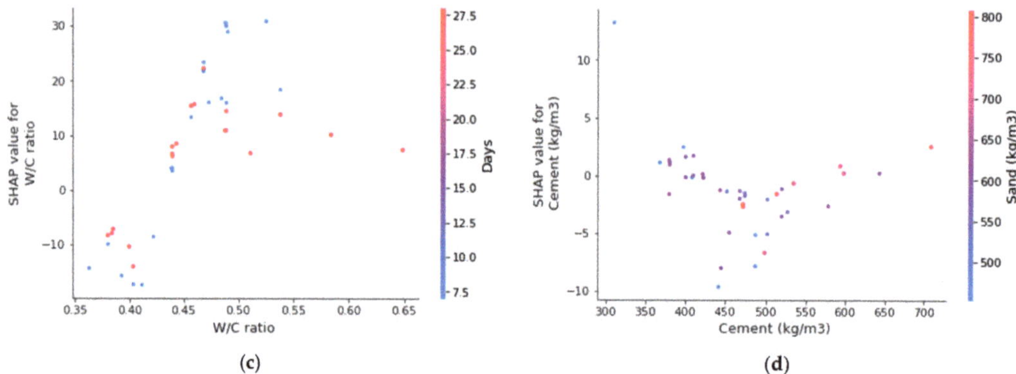

Figure 17. Interaction plots of various parameters: (**a**) marble dust; (**b**) days; (**c**) w/c ratio; and (**d**) cement.

5. Conclusions

The incorporation of marble waste dust into concrete can be an efficient way to improve the environment and reduce landfill pollution. To achieve this, waste marble dust was used in concrete. Additionally, soft computing techniques were compared to predict waste marble dust concrete (WMDC) characteristics. Based on the conducted research, the following conclusions were drawn:

- An amount of 10% marble dust in concrete influenced the ultrasonic pulse velocity. The ultrasonic pulse velocity increased due to the reduced porosity of concrete with marble dust. In this case, waste marble dust concrete with 10% marble dust (as a replacement) showed improved UPV compared to the control mix with 0% marble dust.
- Due to its greater R^2 and lower error levels, the Random Forest model outperformed AdaBoost, Bagging, and MLPNN techniques in terms of prediction. The MLPNN, Bagging, AdaBoost, and Random Forest models had R^2 values of 0.88, 0.94, 0.91, and 0.97, respectively. However, the ensemble model results for Random Forest, followed by Bagging and AdaBoost, were acceptable.
- A k-fold cross-validation technique and statistical analyses revealed adequate Random Forest, AdaBoost, and Bagging outcomes. These tests also showed that the Random Forest model outperformed the MLPNN, AdaBoost, and Bagging models.
- The study validated the application of ultrasonic pulse velocity for forecasting the ultrasonic pulse velocity of sustainable cementitious composite. Therefore, the presented techniques using artificial intelligence seemed reliable for predicting waste marble dust concrete properties.
- A higher SHAP-plot value depicted the positive relation of marble dust content with the UPV of marble dust concrete.
- The feature interaction plot represented that marble dust and curing days positively interacted with cement content and improved the UPV of concrete.

This study was limited to the prediction of the UPV of waste marble dust concrete with limited input parameters and machine learning algorithms (an MLPNN-based approach and decision-tree-based approaches). It is suggested that more comprehensive research on waste marble dust needs to be conducted with more criteria included. Adding additional input factors and expanding the database can produce more trustworthy findings and provide a more comprehensive expression. These parameters should include, in the future, compressive strength, temperature effect, acid attack resistance, chlorine resistance, sulphate resistance, and corrosion. Advanced technologies such as particle swarm optimization (PSO) and M5P trees can be used to make more accurate predictions. However, for better results, machine learning approaches can be coupled with heuristic methods, such as the whale optimization algorithm and ant colony optimization, and then compared with the current study. Further studies should be carried out to investigate the chemical properties of waste marble dust, as well as all other mechanical properties that are key to any application in concrete.

Author Contributions: D.Y.: Conceptualization, Investigation, Methodology, Formal Analysis, Writing—Original Draft. J.Z.: Investigation, Methodology, Data Curation, Writing—Review and Editing. S.A.S.: Formal Analysis, Software, Validation, Data Curation, Writing—Review and Editing. W.A.: Data Curation, Writing—Review and Editing, Supervision. P.K.: Funding Acquisition, Investigation, Resources, Validation, Writing—Review and Editing. A.D.: Data Curation, Software, Methodology, Formal Analysis, Writing—Review and Editing. A.S.: Visualization, Project Administration, Writing—Review and Editing. A.M.: Visualization, Software, Data Curation, Writing—Review and Editing. All authors have read and agreed to the published version of the manuscript.

Funding: This study received no external funding.

Institutional Review Board Statement: Not applicable.

Informed Consent Statement: Not applicable.

Data Availability Statement: Data available on request from the corresponding authors.

Acknowledgments: The authors acknowledge the supportive roles of their respective institutions.

Conflicts of Interest: The authors declare no conflict of interest.

References

1. Assi, L.N.; Carter, K.; Deaver, E.; Ziehl, P. Review of availability of source materials for geopolymer/sustainable concrete. *J. Clean. Prod.* **2020**, *263*, 121477. [CrossRef]
2. Ahmad, W.; Ahmad, A.; Ostrowski, K.A.; Aslam, F.; Joyklad, P. A scientometric review of waste material utilization in concrete for sustainable construction. *Case Stud. Constr. Mater.* **2021**, *15*, e00683. [CrossRef]
3. Khan, M.; Cao, M.; Ai, H.; Hussain, A. Basalt Fibers in Modified Whisker Reinforced Cementitious Composites. *Period. Polytech. Civ. Eng.* **2022**, *66*, 344–354. [CrossRef]
4. De Azevedo, A.R.; Alexandre, J.; Xavier, G.D.C.; Pedroti, L.G. Recycling paper industry effluent sludge for use in mortars: A sustainability perspective. *J. Clean. Prod.* **2018**, *192*, 335–346. [CrossRef]
5. Khan, M.; Ali, M. Improvement in concrete behavior with fly ash, silica-fume and coconut fibres. *Constr. Build. Mater.* **2019**, *203*, 174–187. [CrossRef]
6. Cao, M.; Xie, C.; Li, L.; Khan, M. Effect of different PVA and steel fiber length and content on mechanical properties of $CaCO_3$ whisker reinforced cementitious composites. *Mater. Construcción* **2019**, *69*, e200. [CrossRef]
7. Cao, M.; Khan, M.; Ahmed, S. Effectiveness of Calcium Carbonate Whisker in Cementitious Composites. *Period. Polytech. Civ. Eng.* **2020**, *64*, 265. [CrossRef]
8. Xie, C.; Cao, M.; Khan, M.; Yin, H.; Guan, J. Review on different testing methods and factors affecting fracture properties of fiber reinforced cementitious composites. *Constr. Build. Mater.* **2021**, *273*, 121766. [CrossRef]
9. Tayeh, B.A. Effects of marble, timber, and glass powder as partial replacements for cement. *J. Civ. Eng. Constr.* **2018**, *7*, 63–71. [CrossRef]
10. Marvila, M.T.; de Azevedo, A.R.; Alexandre, J.; Colorado, H.; Pereira Antunes, M.L.; Vieira, C.M. Circular economy in cementitious ceramics: Replacement of hydrated lime with a stoichiometric balanced combination of clay and marble waste. *Int. J. Appl. Ceram. Technol.* **2021**, *18*, 192–202. [CrossRef]
11. Kore, S.D.; Vyas, A. Impact of marble waste as coarse aggregate on properties of lean cement concrete. *Case Stud. Constr. Mater.* **2016**, *4*, 85–92. [CrossRef]
12. Marvila, M.T.; Alexandre, J.; de Azevedo, A.R.; Zanelato, E.B. Evaluation of the use of marble waste in hydrated lime cement mortar based. *J. Mater. Cycles Waste Manag.* **2019**, *21*, 1250–1261. [CrossRef]
13. Rajamony Laila, L.; Gurupatham, B.G.A.; Roy, K.; Lim, J.B.P. Effect of super absorbent polymer on microstructural and mechanical properties of concrete blends using granite pulver. *Struct. Concr.* **2021**, *22*, E898–E915. [CrossRef]
14. Rajamony Laila, L.; Gurupatham, B.G.A.; Roy, K.; Lim, J.B. Influence of super absorbent polymer on mechanical, rheological, durability, and microstructural properties of self-compacting concrete using non-biodegradable granite pulver. *Struct. Concr.* **2021**, *22*, E1093–E1116. [CrossRef]
15. De Azevedo, A.R.G.; Alexandre, J.; Zanelato, E.B.; Marvila, M.T. Influence of incorporation of glass waste on the rheological properties of adhesive mortar. *Constr. Build. Mater.* **2017**, *148*, 359–368. [CrossRef]
16. De Azevedo, A.R.; Teixeira Marvila, M.; Barbosa de Oliveira, L.; Macario Ferreira, W.; Colorado, H.; Rainho Teixeira, S.; Mauricio Fontes Vieira, C. Circular economy and durability in geopolymers ceramics pieces obtained from glass polishing waste. *Int. J. Appl. Ceram. Technol.* **2021**, *18*, 1891–1900. [CrossRef]

17. De Azevedo, A.R.; Marvila, M.T.; Ali, M.; Khan, M.I.; Masood, F.; Vieira, C.M.F. Effect of the addition and processing of glass polishing waste on the durability of geopolymeric mortars. *Case Stud. Constr. Mater.* **2021**, *15*, e00662. [CrossRef]
18. Marvila, M.T.; Azevedo, A.R.G.D.; Matos, P.R.d.; Monteiro, S.N.; Vieira, C.M.F. Rheological and the fresh state properties of alkali-activated mortars by blast furnace slag. *Materials* **2021**, *14*, 2069. [CrossRef]
19. Ahmad, W.; Ahmad, A.; Ostrowski, K.A.; Aslam, F.; Joyklad, P.; Zajdel, P. Sustainable approach of using sugarcane bagasse ash in cement-based composites: A systematic review. *Case Stud. Constr. Mater.* **2021**, *15*, e00698. [CrossRef]
20. Alaloul, W.S.; Musarat, M.A.; Tayeh, B.A.; Sivalingam, S.; Rosli, M.F.B.; Haruna, S.; Khan, M.I. Mechanical and deformation properties of rubberized engineered cementitious composite (ECC). *Case Stud. Constr. Mater.* **2020**, *13*, e00385. [CrossRef]
21. Alyousef, R.; Ahmad, W.; Ahmad, A.; Aslam, F.; Joyklad, P.; Alabduljabbar, H. Potential use of recycled plastic and rubber aggregate in cementitious materials for sustainable construction: A review. *J. Clean. Prod.* **2021**, *329*, 129736. [CrossRef]
22. Mustafa, M.A.-T.; Hanafi, I.; Mahmoud, R.; Tayeh, B.A. *Effect of Partial Replacement of Sand by Plastic Waste on Impact Resistance of Concrete: Experiment And Simulation*; Elsevier: Amsterdam, The Netherlands, 2019; pp. 519–526.
23. Tawfik, T.A.; Metwally, K.A.; El-Beshlawy, S.A.; Al Saffar, D.M.; Tayeh, B.A.; Hassan, H.S. Exploitation of the nanowaste ceramic incorporated with nano silica to improve concrete properties. *J. King Saud Univ.-Eng. Sci.* **2021**, *33*, 581–588. [CrossRef]
24. Amin, M.; Zeyad, A.M.; Tayeh, B.A.; Agwa, I.S. Engineering properties of self-cured normal and high strength concrete produced using polyethylene glycol and porous ceramic waste as coarse aggregate. *Constr. Build. Mater.* **2021**, *299*, 124243. [CrossRef]
25. Khan, M.; Cao, M.; Xie, C.; Ali, M. Efficiency of basalt fiber length and content on mechanical and microstructural properties of hybrid fiber concrete. *Fatigue Fract. Eng. Mater. Struct.* **2021**, *44*, 2135–2152. [CrossRef]
26. Khan, M.; Cao, M.; Xie, C.; Ali, M. Hybrid fiber concrete with different basalt fiber length and content. *Struct. Concr.* **2021**, *23*, 346–364. [CrossRef]
27. Khan, M.; Cao, M.; Hussain, A.; Chu, S. Effect of silica-fume content on performance of $CaCO_3$ whisker and basalt fiber at matrix interface in cement-based composites. *Constr. Build. Mater.* **2021**, *300*, 124046. [CrossRef]
28. Arshad, S.; Sharif, M.B.; Irfan-ul-Hassan, M.; Khan, M.; Zhang, J.-L. Efficiency of supplementary cementitious materials and natural fiber on mechanical performance of concrete. *Arab. J. Sci. Eng.* **2020**, *45*, 8577–8589. [CrossRef]
29. Arafa, M.; Tayeh, B.A.; Alqedra, M.; Shihada, S.; Hanoona, H. Investigating the effect of sulfate attack on compressive strength of recycled aggregate concrete. *J. Eng. Res. Technol.* **2017**, *4*, 1–7.
30. Elsayed, M.; Tayeh, B.A.; Mohamed, M.; Elymany, M.; Mansi, A.H. Punching shear behaviour of RC flat slabs incorporating recycled coarse aggregates and crumb rubber. *J. Build. Eng.* **2021**, *44*, 103363. [CrossRef]
31. Taher, S.M.S.; Saadullah, S.T.; Haido, J.H.; Tayeh, B.A. Behavior of geopolymer concrete deep beams containing waste aggregate of glass and limestone as a partial replacement of natural sand. *Case Stud. Constr. Mater.* **2021**, *15*, e00744. [CrossRef]
32. Amin, M.; Zeyad, A.M.; Tayeh, B.A.; Agwa, I.S. Effects of nano cotton stalk and palm leaf ashes on ultrahigh-performance concrete properties incorporating recycled concrete aggregates. *Constr. Build. Mater.* **2021**, *302*, 124196. [CrossRef]
33. Aliabdo, A.A.; Abd Elmoaty, M.; Auda, E.M. Re-use of waste marble dust in the production of cement and concrete. *Constr. Build. Mater.* **2014**, *50*, 28–41. [CrossRef]
34. Keleştemur, O.; Arıcı, E.; Yıldız, S.; Gökçer, B. Performance evaluation of cement mortars containing marble dust and glass fiber exposed to high temperature by using Taguchi method. *Constr. Build. Mater.* **2014**, *60*, 17–24. [CrossRef]
35. Rodrigues, R.D.; De Brito, J.; Sardinha, M. Mechanical properties of structural concrete containing very fine aggregates from marble cutting sludge. *Constr. Build. Mater.* **2015**, *77*, 349–356. [CrossRef]
36. Aruntaş, H.Y.; Gürü, M.; Dayı, M.; Tekin, I. Utilization of waste marble dust as an additive in cement production. *Mater. Des.* **2010**, *31*, 4039–4042. [CrossRef]
37. Arel, H.Ş. Recyclability of waste marble in concrete production. *J. Clean. Prod.* **2016**, *131*, 179–188. [CrossRef]
38. Badurdeen, F.; Aydin, R.; Brown, A. A multiple lifecycle-based approach to sustainable product configuration design. *J. Clean. Prod.* **2018**, *200*, 756–769. [CrossRef]
39. Uysal, M.; Yilmaz, K. Effect of mineral admixtures on properties of self-compacting concrete. *Cem. Concr. Compos.* **2011**, *33*, 771–776. [CrossRef]
40. Belaidi, A.; Azzouz, L.; Kadri, E.; Kenai, S. Effect of natural pozzolana and marble powder on the properties of self-compacting concrete. *Constr. Build. Mater.* **2012**, *31*, 251–257. [CrossRef]
41. Gesoğlu, M.; Güneyisi, E.; Kocabağ, M.E.; Bayram, V.; Mermerdaş, K. Fresh and hardened characteristics of self compacting concretes made with combined use of marble powder, limestone filler, and fly ash. *Constr. Build. Mater.* **2012**, *37*, 160–170. [CrossRef]
42. Gencel, O.; Ozel, C.; Koksal, F.; Erdogmus, E.; Martínez-Barrera, G.; Brostow, W. Properties of concrete paving blocks made with waste marble. *J. Clean. Prod.* **2012**, *21*, 62–70. [CrossRef]
43. Li, L.; Huang, Z.; Tan, Y.; Kwan, A.; Chen, H. Recycling of marble dust as paste replacement for improving strength, microstructure and eco-friendliness of mortar. *J. Clean. Prod.* **2019**, *210*, 55–65. [CrossRef]
44. Li, L.; Huang, Z.; Tan, Y.; Kwan, A.; Liu, F. Use of marble dust as paste replacement for recycling waste and improving durability and dimensional stability of mortar. *Constr. Build. Mater.* **2018**, *166*, 423–432. [CrossRef]
45. Feng, D.-C.; Liu, Z.-T.; Wang, X.-D.; Chen, Y.; Chang, J.-Q.; Wei, D.-F.; Jiang, Z.-M. Machine learning-based compressive strength prediction for concrete: An adaptive boosting approach. *Constr. Build. Mater.* **2020**, *230*, 117000. [CrossRef]

46. Amlashi, A.T.; Abdollahi, S.M.; Goodarzi, S.; Ghanizadeh, A.R. Soft computing based formulations for slump, compressive strength, and elastic modulus of bentonite plastic concrete. *J. Clean. Prod.* **2019**, *230*, 1197–1216. [CrossRef]
47. Farooq, F.; Ahmed, W.; Akbar, A.; Aslam, F.; Alyousef, R. Predictive modeling for sustainable high-performance concrete from industrial wastes: A comparison and optimization of models using ensemble learners. *J. Clean. Prod.* **2021**, *292*, 126032. [CrossRef]
48. Chaabene, W.B.; Flah, M.; Nehdi, M.L. Machine learning prediction of mechanical properties of concrete: Critical review. *Constr. Build. Mater.* **2020**, *260*, 119889. [CrossRef]
49. Tiwari, N.; Satyam, N. Coupling effect of pond ash and polypropylene fiber on strength and durability of expansive soil subgrades: An integrated experimental and machine learning approach. *J. Rock Mech. Geotech. Eng.* **2021**, *13*, 1101–1112. [CrossRef]
50. Ahmad, W.; Ahmad, A.; Ostrowski, K.A.; Aslam, F.; Joyklad, P.; Zajdel, P. Application of Advanced Machine Learning Approaches to Predict the Compressive Strength of Concrete Containing Supplementary Cementitious Materials. *Materials* **2021**, *14*, 5762. [CrossRef]
51. Song, H.; Ahmad, A.; Ostrowski, K.A.; Dudek, M. Analyzing the compressive strength of ceramic waste-based concrete using experiment and artificial neural network (ANN) approach. *Materials* **2021**, *14*, 4518. [CrossRef]
52. Amini, K.; Jalalpour, M.; Delatte, N. Advancing concrete strength prediction using non-destructive testing: Development and verification of a generalizable model. *Constr. Build. Mater.* **2016**, *102*, 762–768. [CrossRef]
53. Turgut, P. Evaluation of the ultrasonic pulse velocity data coming on the field. *Ibis* **2004**, *6*, 573–578.
54. *CEN EN 12504–4:2006*; Testing Concrete–Part 4: Determination of Ultrasonic Pulse Velocity. British Standards Institution: London, UK, 2004.
55. British Standards Institution. *Testing Concrete in Structures: Part 2: Non-Destructive Testing-Determination of Rebound Number*; British Standards Institution: London, UK, 2001.
56. Lundberg, S.M.; Erion, G.; Chen, H.; DeGrave, A.; Prutkin, J.M.; Nair, B.; Katz, R.; Himmelfarb, J.; Bansal, N.; Lee, S.-I. From local explanations to global understanding with explainable AI for trees. *Nat. Mach. Intell.* **2020**, *2*, 56–67. [CrossRef] [PubMed]
57. Ahmad, A.; Chaiyasarn, K.; Farooq, F.; Ahmad, W.; Suparp, S.; Aslam, F. Compressive strength prediction via gene expression programming (GEP) and artificial neural network (ANN) for concrete containing RCA. *Buildings* **2021**, *11*, 324. [CrossRef]
58. Aslam, F.; Farooq, F.; Amin, M.N.; Khan, K.; Waheed, A.; Akbar, A.; Javed, M.F.; Alyousef, R.; Alabdulijabbar, H. Applications of gene expression programming for estimating compressive strength of high-strength concrete. *Adv. Civ. Eng.* **2020**, *2020*, 8850535. [CrossRef]
59. Feng, D.-C.; Liu, Z.-T.; Wang, X.-D.; Jiang, Z.-M.; Liang, S.-X. Failure mode classification and bearing capacity prediction for reinforced concrete columns based on ensemble machine learning algorithm. *Adv. Eng. Inform.* **2020**, *45*, 101126. [CrossRef]
60. Farooq, F.; Nasir Amin, M.; Khan, K.; Rehan Sadiq, M.; Faisal Javed, M.; Aslam, F.; Alyousef, R. A comparative study of random forest and genetic engineering programming for the prediction of compressive strength of high strength concrete (HSC). *Appl. Sci.* **2020**, *10*, 7330. [CrossRef]
61. Ahmad, A.; Ahmad, W.; Chaiyasarn, K.; Ostrowski, K.A.; Aslam, F.; Zajdel, P.; Joyklad, P. Prediction of geopolymer concrete compressive strength using novel machine learning algorithms. *Polymers* **2021**, *13*, 3389. [CrossRef]
62. Lundberg, S.M.; Erion, G.; Chen, H.; DeGrave, A.; Prutkin, J.M.; Nair, B.; Katz, R.; Himmelfarb, J.; Bansal, N.; Lee, S.-I. Explainable AI for trees: From local explanations to global understanding. *arXiv* **2019**, arXiv:1905.04610. [CrossRef]

materials

MDPI

Article

Exploring the Use of Waste Marble Powder in Concrete and Predicting Its Strength with Different Advanced Algorithms

Kaffayatullah Khan [1,*], Waqas Ahmad [2], Muhammad Nasir Amin [1], Ayaz Ahmad [3], Sohaib Nazar [2], Anas Abdulalim Alabdullah [1] and Abdullah Mohammad Abu Arab [1]

1 Department of Civil and Environmental Engineering, College of Engineering, King Faisal University, Al-Ahsa 31982, Saudi Arabia; mgadir@kfu.edu.sa (M.N.A.); 218038024@student.kfu.edu.sa (A.A.A.); 219041496@student.kfu.edu.sa (A.M.A.A.)

2 Department of Civil Engineering, COMSATS University Islamabad, Abbottabad 22060, Pakistan; waqasahmad@cuiatd.edu.pk (W.A.); sohaibnazar@cuiatd.edu.pk (S.N.)

3 MaREI Centre, Ryan Institute, School of Engineering, College of Science and Engineering, National University of Ireland Galway, H91 HX31 Galway, Ireland; a.ahmad8@nuigalway.ie

* Correspondence: kkhan@kfu.edu.sa

Abstract: Recently, the high demand for marble stones has progressed in the construction industry, ultimately resulting in waste marble production. Thus, environmental degradation is unavoidable because of waste generated from quarry drilling, cutting, and blasting methods. Marble waste is produced in an enormous amount in the form of odd blocks and unwanted rock fragments. Absence of a systematic way to dispose of these marble waste massive mounds results in environmental pollution and landfills. To reduce this risk, an effort has been made for the incorporation of waste marble powder into concrete for sustainable construction. Different proportions of marble powder are considered as a partial substitute in concrete. A total of 40 mixes are prepared. The effectiveness of marble in concrete is assessed by comparing the compressive strength with the plain mix. Supervised machine learning algorithms, bagging (Bg), random forest (RF), AdaBoost (AdB), and decision tree (DT) are used in this study to forecast the compressive strength of waste marble powder concrete. The models' performance is evaluated using correlation coefficient (R^2), root mean square error, and mean absolute error and mean square error. The achieved performance is then validated by using the k-fold cross-validation technique. The RF model, having an R^2 value of 0.97, has more accurate prediction results than Bg, AdB, and DT models. The higher R^2 values and lesser error (RMSE, MAE, and MSE) values are the indicators for better performance of RF model among all individual and ensemble models. The implementation of machine learning techniques for predicting the mechanical properties of concrete would be a practical addition to the civil engineering domain by saving effort, resources, and time.

Keywords: waste; concrete; marble powder; compressive strength; machine learning algorithms

Citation: Khan, K.; Ahmad, W.; Amin, M.N.; Ahmad, A.; Nazar, S.; Alabdullah, A.A.; Arab, A.M.A. Exploring the Use of Waste Marble Powder in Concrete and Predicting Its Strength with Different Advanced Algorithms. *Materials* **2022**, *15*, 4108. https://doi.org/10.3390/ma15124108

Academic Editor: Jorge Otero

Received: 12 May 2022
Accepted: 1 June 2022
Published: 9 June 2022

Publisher's Note: MDPI stays neutral with regard to jurisdictional claims in published maps and institutional affiliations.

Copyright: © 2022 by the authors. Licensee MDPI, Basel, Switzerland. This article is an open access article distributed under the terms and conditions of the Creative Commons Attribution (CC BY) license (https://creativecommons.org/licenses/by/4.0/).

1. Introduction

Iran, Italy, China, Turkey, India, Egypt, Spain, Brazil, Algeria, Sweden, and France are the main marble-producing countries [1–4]. India is the third most marble-producing country around the globe, and almost 10% of the worldwide marble powder is quarried here [5]. In addition, the import and processing of stone are majorly done in countries such as Pakistan, the United States, Egypt, Saudi Arabia, Portugal, Germany, France, Norway, and Greece [6]. During different stages of stone mining and processing procedures, a bulk quantity of marble waste is generated. Out of which, up to 60% is generated as a result of marble quarrying only [7]. Marble dust in finer form that is produced as a result of its sawing and cutting can cause harmful health issues. Furthermore, the dumping of this marble dust can result in poor soil properties and the fertility reduction of respective land [8]. Almost 30% of marble waste is produced during the working of marble stone [9].

The global annual production of marble and granite was nearly 140 million tonnes in 2014, as per USGS [10]. There were approximately 2 billion tonnes of marble resources in India only, as of April 2015, as per the UNFC system. Only 0.23% were reserved resources, and 99.77% were under the remaining resources category [11]. In 2015, China produced around 350 million sq. meters of marble planks, depicting China as World's largest marble producer [12]. Chauhdary [13] reported the availability of almost 160 million tonnes of marble reserves and around 2 billion M.T granite reserves in Pakistan as of 2006. In the mining industry of Iran, there were approximately 4.8 million tonnes of raw and/or semi-processed stone in the year 2012–2013 from a total of 473 quarries of marble stone [14]. Egypt used to export nearly 13 lac tonnes of stones annually as unprocessed and processed stones. From Shaq Al–Thoban industrial/site areas of Egypt, nearly 7 lac tonnes waste is generated annually [15]. As far as the marble reserves of Turkey are concerned, these are around 3.8 billion cubic meters [16]. In Turkey, Binici, et al. [17] reported an emerging threat to agriculture and health in the form of marble wastes usually left in situ or settled by sedimentation. Approximately 47 thousand tonnes of solid waste powder is collected annually from quarries in Jordan every year [18]. The same is the case with Spain and some other countries [19]. In past years, the marble powder is usually used in mortar, concrete, tiles, cement, embankments, and pavements [20], in addition to the desulfurization process, soil stabilization, ceramics, and asphalt and polymer-based composites [21]. In Italy, a group of researchers also developed a consortium to rehabilitate and restore the Oresei marble chain in Sardinia. This chain was being exploited for quarrying and landfilling [22]. As per the definition of sustainable development by Brundtland [23], keeping in mind the environmental perspective, the addition of mineral admixtures and different waste materials has gained much importance with the aim to reduce the consumption of natural resources. However, the natural resources consumption for the production of concrete is still inevitable. In addition, the extraction of local natural resources within limited surrounding region is unable to meet the said needs; thus becoming un-sustainable in near future. Accordingly, the usage of waste materials in concrete production should be promoted in construction sector. In addition, the alternative sustainable approaches should also be introduced for reducing the consumption of natural materials at national as well as international level [24–27]. Whereas, at local level, recycled aggregates are usually used for road materials stabilization. This is a rare approach due to the less feasible crushing process with respect to traditional approach. The extraction of natural resources is required in traditional approach. Bottom ash and marble dust (MD) are some locally and abundantly available by-products that are usually treated as waste materials and thus ultimately causing environmental pollution.

On a rough estimate, the global annual concrete production is approximately 25 billion tons. Concrete has a very low embodied energy and carbon footprint compared to other building materials. However, due to its wide use in many applications, concrete production has a considerable carbon footprint, contributing to 8% of global carbon dioxide emissions [28,29]. Globally, concrete production accounts for 7.8% of nitrogen oxide emissions, 4.8% of sulfur oxide emissions, 5.2% of particulate matter emissions smaller than 10 mm, and 6.4% of particulate matter emissions smaller than 2.5 microns [30]. It is worth noting that only half of the cement is used in concrete [31], and the remaining is used in blocks, mortar, and plaster [32]. Nonetheless, due to the widespread use of concrete in modern civilization, concrete production accounts for a significant portion of global CO_2 emissions through construction [32]. Aiming toward sustainable development, the usage of environment-friendly by-products is considered an effective strategy toward reducing CO_2 emissions [31,33–35]. Marble dust (MD), having abundant availability in Turkey, China, Iran, Italy, and India, is also an alternative which can be used as a replacement for cement in the production of concrete. Marble, due to its durable properties, is usually used in multiple non-structural applications such as cladding, floors, architectural decoration for indoors and sculpture etc. Considerable waste is generated during the shaping and cutting processes of various marble applications in the form of dust particles. These materials are contaminating the natural resources in terms of environmental damage. Partial replacement of cement and other constituents of concrete has already been made extensively by industrial by-products in various studies [36–44]. The reuse of MD, due to its chemical nature, in the production of concrete came out to be an alternative sustainable approach. The use of MD, either as a natural aggregate [9,45,46] or as a replacement for Portland cement (PC) [16,47–49], has been studied in various research. Generally, MD has been used as up to 60% replacement in different forms. Gesoğlu, et al. [50] reported a 20% decreased slump due to MD as a PC replacement. Concrete having MD showed similar consistency with respect to reference mix as reported by Seghir, Mellas, Sadowski, and Żak [4]. Contrary to this, Alyamac, Ghafari, and Ince [19] stated that the incorporation of MD in concrete improved its fresh properties. In addition, the strength of concrete having MD is still questionable. Topcu, et al. [51] reported the decreased compressive strength with an increase in MD content. The same behavior was also reported by Gencel, et al. [52]. The 5% of MD replacement in concrete production came out to be an optimum content for compressive strength, as reported in several studies [50,53,54]. However, Li, et al. [55] reported the same with 10% MD replacement in concrete. Li, Huang, Tan, Kwan and Liu [12] and Li, Huang, Tan, Kwan and Chen [55] also proposed a paste replacement method for reducing significant (i.e., 33%) cement content and enhancing the utilization of MD waste, having enhanced durability and strength. Seghir, Mellas, Sadowski and Żak [4] reported an enhancement of marble powder porosity by 15% in result of reduced hydration products. The major focus of existing studies is on replacement of cement with alternative sustainable materials for reduction in emissions, caused by PC. Marble waste is used as cement replacement in concrete by various researchers [9,46,50,52,54,56]. Rodrigues, De Brito and Sardinha [46] investigated the incorporation of marble dust having 5, 10, and 20% content as cement replacement in concrete. The study reported positive effect on compressive strength of concrete with cement replaced up to 10% marble dust; however, reduced compressive strength is observed in concrete having 25% of marble dust. The compressive strength is reduced by 13.46% with 20% marble dust content, as reported by Gesoğlu, Güneyisi, Kocabağ, Bayram and Mermerdaş [50]. Another study reported decrement in compressive strengths by 91%, 86%, and 76% having cement replaced by 20%, 30%, and 40% marble dust contents, respectively [52]. Şanal [57] reported enhancement of pore structure due to an increase in the capillary structure of concrete by adding 10% marble dust as cement replacement, ultimately resulting in reduced mechanical properties of concrete.

Concrete is the second most widely used commodity around the globe [58]. Due to its multiple properties such as strength, stiffness, density, fire/thermal resistance, porosity, and durability, concrete is being most commonly used as a building material all around the world. Compressive strength is the most dominating factor among all these, as it directly affects the durability of concrete [59,60]. Concrete is a heterogeneous material constituted by cement, sand, aggregates, and water, as it has different compressive strength values [61]. All the ingredients mentioned above and respective mixtures affect the compressive strength of concrete in terms of water/binder ratio, aggregate size, binder type, or waste composition [62]. The compressive strength of concrete is hard to predict precisely due to its complicated mixture. The determination of concrete compressive strength can be made in the laboratory by crushing standardized cylinders/cubes after specified curing post to the casting of samples [63]. This is globally a standardized method. However, as a result of advancements in technological development, laboratory tests are now insufficient and uneconomical due to the involved time and cost. Nowadays, due to the artificial intelligence (AI) evolution, mechanical properties of concrete can also be predicted by using machine learning (ML) algorithms [64–66]. ML techniques such as classification, clustering, and regression, can be used to estimate various parameters along with varied efficiency and can also help in predetermining the accurately précised compressive strength of concrete.

The performance prediction of various parameters using machine learning algorithms is known for many years. As far as the field of civil engineering is concerned, this trend is increased significantly in the past few years. It is because of the highly accurate prediction of mechanical properties (Table 1). The working principle of machine learning is the same as that of conventional algorithms high accuracy of nonlinear behavior with respect to the linear one. Artificial neural networks (ANN), support vector machines (SVM), decision trees (DT), gene expression programming (GEP), random forest (RF), and deep learning (DL) are widely used prediction techniques in case of mechanical properties of concrete [67]. The shear strength of steel fibers reinforced concrete beams was predicted with the help of eleven algorithms by Rahman, et al. [68]. ANN with optimizer as multi-objective grey wolves (MOGW) was used by Behnood and Golafshani [69] for predicting the static properties of silica fume modified concrete. Güçlüer, et al. [70] used ANN, DT, LR, and SVR to predict the compressive strength of concrete. The tensile strength and compressive strength of waste concrete were predicted with ANN algorithm by Getahun, et al. [71]. Ling, et al. [72] used SVM to predict concrete compressive strength in marine and the results were compared with that of DT and ANN models. Yaseen, et al. [73] also used different ML approaches for the prediction of load carrying capacity, under compression, of light-weight foamed concrete. A machine learning algorithm was also used by Taffese and Sistonen [74] for assessing reinforced concrete structures' durability. Yokoyama and Matsumoto [75] developed an automatic crack detector for concrete structures using machine learning. Concrete samples photographs were used for learning data, whereas deep learning was applied for crack detection. The accuracy level of ML models was determined by Chaabene, et al. [76]. Ahmad, et al. [77] performed ensembled machine learning (EML) and standalone techniques for the prediction of concrete's compressive strength and accuracy comparison. It is reported that the outcome predicted from EML techniques has more accuracy than that by standalone technique. However, the range of standalone technique results was also acceptable. Song, et al. [78] determined the compressive strength of ceramic waste modified concrete both experimentally and with standalone techniques. Marginal variation in experimental results and prediction model's outcomes was reported. Neural networks and decision trees, which are also called classification trees, are two popular ways to model data. These two models have different ways of modeling data and finding relationships between variables. The nodes in the neural network make it look like the human brain and very complex structure is formed. While the decision tree is an easy way to look at data from the top down. Decision trees have a natural flow that is easy to understand and are also easy for computer systems to program. The data point in decision tree models at the top of the tree has the most effect on the response variable in the model. On the other hand, the visual representation of neural network models does not make it easy to understand the working. For neural network model, it is hard to make computer systems, and it is almost impossible to make an explanation because of complex structure. Therefore, decision tree-based algorithms (AdaBoost and bagging) are considered in the study because these trees are so easy to understand, they are very useful for modeling and showing the data visually without any complex structure. Accordingly, the current study aims the usage of advanced techniques for forecasting the concrete properties.

Table 1. Machine learning algorithms in the literature.

Algorithm Name	Notation	Prediction Properties	Year	Waste Material Used	Ref.
Individual (decision tree) and ensemble algorithm (bagging)	DT and Bg	Compressive Strength	2021	FA	[79]
Ensemble modelling (bagging and boosting)	Bg and AdB	Compressive strength	2021	FA	[22]
Individual Algorithms (decision tree)	DT	Chloride Concentration	2021	FA	[18]
Data Envelopment Analysis	DEA	Compressive strength Slump test L-box test V-funnel test	2021	FA	[80]
Multivariate	MV	Compressive strength	2020	Crumb rubber with SF	[81]
Support vector machine	SVM	Slump test L-box test V-funnel test Compressive strength	2020	FA	[82]
Adaptive neuro fuzzy inference system	ANFIS with ANN	Compressive strength	2020	POFA	[83]
Random forest	RF	Compressive strength	2020	-	[84]
Intelligent rule-based enhanced multiclass support vector machine and fuzzy rules	IREMSVM-FR with RSM	Compressive strength	2019	FA	[85]
Random forest	RF	Compressive strength	2019	FA GGBFS FA	[86]
Decision tree	DT	Compressive strength	2021	Ceramic waste	[62]

2. Research Significance

The incorporation of waste materials in concrete to improve its mechanical characteristics has been done in various studies. However, the stepwise laboratory procedure, i.e., casting of specimens, curing for a specified time, and testing is still a concern in terms of cost and time. Novel machine learning techniques are being introduced for forecasting the behavior of waste concrete in terms of mechanical properties to overcome the issues mentioned above, i.e., the excessive consumption of time and cost. However, the results of different machine learning models are still inconsistent depending on the type of material, data set, and other contributing input/output parameters. Therefore, this paper aims to investigate marble dust concrete with the intention of marble dust waste management and identify the optimal machine learning technique. The novelty and significance of the current study are to conduct experimentation on waste marble (powder-based) concrete (WMC) and development of WMC prediction model by computational methods. Additionally, this study is focused on predicting and comparing the compressive strength of WMC through supervised ML approaches. The AdB, RF, Bg, and DT approaches are employed to predict and compare outcomes against actual results. Twenty sub-models are developed in EML modelling to have more accuracy in R^2 value for the optimization. Prediction performance of each technique is done by using these applications. This research is significant for understanding the input parameter's role and accuracy for the outcomes obtained through ML algorithms. Individual ML and ensemble approaches are also compared against the results obtained from experimental work. The k-fold cross-validation and statistical checks are also used to evaluate the performance of each model. A discussion on the use of marble for sustainable construction is made.

3. Experimentation and Data Description

Cement, marble powder, and fine and coarse aggregates are used to prepare 40 mixes. Type-I Ordinary Portland Cement (OPC) is used. ASTM C150 is used to conduct the entire investigation in this research. The chemical composition of used marble and cement is listed in Table 2. The properties of fine aggregate are also determined as per the ASTM standard. Locally available coarse aggregates having a maximum nominal size of 25.4 mm are being used. Furthermore, the physical properties of fine and coarse aggregate can also be seen in Table 3. Marble powder, collected from a local company, is used in this study, as shown in Figure 1. The Blaine fineness value was 2196 m^2/kg, and the relative density was 2.43 g/cm^3. The marble powder has a large specific surface area, suggesting that adding it to concretes would improve their cohesiveness.

Table 2. Chemical composition of cement and marble powder.

Components Details	Cement	Marble Powder
Calcium Oxide (CaO)	61.81	42.14
Magnesium Oxide (MgO)	1.96	2.77
Silica (SiO_2)	22.07	0.79
Potassium Oxide (K_2O)	0.46	0.63
Alumina (Al_2O_3)	6.96	2.69
Sodium Oxide (Na_2O)	0.11	0.61
Iron Oxide (Fe_2O_3)	3.62	1.94
Sulfur Trioxide (SO_3)	2.14	0.042
LOI	1.2	42.28

Table 3. Physical properties of sand and aggregates.

Property	Dry Rodded Bulk Density	Bulk Specific Gravity	Moisture Content	Water Absorption	Fineness Modulus	Nominal Maximum Size
	kg/m^3	-	%	%	-	mm
Sand	1800	2.61	1.57	2	2.72	-
Aggregate	1601	2.51	1.49	1.65	-	25.4
Followed Standards	ASTM C29	ASTM C128/C127		ASTM C566	ASTM C136	-

Figure 1. Waste marble powder.

In this study, two different mix designs are considered. Twenty mixes for controlled concrete and twenty for marble replaced concrete are prepared at every 7 days and 28 days. A total of 40 combinations with 240 specimens are prepared (120 in number for each respective day) with a size of 150 mm^3. De-molding of specimens is done after 24 h, followed by 28 days of water curing. The compression test is performed afterwards, as per ASTM C39, to determine compressive strength. The dataset includes six inputs, i.e., i. cement, ii. marble powder, iii. w/c ratio, iv. coarse aggregates, v. sand and vi. Days for single output, i.e., compressive strength of concrete (refer Table S1 in supplementary materials). The description of statistical analysis regarding input parameters is given in Tables 4 and 5. Table 4 shows the mean value, the average of the numbers by adding up, and then dividing by total number of values in a dataset. All the parameters are considered in weight units, i.e., kg/m^3, except for age, which is being considered in days. Brief descriptive coefficients are collected to summarize descriptive statistics to produce a result. Descriptive analysis results are based on input variables data reflecting various information. The minimum and maximum values and ranges for each variable that is used to run the model are also given in tables. However, other analysis parameters, such as standard deviation, mean, mode, and summation of all data points against each variable, are also used for depicting relevant values. Frequency dispersion for every factor that is being utilized in mixes is shown in Figure 2. It has a close connection with distribution probability, a widely used statistics. A relative frequency distribution shows the total observations associated with a class of values or every single value.

Table 4. Input parameters description analysis.

Parameters	Input Variables					
	Cement (kg/m^3)	Marble Powder (kg/m^3)	Sand (kg/m^3)	Aggregate (kg/m^3)	W/C Ratio	Days
Mean	484.396	25.4941	618.7	1202.28	0.45045	17.5
Standard Error	9.72503	2.95033	19.5542	33.6188	0.00637	1.18
Median	472.838	17.238	615.264	1116.36	0.43994	17.5
Mode	486.948	0	620.058	1201.29	0.37997	7
Standard Deviation	86.9833	26.3886	174.898	300.695	0.05696	10.56
Range	398.65	70.89	891.174	1091.64	0.28578	21

Table 5. Input and output variables range.

	Parameters	Abbreviation	Unit	Minimum Value	Maximum Value
	Cement	C	kg/m^3	310.148	708.798
	Marble powder	MP	kg/m^3	0	70.89
Input	Sand	S	kg/m^3	129.472	1020.65
	Aggregate	A	kg/m^3	659.328	1750.97
	Water to cement ratio	W/C	kg/m^3	0.36273	0.64851
	Days	D	Days	7	28
Output	Compressive Strength	C.S	MPa	9.49	72.11

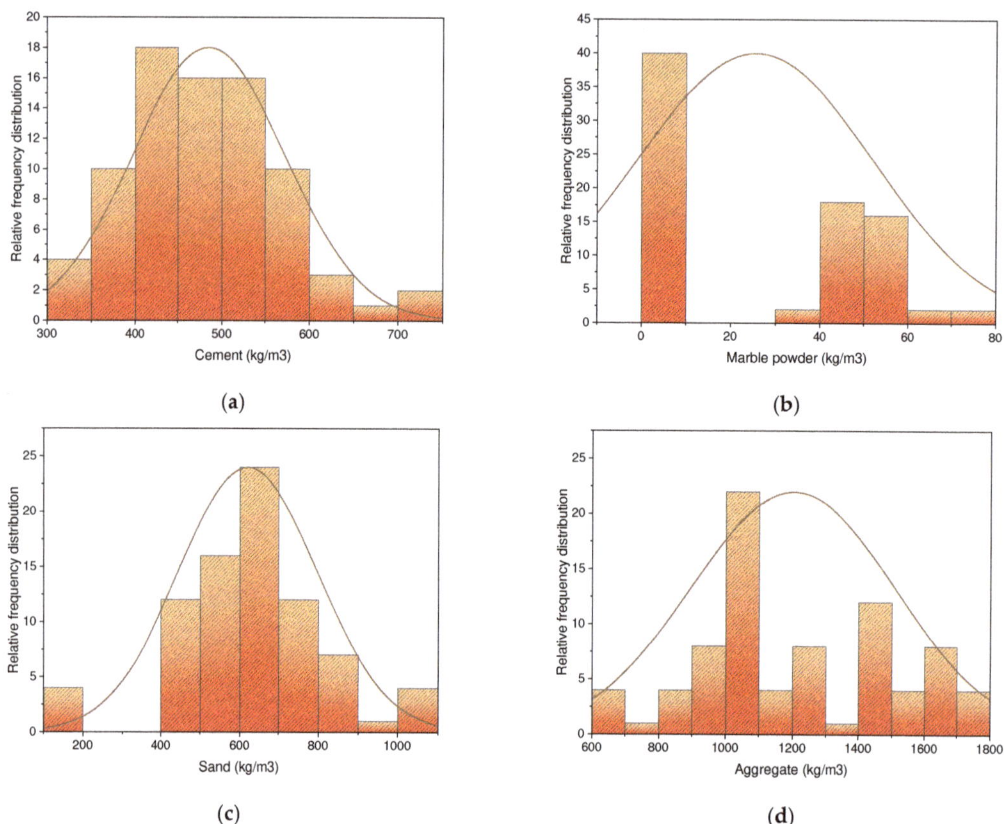

Figure 2. Input parameters relative frequency distribution: (**a**) cement; (**b**) marble powder; (**c**) sand; (**d**) aggregate.

4. Modelling Techniques Description

Concrete compressive strength prediction algorithms are described in this section. Individual ML (DT) and ensembled ML techniques (i.e., bagging models, random forest and AdaBoost) are employed over Anaconda software by using Python code. Spyder (version 4.3.5) of Anaconda navigator is opted for running the random forest, bagging models and AdaBoost. Such algorithms are usually used to predict required outcomes as per input variables. Six input parameters against one output parameter (i.e., compressive strength) are used for all techniques during the modelling phase. R^2 values demonstrate the accuracy/validity of all the models. The R^2 statistic (also named determination coefficients) evaluates the variance response variable as demonstrated by the model fitted against the mean response. It can also be stated as the measurement of how well a model fits this data. 0 value implies the comparison of fitting the mean and model, whereas 1 depicts a perfect fit among data and model. C.S prediction is made with individual, i.e., decision tree, and ensemble algorithm, i.e., bagging models, random forest, and AdaBoost. Figure 3 shows a detailed flowchart of the used algorithm. It may be noted that 50% of data is used for training, and rest of the 50% is used for testing and validation. The error between the experimental and predicted values is also reported for each algorithm, and a discussion is made in Section 6.

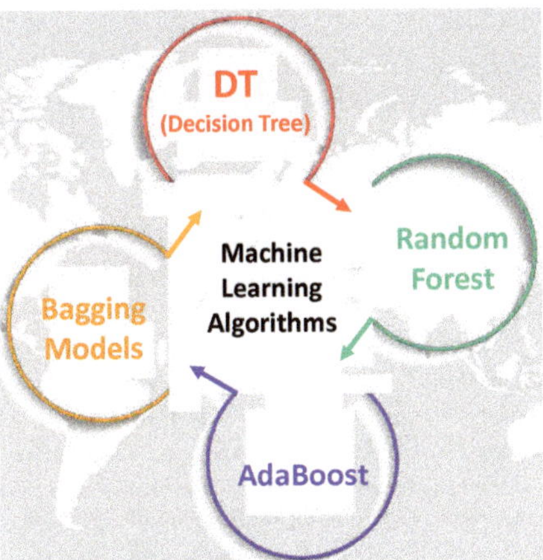

Figure 3. Algorithm flowchart.

4.1. Decision Tree Algorithm

DT is widely utilized to categorize regression problems and classify difficulties [87]. There are classes within a tree. However, the regression technique is used to predict outcome-independent variables in case of the non-existence of any class [88]. In DT, database attributes are represented by inner nodes. Conclusion rules are denoted by branches, whereas the leaf nodes represent the result. Two nodes, i.e., the decision node and leaf node, are the composition of a DT. Several branches of decision nodes can make a decision, and leaf nodes depicts. Leaf nodes depict the decision's output, lacking branches. It is named a decision tree as it resembles a tree-like structure that begins with grows as per the number of branches based on a root node [76]. Data samples are bifurcated in multiple segments by DT. An executed algorithm determines the difference between forecasted values and goal at each division point. Errors are also calculated at each division point, and the lowest value variable is selected as a split point for the fitness function, and the same procedure/method is repeated. Figure 4 depicts the decision tree schematic diagram.

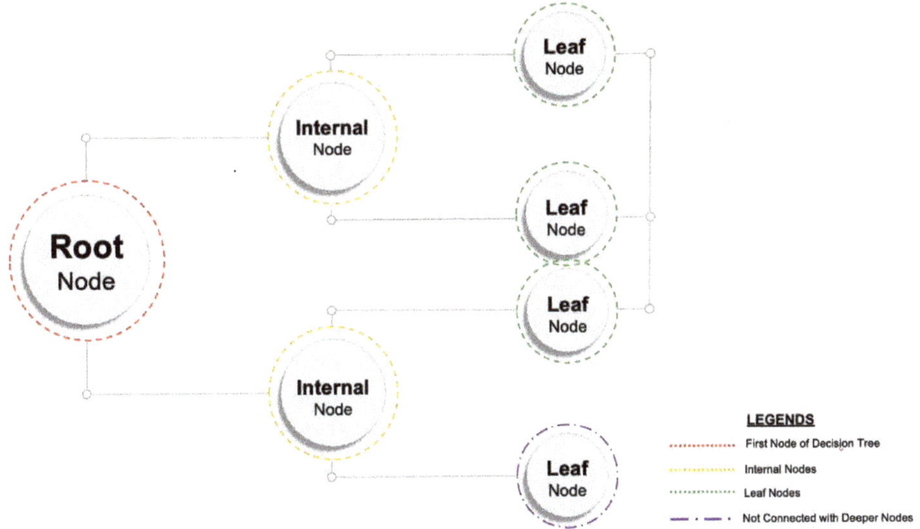

Figure 4. Decision tree schematic diagram.

4.2. Random Forest Algorithm

The random forest model is a regression and classification-based approach that has been studied by various researchers till now [86,89]. The compressive strength of concrete is predicted by using the RF model, as done by Shaqadan [90]. The prime difference between RF and DT is the number of trees as shown in Figure 5. A single tree is developed in DT; however, in RF, multiple trees are built that are known as forest. Dissimilar data are selected arbitrarily and accordingly, allocated to respective trees. Each tree has data in rows and columns, and different dimensions of rows and columns are selected. Following steps are carried out for the growth of each tree; the data frame comprises 2/3rd of the whole data that is randomly selected for each tree. This method is known as bagging. Random selection is made for prediction variables, and the node splitting is done by finely splitting these variables. For all trees, the remaining data are utilized to estimate out-of-bag error. Accordingly, the final out-of-bag error rate is assessed by combining errors from each tree. Each tree provides regression, and among all forest trees, the forest with greater votes is selected for the model. The value of votes can either be 1's or 0's. Prediction probability is specified by the obtained proportion of 1's. Among all ensemble algorithms, random forest (RF) is the most sophisticated one. It includes desirable features for variable importance measures (VIMs) with robust overfitting resistance and fewer model parameters. DT is used as a base predictor for RF. Acceptable results can be produced by RF models with default parameter settings [91]. As allowed by RF combinations of parameter settings, and base predictors can be reduced to one.

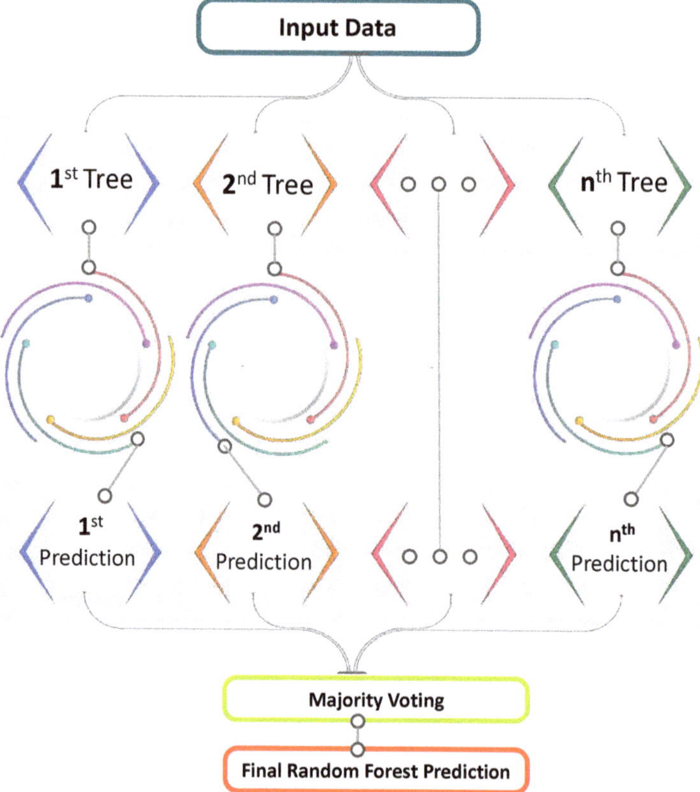

Figure 5. Random forest schematic diagram.

4.3. AdaBoost Algorithm

Figure 6 shows the entire process of forecasting the AR algorithm outcome. The Ensemble technique is a concept of ML that is utilized for training various models by using a learning algorithm of the same kind [92]. Multiple algorithms are collected, as multi-classifiers, for making an ensemble. A group comprises almost a thousand learners working with the same objective of resolving the issue. Ensemble learning is employed by an AdaBoost algorithm, which is a supervised ML technique. It can also be referred to as adaptive boosting, as weights are re-linked to every instance, with higher weights linked to wrongly classified instances. Boosting techniques are widely utilized to minimize variance and bias in supervised ML. Weak learners can be strengthened by using the said ensemble techniques. Infinite no. of DTs are employed by it for the input data during a training phase. During constructing the initial DT, the erroneously categorized recorded data are prioritized throughout the initial model. Same data records are used only as an input for different other models. The technique mentioned above is repeated till the creation of specified base learners. AdaBoost optimizes enhancement of DTs performance on binary classification issues. In addition, it is also used for enhancing ML algorithms performance. It is specifically effective when it is used with slow learners. These ensemble algorithms are very prevalent in the civil engineering field, especially for predicting concrete mechanical properties.

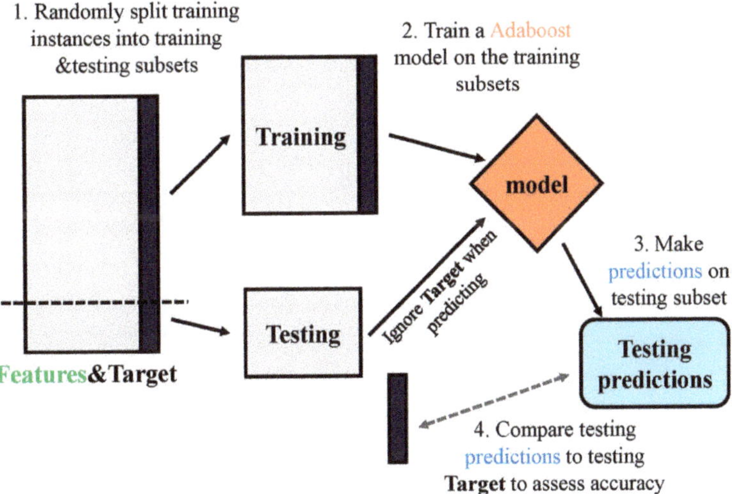

Figure 6. Complete process of prediction via AdaBoost algorithm [93].

4.4. Bagging Algorithm

The detailed procedural flow chart of the bagging algorithm is shown in Figure 7. It is basically an equivalent ensemble method that describes the prediction model variance by supplementation with additional data throughout the training stage. The technique of irregular sampling includes the data replacement from a primary set. Employing replaced sampling, every new training dataset can duplicate specific observations. In the procedure of bagging, for every component, there is an equal possibility of appearing in a new dataset. The training set size is not dependent on predictive force. Furthermore, variance may be remarkably declined by precisely tuning the prediction of the desired outcome. Additional models are trained by using these data sets. The mean of predictions by all models is used for this ensemble. In regression, the average of various models' predictions can be a forecast [94]. A total of twenty sub-models are being utilized for tweaking the bagging algorithm with DT to find the optimized value which produces firm output.

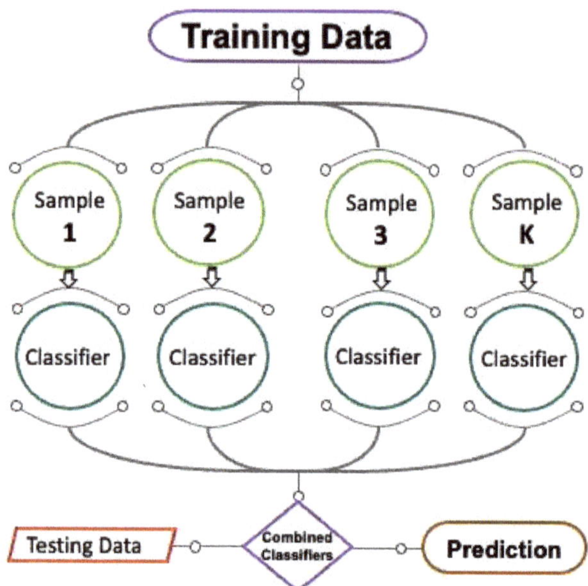

Figure 7. Bagging algorithm flow chart indicating the step-by-step procedure of prediction.

The flowchart depicting the research approach is shown in Figure 8. Given the three algorithms mentioned above anomaly, further to DT, a combination of ensembles (i.e., AdaBoost, bagging models, and random forest) algorithms is employed in this study for maximizing the respective benefits. Twenty sub-models are employed by ensembled strategies for the determination of ideal value, which develops a firm output. In addition, error evaluations such as mean square error (MSE), mean absolute error (MAE), k-fold cross-validation and root mean square error (RMSE), and statistical checks are made for evaluating the model's performance. Finally, the comparison of different machine learning models is made, as well as the suitability of waste marble powder in concrete for sustainable construction.

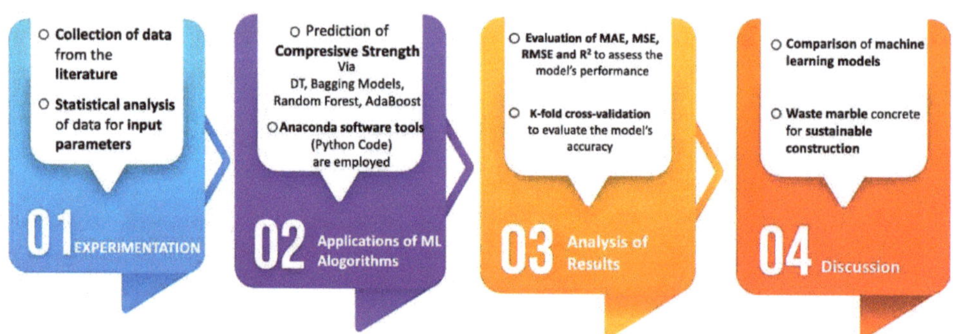

Figure 8. Research methodology.

5. Experimental Compressive Strength Test Results

From the compressive strength test results, it is identified that a decrement in compressive strength is observed with an increase in the content of marble powder in bricks (Figure 9). The highest C.S at 7-days and 28-days of 34.13 MPa and 41.03 MPa is obtained by M18, which contained 0% marble powder content. Specimens of waste marble powder group achieved a maximum compressive strength of 31.06 and 37.83 MPa at 7 and 28-days, respectively. The maximum decrease in waste marble concrete range is 9.97–48.14%, as compared to 7 days of plain mix. The maximum decrease in waste marble concrete range is 2.9–46.9%, compared to 28 days of plain mix. The increased porosity level with the increase in marble powder content in concrete, and hence the compressive strength is decreased. Şanal [57] reported enhancement of pore structure due to an increase in the capillary structure of concrete by adding 10% marble dust as cement replacement, ultimately resulting in reduced mechanical properties of concrete. This can be caused by the dissimilar C_3A—tricalcium aluminate content in cement due to its replacement by marble dust [50]. However, in the current study, the worst mechanical property was observed that might result from the increase in the capillary structure of the pores with the addition of marble dust, as reported in the previous study [57].

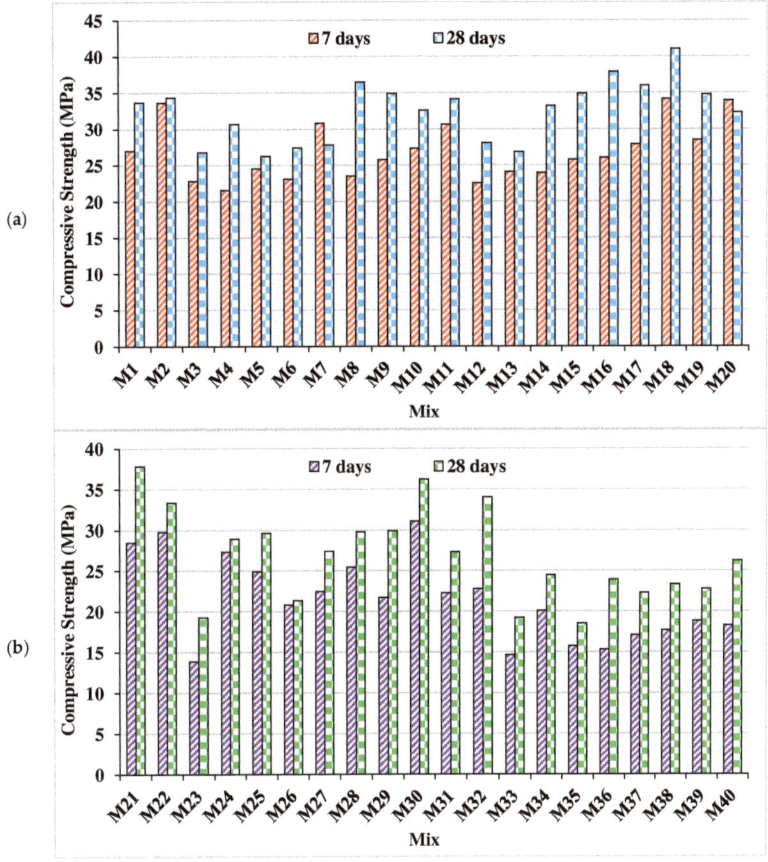

Figure 9. Experimental compressive strength; (**a**) plain concrete; (**b**) marble powder concrete.

6. Analysis and Modelling Results

6.1. Prediction of Compressive Strength by Different Models

i.   Decision tree modelling

Figure 10 depicts a statistical analysis of projected and actual results regarding C.S of WMC for DT modelling. A reasonably précised output and a very low variation between anticipated and actual values can be obtained by DT technique. The accuracy of predicting results can be assessed by having a 0.86 R^2 value. The blue line represents the correlation between the experimental and predicted values, as evident by the R^2 value. The higher R^2 denotes the higher accuracy of the model. The dispersion for predicted and experimental values (targets) and DT model errors is shown in Figure 11. The average, highest, and lowest values of the training set are 6.20, 20.7, and 0.07 MPa, respectively. Whereas 12.5% error values are less than 1 MPa, 37.5% are from 2 to 5 MPa, 32.5% are from 6 to 10 MPa, and 17.5% are higher than 5 MPa.

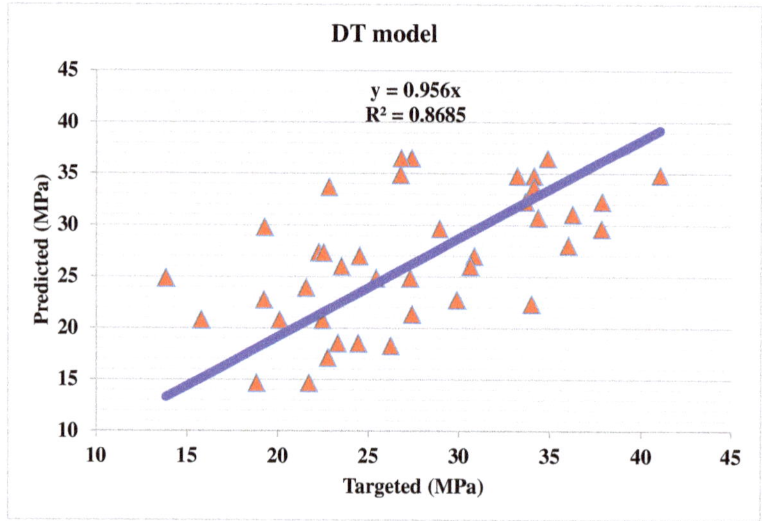

Figure 10. Predicted and actual results of DT model.

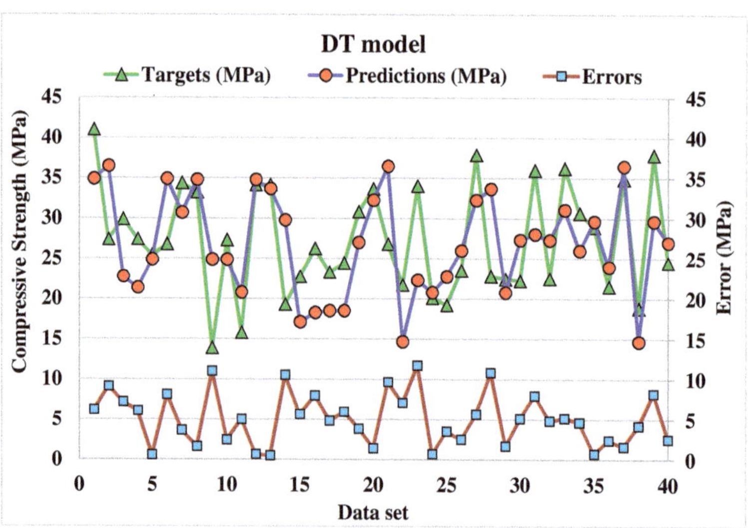

Figure 11. Dispersion of predicted and actual values along with errors for DT model.

ii.   Random forest modelling

The correlation between projected and actual results of RF model is shown in Figure 12. The R^2 value for the RF model comes out to be 0.97, which represents the highly precise and more accurate of RF w.r.t Bg, DT, and AdB models. Furthermore, the dispersion of projected values, actual targeted values and errors for RF model is shown in Figure 13. The minimum, maximum, and average error values are 0.07, 10.9 and 3.93 MPa. It is noted that 15% of error data are below 1 MPa, 57.5% from 2 to 5 MPa, 22.5% from 6 to10 MPa, and only 5% higher than 10 MPa. This analysis reveals the higher accuracy of RF model w.r.t AdB, DT, and Bg models. It can also be depicted from lower error and greater R^2 values. In addition, twenty sub-models are employed by EML (Bg, DT, and AdB) to get the optimized value that produces a firm output.

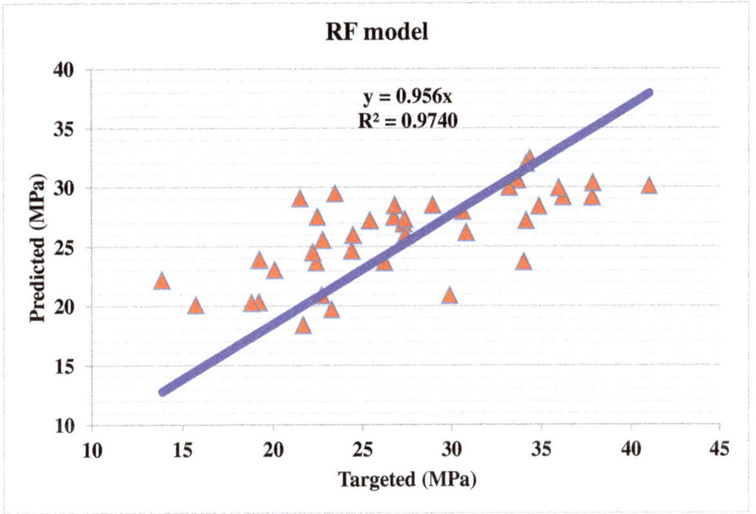

Figure 12. Predicted and actual results for RF model.

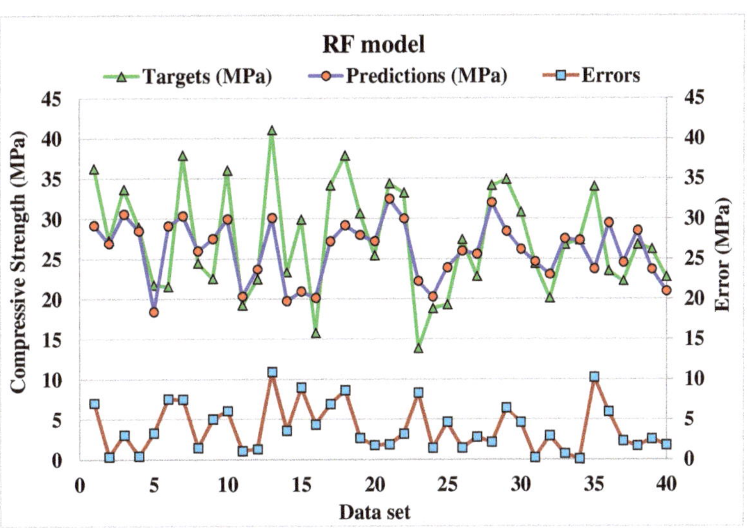

Figure 13. Dispersion of predicted and actual values along with errors for RF model.

iii. AdaBoost modelling

A comparison of projected and actual outputs of AdB model is shown in Figures 14 and 15. The correlation between them is illustrated in Figure 14. The R^2 value is 0.91, which shows better outcomes when compared to the DT model. The dispersion of actual and predicted values along with errors for AdB model is illustrated in Figure 15. 19.7, 0.15, and 6.34 MPa are the maximum, minimum, and average values for the training set. Whereas 27.5% of error values are below 1 MPa, 20% range from 2 to 5 MPa, 30% range from 6 to 10 MPa, and only 22.5% are higher than 10 MPa. The higher accuracy of AdB model in comparison with the DT model is also depicted by lower error values.

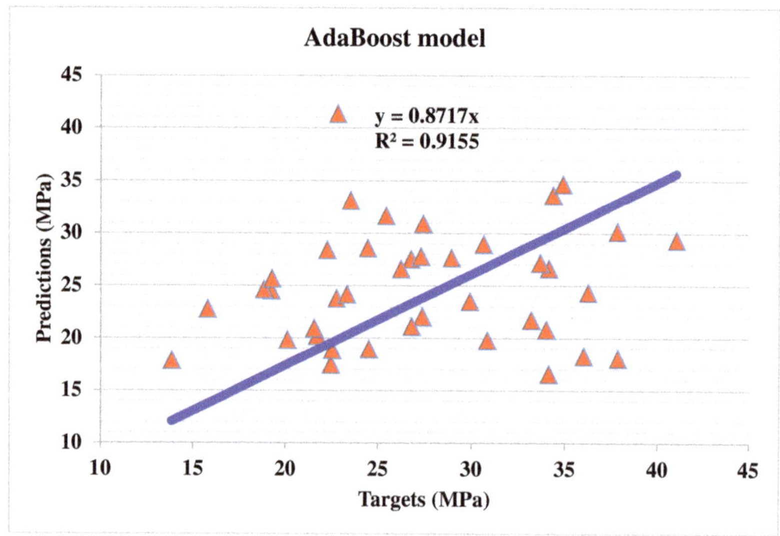

Figure 14. Predicted and actual results for AdaBoost model.

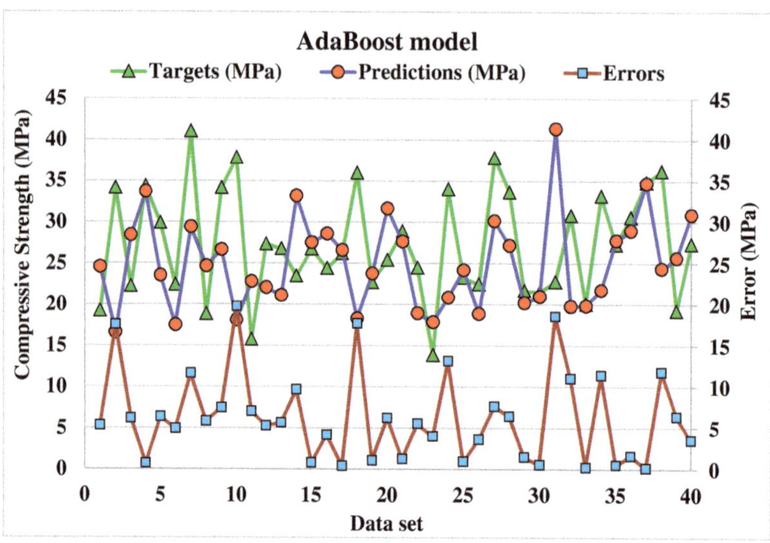

Figure 15. Dispersion of predicted and actual values along with errors for AdaBoost model.

iv.   Bagging modelling

The correlation between predicted and actual output values for Bg model is provided in Figure 16. The R^2 value for this model comes out to be 0.95, showing considerable accuracy as compared to that of DT and AdB models. The dispersion of actual and predicted values and errors for the Bg model is shown in Figure 17. The maximum, average, and minimum in the training set are 11.07, 3.96, and 0.01 MPa, respectively. Whereas only 25% of error values are below 1 MPa, 45% of values range from 2 to 5 MPa, and 27.5% values range from 6 to 10 MPa. The error distribution and R^2 are more accurate than that of DT and AdB models for the C.S prediction of WMC. Whereas the R^2 and error values obtained from all considered ensembled ML models are in an acceptable range, thus depicting better prediction outcomes. Hence, it is observed in this study that EML techniques (RF, AdB and Bg) can predict high accuracy outcomes when compared to standalone DT techniques.

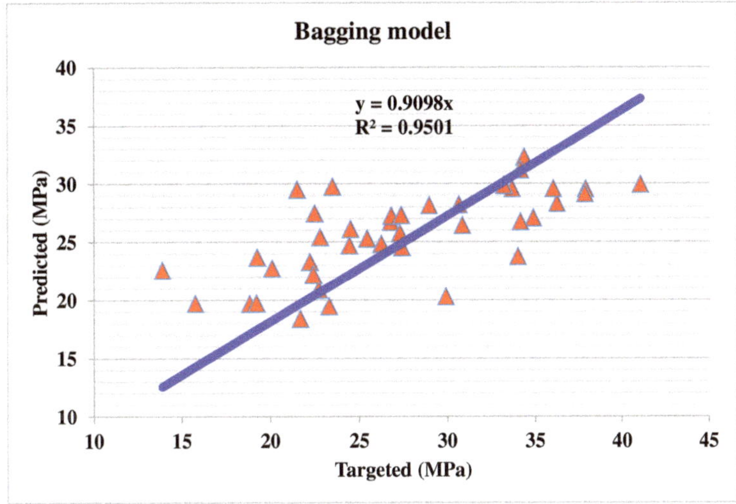

Figure 16. Predicted and actual results for bagging model.

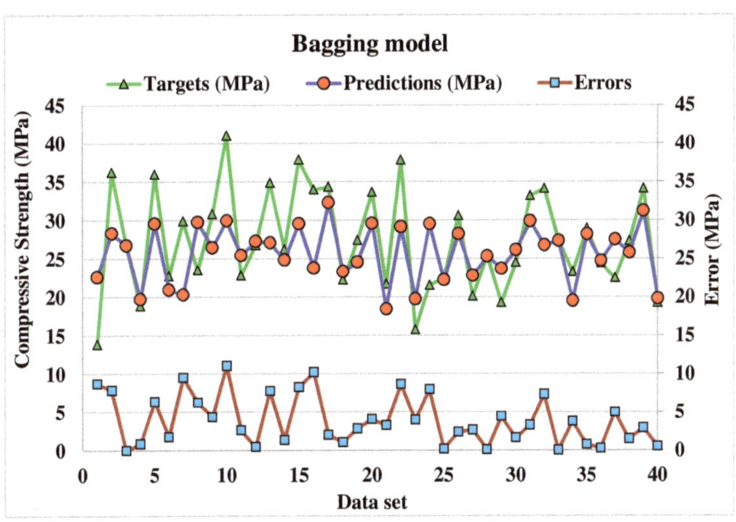

Figure 17. Dispersion of predicted and actual values along with errors for bagging model.

6.2. K-Fold Cross Validation Checks

Statistical analysis with Equations (1)–(3) is utilized to predict the model's response. The model's legitimacy is evaluated by utilizing the k-fold cross-validation approach during execution [95–97]. Usually, the validity of the model is done with a k-fold cross validation process [92], in which random dispersion is done by splitting it into ten groups. The greater the R^2 value and less the errors (RMSE and MAE), the more a model's accuracy is. Furthermore, this process should be repeated multiple (i.e., 10) times for a satisfactory result. The exceptional precision of the model can be achieved by using this comprehensive approach. In addition, statistical analysis (i.e., RMSE and MSE) is also performed for all the models (Table 6). The RF model accuracy (inversely related to error values) compared to AdB, Bg, and DT models is also supported by these checks. Statistical analysis, as reported in the literature [98–100], is used to assess the model's response to the prediction. The k fold cross validation is assessed by utilizing R^2, MSE, and MAE. Respective dispersions for the decision tree, random forest, AdaBoost, and bagging models are presented in Figures 18–21. Minimum, average, and maximum values of R^2 for the decision tree are 0.52, 0.68, and 0.86, respectively (refer to Figure 18). Whereas the maximum, average and minimum values of R^2 for random forest are 0.97, 0.78, and 0.66, respectively (see Figure 19). Contrary to it, the maximum, minimum, and average R^2 values of the AdaBoost model are 0.91, 0.53, and 0.71, respectively, as portrayed in Figure 20. The maximum, average, and minimum values of R^2 for Bg model are 0.95, 0.78, and 0.64, respectively are shown in Figure 21. Upon comparing error values (MSE and MAE), the average MSE and MAE values for DT model are 11.58 and 9.45, respectively. Whereas, average MSE and MAE values for AdaBoost model are 10.08 and 8.45, respectively, and average MSE and MAE values for the Bg model are 7.65 and 7.03, respectively. The RF model with the lowest error and higher R^2 value performs better for results prediction.

$$\text{MAE} = \frac{1}{n} \sum_{i=1}^{n} |x_i - x| \tag{1}$$

$$\text{MSE} = \frac{1}{n} \sum_{i=1}^{n} \left(y_{pred} - y_{ref} \right)^2 \tag{2}$$

$$\text{RMSE} = \sqrt{\sum \frac{\left(y_{pred} - y_{ref} \right)^2}{N}} \tag{3}$$

where:

n = Total data samples,

x, y_{ref} = data sample reference values,

x_i, y_{pred} = model prediction values.

Table 6. Statistical checks of decision tree, random forest, AdaBoost, and bagging models.

Models	Mean Absolute Error (MPa)	Mean Square Error (MPa)	Root Mean Square Error (MPa)
Decision tree	6.204	26.173	5.116
Random forest	3.937	24.197	4.919
AdaBoost	4.665	30.947	5.563
Bagging	3.969	62.679	7.917

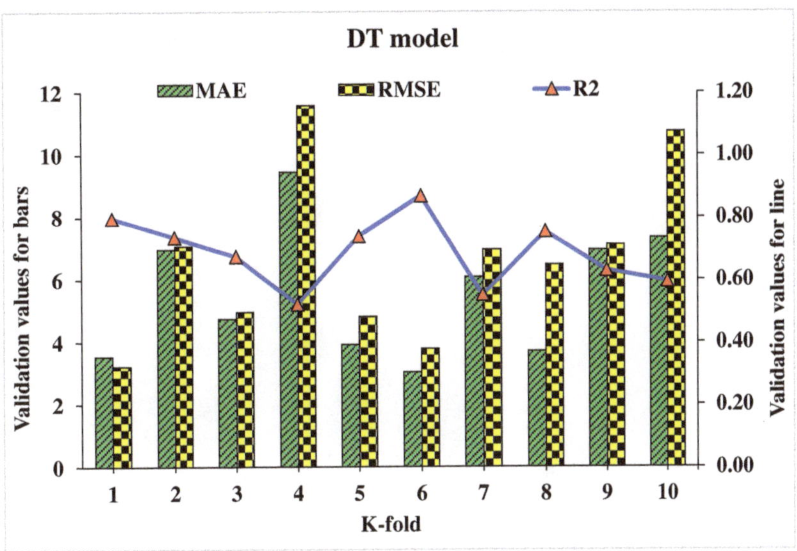

Figure 18. Statistical analysis of DT model for K-fold cross-validation.

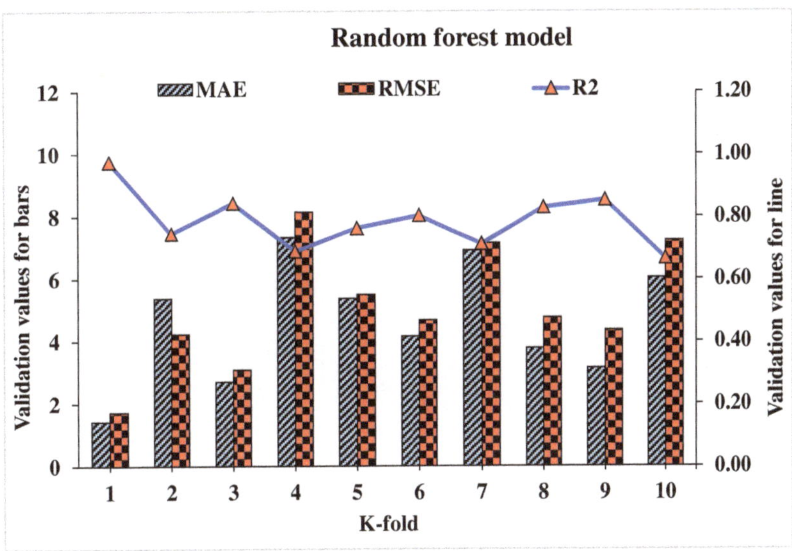

Figure 19. Statistical analysis of RF model for K-fold cross-validation.

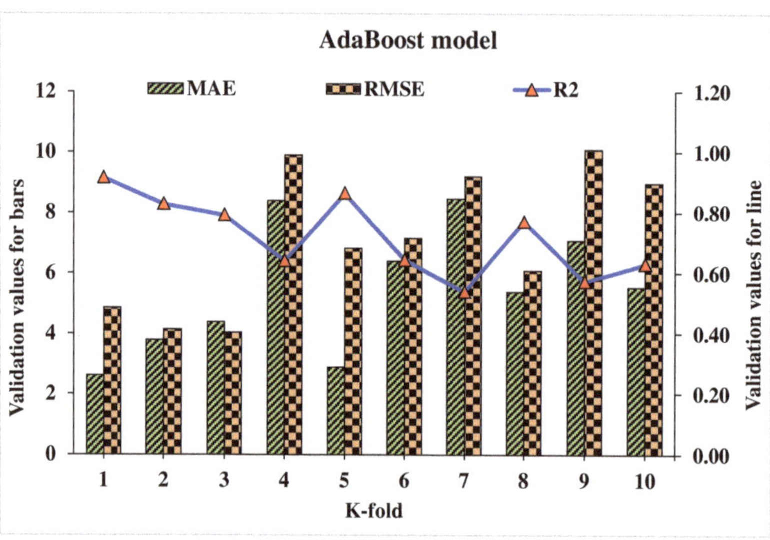

Figure 20. Statistical analysis of AdaBoost model for K-fold cross-validation.

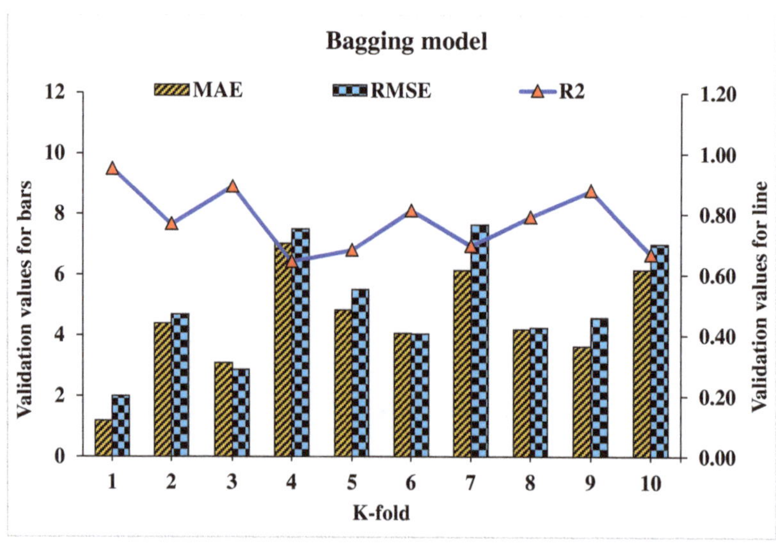

Figure 21. Statistical analysis of bagging model for K-fold cross-validation.

7. Discussion

7.1. Comparison of Machine Learning Models

Ensembled ML and individual approaches are explored in this study to estimate WMC with the aim of sustainable development in environment-friendly construction materials. RF, Bg, AdB, and DT machine learning techniques are used in this study to predict the compressive strength of WMC. The DT algorithm's goal is to develop a model that can predict the target variable accurately, for which a tree like structure, i.e., a decision tree, is developed for problem-solving. In DT, the class label is represented by a leaf node and attributes are represented by interior node. Both variance and bias are reduced by boosting supervised learning. Learners develop this idea sequentially on which it is based. The growth of all subsequent learners is based on prior learners, except for the initial one. In this way, strong learners are formed from weak ones. Whereas, in bagging technique, a random sample is selected for data from the training set; i.e., the selection of individual data points can be made multiple times. Individual training of said weak models is done in pursuance of numerous data samples generation and based on task type like; classification or regression, the average and/or majority of these predictions give an estimate with high accuracy. To establish the algorithm's prediction superiority, employed algorithms are compared for targeted performance. The output of the random forest model comes out to be more accurate, having a 0.97 R^2 value, compared to bagging with 0.95 R^2, AdB with 0.91 R^2, and DT with 0.86 R^2. Furthermore, the performance of AdB, RF, DT, and Bg models is also evaluated by utilizing the k-fold cross-validation technique and statistical analysis. The performance of the model is higher with low error levels. But it is tough to assess optimized machine learning regressors to forecast results from a wide range of topics because the model's performance is very much dependable on data points and the model's input parameters. On the other hand, in ensemble ML techniques, sub-models are generated to leverage the weak learner that can be optimized and trained on data for achieving the higher value of R^2. Dispersion of values for the determinant coefficient of AdB, bg, and RF sub-models is shown in Figure 22. The values of R^2 for all sub-models of RF are greater than 0.76, as shown in Figure 22a, while most values of R^2 in the case of sub-models for AdB and Bg are less than 0.51 (Figure 22b) and 0.66 (Figure 22c), respectively. It depicts higher accuracy of RF technique for results prediction having a maximum value of R^2, i.e., 0.97. Therefore, the RF model is suggested to predict the compressive strength of waste materials such as marble powder.

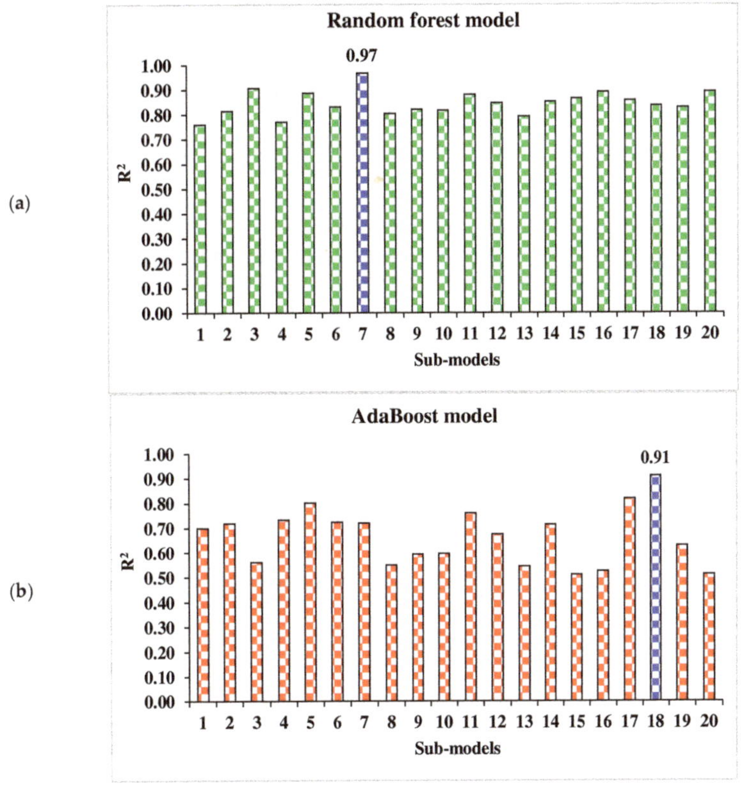

Figure 22. *Cont.*

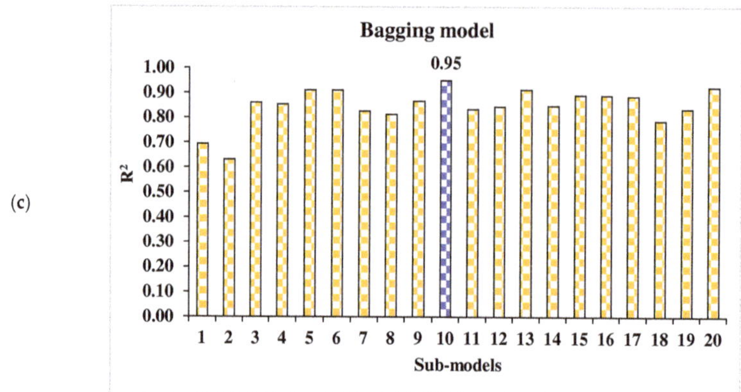

Figure 22. Determination coefficient (R^2) values for sub-models. (**a**) Random forest; (**b**) AdaBoost; (**c**) Bagging.

7.2. Waste Marble Concrete for Sustainable Construction

Planet earth is facing destruction of the ecosystem in terms of ground contamination, water pollution, and air quality. These are the leading causes of severe diseases leading to mortality. In addition to health issues, pollution is also the main hindrance to achieving sustainability. A substantial expense for society and the economy is imposed by high levels of environmental pollution, i.e., air, water, and land treatment. Construction wastes are a major contributor to environmental pollution. Singh, et al. [101] reported that 30% of marble is wasted during processing because of its uneven shape or smaller size. In the case of semi-processed slabs, the quantity of waste is 2–5%. In a vertical/horizontal cutter, one ton of processed marble stone produces nearly one ton of slurry with 35–45% water content. Construction industries are expanding too quickly, resulting in a massive amount of waste, wreaking havoc on the environment in terms of air pollution, water pollution, and soil deterioration, such as waste generated by marble industries. To address this major challenge, strong strategy action is required. Researchers/engineers are more focused on the effective usage of waste materials in the construction industry to minimize the challenge mentioned above. The incorporation of waste materials, such as marble powder, is among the effective steps toward sustainability as it would not only reduce the impact on the environment, but would also save natural resources and lower the project's overall cost, ultimately bringing economic value for waste materials. According to this viewpoint, the building sector is the primary focus for the reuse of waste products such as waste marble and granite, natural waste fibers, aggregate, and mortar wastes, etc. These wastes may be used in large-scale concrete production, whereas renewable resources such as natural sand may last longer and minimize cement usage, resulting in more productive fields, lower project costs, and reduced environmental contamination risk. In the current research, waste marble powder usage is pointed out for concrete manufacturing to reduce waste disposal problems as shown in Figure 23. The concrete blocks are mostly used in the interior and the exterior of buildings. Blocks are used for partition as non-load bearing walls when used in frame structures that are constructed with reinforced cement concrete (RCC). The waste marble powder concrete blocks can deliver several flexible choices that can be used to customize one's home aesthetics with minimum effort. Because of this functionality, concrete blocks allow design ideas for innovation in the street and building floors. Sustainable concrete blocks are readily recyclable, thus reducing the overall cost of building construction, ultimately eliminating potential pollution. Marble powder is added to concrete to make these blocks which can be used in the construction of roadside walkways. C.S of concrete is reduced by adding waste marble powder to it, as reported in the current study, allowing its application in emergency light-weight structures such as shelter homes, hospitals after earthquakes and flooding, and restrooms for passengers on highways and in railway/bus stations. In this scenario, waste marble powder concrete blocks are proposed to be used as sustainable construction material.

Figure 23. Marble powder disposal problems.

8. Conclusions

Marble stone waste materials are a major concern for the construction industry. Accordingly, the incorporation of marble waste powder in concrete composite during its manufacturing could be an effective addition to the category of sustainable construction materials and an effective effort to improve the surrounding environment. For this purpose, an approach has been made to use marble powder with different proportions in concrete. Additionally, this study aims to explore the usage of ensembles machine learning (ML) and individual approaches for the prediction of compressive strength (C.S) of waste marble concrete (WMC). Forecasting the compressive strength of waste marble concrete is achieved by utilizing random forest (RF), AdaBoost (AdB), bagging (Bg), and decision tree (DT) techniques. The conclusions are as follows:

- Bricks manufactured of 10% marble powder as a substitute had the highest compressive strengths of 37.8 MPa at 28 days. Such type of waste marble concrete may be used in the form of blocks for emergency light-weight structures such as hospitals and refuge homes during earthquakes and flooding. In this scenario, WMC having a 10% marble powder content (as a substitute) is proposed to be used as construction material.
- The random forest model has come out to be most effective in terms of prediction with respect to AdaBoost, bagging, and decision tree approaches due to higher values of R^2 with lower error values. Decision tree, random forest, AdaBoost, and bagging models have R^2 values of 0.86, 0.97, 0.91, and 0.95, respectively. However, the findings of ensembled models (RF, AdaBoost, and bagging) are within an acceptable range.
- Satisfactory outputs of random forest, AdaBoost, and bagging are also demonstrated by the k-fold cross-validation approach and statistical analysis. In addition, the higher performance of the random forest model with respect to the decision tree, AdaBoost, and bagging models is also established through these checks.
- ML can achieve more accurate prediction of material strength properties approaches without putting additional effort and time for sampling, casting, curing, and testing.

This study evaluated the compressive strength of waste marble concrete considering limited mix proportions with limited input parameters. However, in the future, increasing the number of datasheets and importing a substantially higher number of mixtures and considering higher input parameters could result in a better applicable model. As a result, experimental work, field tests, and numerical analysis employing a variety of methodologies should be used to increase the quantity of data points and outcomes in future investigations (e.g., the Monte Carlo simulation).

Supplementary Materials: The following supporting information can be downloaded at: https://www.mdpi.com/article/10.3390/ma15124108/s1, Table S1: Data used for modeling.

Author Contributions: K.K.: conceptualization, funding acquisition, resources, project administration, supervision, writing, reviewing, and editing. W.A.: conceptualization, data curation, software, methodology, investigation, validation, writing—original draft. M.N.A.: methodology, investigation, writing, reviewing, and editing. A.A.: resources, visualization, writing, reviewing, and editing. S.N.: data curation, methodology, writing, reviewing, and editing. S.N.: visualization, writing, reviewing, and editing. A.A.A.: Software, resources, validation, visualization, writing, reviewing, and editing. A.M.A.A.: investigation, resources, validation, visualization, writing, reviewing, and editing. All authors have read and agreed to the published version of the manuscript.

Funding: This work was supported by the Deanship of Scientific Research, Vice Presidency for Graduate Studies and Scientific Research, King Faisal University, Saudi Arabia [Project No. GRANT706].

Institutional Review Board Statement: Not applicable.

Informed Consent Statement: Not applicable.

Data Availability Statement: The data used in this research has been properly cited and reported in the main text.

Acknowledgments: The authors acknowledge the Deanship of Scientific Research, Vice Presidency for Graduate Studies and Scientific Research, King Faisal University, Saudi Arabia [Project No. GRANT706].

Conflicts of Interest: The authors declare no conflict of interest.

References

1. Rana, A.; Kalla, P.; Csetenyi, L.J. Recycling of dimension limestone industry waste in concrete. *Int. J. Min. Reclam. Environ.* **2017**, *31*, 231–250. [CrossRef]
2. Rana, A.; Kalla, P.; Verma, H.; Mohnot, J. Recycling of dimensional stone waste in concrete: A review. *J. Clean. Prod.* **2016**, *135*, 312–331. [CrossRef]
3. Munir, M.J.; Kazmi, S.M.S.; Wu, Y.-F.; Hanif, A.; Khan, M.U.A. Thermally efficient fired clay bricks incorporating waste marble sludge: An industrial-scale study. *J. Clean. Prod.* **2018**, *174*, 1122–1135. [CrossRef]
4. Seghir, N.T.; Mellas, M.; Sadowski, Ł.; Żak, A. Effects of marble powder on the properties of the air-cured blended cement paste. *J. Clean. Prod.* **2018**, *183*, 858–868. [CrossRef]
5. Ashish, D.K. Concrete made with waste marble powder and supplementary cementitious material for sustainable development. *J. Clean. Prod.* **2019**, *211*, 716–729. [CrossRef]
6. Demirel, B.; Alyamaç, K.E. Waste marble powder/dust. In *Waste and Supplementary Cementitious Materials in Concrete*; Elsevier: Amsterdam, The Netherlands, 2018; pp. 181–197.
7. MSME. *Status Report on Commercial Utilization of Marble Slurry in Rajasthan*; Ministry of Micro Small and Medium Enterprisese Development Institute, Govt. of India, Ministry of Micro, Small & Medium Enterprises: Jaipur, India, 2016.
8. Sarkar, R.; Das, S.K.; Mandal, P.K.; Maiti, H.S. Phase and microstructure evolution during hydrothermal solidification of clay–quartz mixture with marble dust source of reactive lime. *J. Eur. Ceram. Soc.* **2006**, *26*, 297–304. [CrossRef]
9. Aliabdo, A.A.; Abd Elmoaty, M.; Auda, E.M. Re-use of waste marble dust in the production of cement and concrete. *Constr. Build. Mater.* **2014**, *50*, 28–41. [CrossRef]
10. USGS. *Stone (Dimension), Dimension Stone Statistics and Information*; National Minerals Information Center: Reston, VA, USA, 2018.
11. IBM. *Indian Minerals Yearbook 2016 (Parte III: Mineral Reviews)*, 55th Edition Marble ed.; Ministry of Mines, Indian Bureau of Mines, Government of India: Nagpur, India, 2018.
12. Li, L.; Huang, Z.; Tan, Y.; Kwan, A.; Liu, F. Use of marble dust as paste replacement for recycling waste and improving durability and dimensional stability of mortar. *Constr. Build. Mater.* **2018**, *166*, 423–432. [CrossRef]

13. Mehdi, A.; Chaudhry, M.A. *Diagnostic Study Marble & Granite Cluster Rawalpindi Pakistan*; UNIDO-SMEDA Cluster Development Programme: Rawalpindi, Pakistan, 2006.
14. Khodabakhshian, A.; Ghalehnovi, M.; De Brito, J.; Shamsabadi, E.A. Durability performance of structural concrete containing silica fume and marble industry waste powder. *J. Clean. Prod.* **2018**, *170*, 42–60. [CrossRef]
15. Mashaly, A.O.; El-Kaliouby, B.A.; Shalaby, B.N.; El–Gohary, A.M.; Rashwan, M.A. Effects of marble sludge incorporation on the properties of cement composites and concrete paving blocks. *J. Clean. Prod.* **2016**, *112*, 731–741. [CrossRef]
16. Arel, H.Ş. Recyclability of waste marble in concrete production. *J. Clean. Prod.* **2016**, *131*, 179–188. [CrossRef]
17. Binici, H.; Kaplan, H.; Yilmaz, S. Influence of marble and limestone dusts as additives on some mechanical properties of concrete. *Sci. Res. Essays* **2007**, *2*, 372–379.
18. Al-Akhras, N.M.; Ababneh, A.; Alaraji, W.A. Using burnt stone slurry in mortar mixes. *Constr. Build. Mater.* **2010**, *24*, 2658–2663. [CrossRef]
19. Alyamac, K.E.; Ghafari, E.; Ince, R. Development of eco-efficient self-compacting concrete with waste marble powder using the response surface method. *J. Clean. Prod.* **2017**, *144*, 192–202. [CrossRef]
20. Shirazi, E.K. Reusing of stone waste in various industrial activities. In Proceedings of the 2011 2nd International Conference on Environmental Science and Development IPCBEE, Singapore, 26–28 February 2011.
21. Sadek, D.M.; El-Attar, M.M.; Ali, H.A. Reusing of marble and granite powders in self-compacting concrete for sustainable development. *J. Clean. Prod.* **2016**, *121*, 19–32. [CrossRef]
22. Ashish, D.K.; Verma, S.K. An overview on mixture design of self-compacting concrete. *Struct. Concr.* **2019**, *20*, 371–395. [CrossRef]
23. Brundtland, G.H. *Report of the World Commission on Environment and Development: "Our Common Future"*; UN: New York, NY, USA, 1987.
24. Levy, S.M.; Helene, P. Durability of recycled aggregates concrete: A safe way to sustainable development. *Cem. Concr. Res.* **2004**, *34*, 1975–1980. [CrossRef]
25. de Azevedo, A.R.; Alexandre, J.; Xavier, G.d.C.; Pedroti, L.G. Recycling paper industry effluent sludge for use in mortars: A sustainability perspective. *J. Clean. Prod.* **2018**, *192*, 335–346. [CrossRef]
26. Li, H.; Xu, W.; Yang, X.; Wu, J. Preparation of Portland cement with sugar filter mud as lime-based raw material. *J. Clean. Prod.* **2014**, *66*, 107–112. [CrossRef]
27. Li, H.; Dong, L.; Jiang, Z.; Yang, X.; Yang, Z. Study on utilization of red brick waste powder in the production of cement-based red decorative plaster for walls. *J. Clean. Prod.* **2016**, *133*, 1017–1026. [CrossRef]
28. Klee, H. *The Cement Sustainability Initiative: Recycling Concrete*; World Business Council for Sustainable Development (WBCSD): Geneva, Switzerland, 2009.
29. Amin, M.N.; Khan, K.; Saleem, M.U.; Khurram, N.; Niazi, M.U.K. Aging and curing temperature effects on compressive strength of mortar containing lime stone quarry dust and industrial granite sludge. *Materials* **2017**, *10*, 642. [CrossRef] [PubMed]
30. Miller, S.A.; Moore, F.C. Climate and health damages from global concrete production. *Nat. Clim. Chang.* **2020**, *10*, 439–443. [CrossRef]
31. Environment, U.; Scrivener, K.L.; John, V.M.; Gartner, E.M. Eco-efficient cements: Potential economically viable solutions for a low-CO_2 cement-based materials industry. *Cem. Concr. Res.* **2018**, *114*, 2–26.
32. Habert, G.; Miller, S.A.; John, V.M.; Provis, J.L.; Favier, A.; Horvath, A.; Scrivener, K.L. Environmental impacts and decarbonization strategies in the cement and concrete industries. *Nat. Rev. Earth Environ.* **2020**, *1*, 559–573. [CrossRef]
33. Thomas, B.S.; Yang, J.; Bahurudeen, A.; Abdalla, J.A.; Hawileh, R.; Hamada, H.M.; Nazar, S.; Jittin, V.; Ashish, D.K. Sugarcane bagasse ash as supplementary cementitious material in concrete–A review. *Mater. Today Sustain.* **2021**, *15*, 100086. [CrossRef]
34. Farooqi, M.U.; Ali, M. Effect of pre-treatment and content of wheat straw on energy absorption capability of concrete. *Constr. Build. Mater.* **2019**, *224*, 572–583. [CrossRef]
35. Farooqi, M.U.; Ali, M. Effect of Fibre Content on Splitting-Tensile Strength of Wheat Straw Reinforced Concrete for Pavement Applications. In *Key Engineering Materials*; Trans Tech Publications Ltd.: Bäch SZ, Switzerland, 2018; pp. 349–354. [CrossRef]
36. Khan, M.; Ali, M. Improvement in concrete behavior with fly ash, silica-fume and coconut fibres. *Constr. Build. Mater.* **2019**, *203*, 174–187. [CrossRef]
37. Cao, M.; Xie, C.; Li, L.; Khan, M. Effect of different PVA and steel fiber length and content on mechanical properties of $CaCO_3$ whisker reinforced cementitious composites. *Mater. De Construcción* **2019**, *69*, e200. [CrossRef]
38. Cao, M.; Khan, M.; Ahmed, S. Effectiveness of Calcium Carbonate Whisker in Cementitious Composites. *Period. Polytechnica. Civ. Eng.* **2020**, *64*, 265. [CrossRef]
39. Xie, C.; Cao, M.; Khan, M.; Yin, H.; Guan, J. Review on different testing methods and factors affecting fracture properties of fiber reinforced cementitious composites. *Constr. Build. Mater.* **2021**, *273*, 121766. [CrossRef]
40. Khan, M.; Cao, M.; Xie, C.; Ali, M. Efficiency of basalt fiber length and content on mechanical and microstructural properties of hybrid fiber concrete. *Fatigue Fract. Eng. Mater. Struct.* **2021**, *44*, 2135–2152. [CrossRef]
41. Khan, M.; Cao, M.; Xie, C.; Ali, M. Hybrid fiber concrete with different basalt fiber length and content. *Struct. Concr.* **2021**, *23*, 346–364. [CrossRef]
42. Khan, M.; Cao, M.; Hussain, A.; Chu, S. Effect of silica-fume content on performance of $CaCO_3$ whisker and basalt fiber at matrix interface in cement-based composites. *Constr. Build. Mater.* **2021**, *300*, 124046. [CrossRef]

43. Arshad, S.; Sharif, M.B.; Irfan-ul-Hassan, M.; Khan, M.; Zhang, J.-L. Efficiency of supplementary cementitious materials and natural fiber on mechanical performance of concrete. *Arab. J. Sci. Eng.* **2020**, *45*, 8577–8589. [CrossRef]
44. Ahmad, W.; Ahmad, A.; Ostrowski, K.A.; Aslam, F.; Joyklad, P.; Zajdel, P. Sustainable approach of using sugarcane bagasse ash in cement-based composites: A systematic review. *Case Stud. Constr. Mater.* **2021**, *15*, e00698. [CrossRef]
45. Keleştemur, O.; Arıcı, E.; Yıldız, S.; Gökçer, B. Performance evaluation of cement mortars containing marble dust and glass fiber exposed to high temperature by using Taguchi method. *Constr. Build. Mater.* **2014**, *60*, 17–24. [CrossRef]
46. Rodrigues, R.d.; De Brito, J.; Sardinha, M. Mechanical properties of structural concrete containing very fine aggregates from marble cutting sludge. *Constr. Build. Mater.* **2015**, *77*, 349–356. [CrossRef]
47. Aruntaş, H.Y.; Gürü, M.; Dayı, M.; Tekin, I. Utilization of waste marble dust as an additive in cement production. *Mater. Des.* **2010**, *31*, 4039–4042. [CrossRef]
48. Badurdeen, F.; Aydin, R.; Brown, A. A multiple lifecycle-based approach to sustainable product configuration design. *J. Clean. Prod.* **2018**, *200*, 756–769. [CrossRef]
49. AYDIN, E. Effects of elevated temperature for the marble cement paste products for better sustainable construction. *Politek. Derg.* **2019**, *22*, 259–267.
50. Gesoğlu, M.; Güneyisi, E.; Kocabağ, M.E.; Bayram, V.; Mermerdaş, K. Fresh and hardened characteristics of self compacting concretes made with combined use of marble powder, limestone filler, and fly ash. *Constr. Build. Mater.* **2012**, *37*, 160–170. [CrossRef]
51. Topcu, I.B.; Bilir, T.; Uygunoğlu, T. Effect of waste marble dust content as filler on properties of self-compacting concrete. *Constr. Build. Mater.* **2009**, *23*, 1947–1953. [CrossRef]
52. Gencel, O.; Ozel, C.; Koksal, F.; Erdogmus, E.; Martínez-Barrera, G.; Brostow, W. Properties of concrete paving blocks made with waste marble. *J. Clean. Prod.* **2012**, *21*, 62–70. [CrossRef]
53. Ergün, A. Effects of the usage of diatomite and waste marble powder as partial replacement of cement on the mechanical properties of concrete. *Constr. Build. Mater.* **2011**, *25*, 806–812. [CrossRef]
54. Uysal, M.; Yilmaz, K. Effect of mineral admixtures on properties of self-compacting concrete. *Cem. Concr. Compos.* **2011**, *33*, 771–776. [CrossRef]
55. Li, L.; Huang, Z.; Tan, Y.; Kwan, A.; Chen, H. Recycling of marble dust as paste replacement for improving strength, microstructure and eco-friendliness of mortar. *J. Clean. Prod.* **2019**, *210*, 55–65. [CrossRef]
56. Belaidi, A.; Azzouz, L.; Kadri, E.; Kenai, S. Effect of natural pozzolana and marble powder on the properties of self-compacting concrete. *Constr. Build. Mater.* **2012**, *31*, 251–257. [CrossRef]
57. Şanal, İ. Significance of concrete production in terms of carbondioxide emissions: Social and environmental impacts. *Politek. Derg.* **2018**, *21*, 369–378. [CrossRef]
58. Farooq, F.; Akbar, A.; Khushnood, R.A.; Muhammad, W.L.B.; Rehman, S.K.U.; Javed, M.F. Experimental investigation of hybrid carbon nanotubes and graphite nanoplatelets on rheology, shrinkage, mechanical, and microstructure of SCCM. *Materials* **2020**, *13*, 230. [CrossRef]
59. Khaloo, A.R.; Dehestani, M.; Rahmatabadi, P. Mechanical properties of concrete containing a high volume of tire–rubber particles. *Waste Manag.* **2008**, *28*, 2472–2482. [CrossRef]
60. Qin, D.; Gao, P.; Aslam, F.; Sufian, M.; Alabduljabbar, H. A comprehensive review on fire damage assessment of reinforced concrete structures. *Case Stud. Constr. Mater.* **2021**, *16*, e00843. [CrossRef]
61. Rashid, M.A.; Mansur, M.A. Considerations in producing high strength concrete. *J. Civ. Eng.* **2009**, *37*, 53–63.
62. Cotsovos, D.; Pavlović, M. Numerical investigation of concrete subjected to compressive impact loading. Part 2: Parametric investigation of factors affecting behaviour at high loading rates. *Comput. Struct.* **2008**, *86*, 164–180. [CrossRef]
63. Li, M.; Hao, H.; Shi, Y.; Hao, Y. Specimen shape and size effects on the concrete compressive strength under static and dynamic tests. *Constr. Build. Mater.* **2018**, *161*, 84–93. [CrossRef]
64. Feng, D.-C.; Liu, Z.-T.; Wang, X.-D.; Chen, Y.; Chang, J.-Q.; Wei, D.-F.; Jiang, Z.-M. Machine learning-based compressive strength prediction for concrete: An adaptive boosting approach. *Constr. Build. Mater.* **2020**, *230*, 117000. [CrossRef]
65. Nafees, A.; Khan, S.; Javed, M.F.; Alrowais, R.; Mohamed, A.M.; Mohamed, A.; Vatin, N.I. Forecasting the Mechanical Properties of Plastic Concrete Employing Experimental Data Using Machine Learning Algorithms: DT, MLPNN, SVM, and RF. *Polymers* **2022**, *14*, 1583. [CrossRef] [PubMed]
66. Nafees, A.; Javed, M.; Khan, S.; Nazir, K.; Farooq, F.; Aslam, F.; Musarat, M.; Vatin, N. Predictive Modeling of Mechanical Properties of Silica Fume-Based Green Concrete Using Artificial Intelligence Approaches: MLPNN, ANFIS, and GEP. *Materials* **2021**, *14*, 7531. [CrossRef]
67. Salehi, H.; Burgueño, R. Emerging artificial intelligence methods in structural engineering. *Eng. Struct.* **2018**, *171*, 170–189. [CrossRef]
68. Rahman, J.; Ahmed, K.S.; Khan, N.I.; Islam, K.; Mangalathu, S. Data-driven shear strength prediction of steel fiber reinforced concrete beams using machine learning approach. *Eng. Struct.* **2021**, *233*, 111743. [CrossRef]
69. Behnood, A.; Golafshani, E.M. Predicting the compressive strength of silica fume concrete using hybrid artificial neural network with multi-objective grey wolves. *J. Clean. Prod.* **2018**, *202*, 54–64. [CrossRef]
70. Güçlüer, K.; Özbeyaz, A.; Göymen, S.; Günaydın, O. A comparative investigation using machine learning methods for concrete compressive strength estimation. *Mater. Today Commun.* **2021**, *27*, 102278. [CrossRef]

71. Getahun, M.A.; Shitote, S.M.; Gariy, Z.C.A. Artificial neural network based modelling approach for strength prediction of concrete incorporating agricultural and construction wastes. *Constr. Build. Mater.* **2018**, *190*, 517–525. [CrossRef]
72. Ling, H.; Qian, C.; Kang, W.; Liang, C.; Chen, H. Combination of Support Vector Machine and K-Fold cross validation to predict compressive strength of concrete in marine environment. *Constr. Build. Mater.* **2019**, *206*, 355–363. [CrossRef]
73. Yaseen, Z.M.; Deo, R.C.; Hilal, A.; Abd, A.M.; Bueno, L.C.; Salcedo-Sanz, S.; Nehdi, M.L. Predicting compressive strength of lightweight foamed concrete using extreme learning machine model. *Adv. Eng. Softw.* **2018**, *115*, 112–125. [CrossRef]
74. Taffese, W.Z.; Sistonen, E. Machine learning for durability and service-life assessment of reinforced concrete structures: Recent advances and future directions. *Autom. Constr.* **2017**, *77*, 1–14. [CrossRef]
75. Yokoyama, S.; Matsumoto, T. Development of an automatic detector of cracks in concrete using machine learning. *Procedia Eng.* **2017**, *171*, 1250–1255. [CrossRef]
76. Chaabene, W.B.; Flah, M.; Nehdi, M.L. Machine learning prediction of mechanical properties of concrete: Critical review. *Constr. Build. Mater.* **2020**, *260*, 119889. [CrossRef]
77. Ahmad, W.; Ahmad, A.; Ostrowski, K.A.; Aslam, F.; Joyklad, P.; Zajdel, P. Application of advanced machine learning approaches to predict the compressive strength of concrete containing supplementary cementitious materials. *Materials* **2021**, *14*, 5762. [CrossRef]
78. Song, H.; Ahmad, A.; Ostrowski, K.A.; Dudek, M. Analyzing the compressive strength of ceramic waste-based concrete using experiment and artificial neural network (ANN) approach. *Materials* **2021**, *14*, 4518. [CrossRef]
79. Ahmad, A.; Farooq, F.; Niewiadomski, P.; Ostrowski, K.; Akbar, A.; Aslam, F.; Alyousef, R. Prediction of compressive strength of fly ash based concrete using individual and ensemble algorithm. *Materials* **2021**, *14*, 794. [CrossRef]
80. Balf, F.R.; Kordkheili, H.M.; Kordkheili, A.M. A New Method for Predicting the Ingredients of Self-Compacting Concrete (SCC) Including Fly Ash (FA) Using Data Envelopment Analysis (DEA). *Arab. J. Sci. Eng.* **2021**, *46*, 4439–4460. [CrossRef]
81. Bušić, R.; Benšić, M.; Miličević, I.; Strukar, K. Prediction models for the mechanical properties of self-compacting concrete with recycled rubber and silica fume. *Materials* **2020**, *13*, 1821. [CrossRef] [PubMed]
82. Saha, P.; Debnath, P.; Thomas, P. Prediction of fresh and hardened properties of self-compacting concrete using support vector regression approach. *Neural Comput. Appl.* **2020**, *32*, 7995–8010. [CrossRef]
83. Al-Mughanam, T.; Aldhyani, T.H.; AlSubari, B.; Al-Yaari, M. Modeling of compressive strength of sustainable self-compacting concrete incorporating treated palm oil fuel ash using artificial neural network. *Sustainability* **2020**, *12*, 9322. [CrossRef]
84. Farooq, F.; Nasir Amin, M.; Khan, K.; Rehan Sadiq, M.; Faisal Javed, M.; Aslam, F.; Alyousef, R. A comparative study of random forest and genetic engineering programming for the prediction of compressive strength of high strength concrete (HSC). *Appl. Sci.* **2020**, *10*, 7330. [CrossRef]
85. Selvaraj, S.; Sivaraman, S. Prediction model for optimized self-compacting concrete with fly ash using response surface method based on fuzzy classification. *Neural Comput. Appl.* **2019**, *31*, 1365–1373. [CrossRef]
86. Zhang, J.; Ma, G.; Huang, Y.; Aslani, F.; Nener, B. Modelling uniaxial compressive strength of lightweight self-compacting concrete using random forest regression. *Constr. Build. Mater.* **2019**, *210*, 713–719. [CrossRef]
87. Karbassi, A.; Mohebi, B.; Rezaee, S.; Lestuzzi, P. Damage prediction for regular reinforced concrete buildings using the decision tree algorithm. *Comput. Struct.* **2014**, *130*, 46–56. [CrossRef]
88. Erdal, H.I. Two-level and hybrid ensembles of decision trees for high performance concrete compressive strength prediction. *Eng. Appl. Artif. Intell.* **2013**, *26*, 1689–1697. [CrossRef]
89. Han, Q.; Gui, C.; Xu, J.; Lacidogna, G. A generalized method to predict the compressive strength of high-performance concrete by improved random forest algorithm. *Constr. Build. Mater.* **2019**, *226*, 734–742. [CrossRef]
90. Shaqadan, A. Prediction of concrete mix strength using random forest model. *Int. J. Appl. Eng. Res.* **2016**, *11*, 11024–11029.
91. Xu, Y.; Ahmad, W.; Ahmad, A.; Ostrowski, K.A.; Dudek, M.; Aslam, F.; Joyklad, P. Computation of High-Performance Concrete Compressive Strength Using Standalone and Ensembled Machine Learning Techniques. *Materials* **2021**, *14*, 7034. [CrossRef] [PubMed]
92. Ribeiro, M.H.D.M.; dos Santos Coelho, L. Ensemble approach based on bagging, boosting and stacking for short-term prediction in agribusiness time series. *Appl. Soft Comput.* **2020**, *86*, 105837. [CrossRef]
93. Wang, C.; Xu, S.; Yang, J. Adaboost Algorithm in Artificial Intelligence for Optimizing the IRI Prediction Accuracy of Asphalt Concrete Pavement. *Sensors* **2021**, *21*, 5682. [CrossRef] [PubMed]
94. Huang, J.; Sun, Y.; Zhang, J. Reduction of computational error by optimizing SVR kernel coefficients to simulate concrete compressive strength through the use of a human learning optimization algorithm. *Eng. Comput.* **2021**, 1–18. [CrossRef]
95. Ahmad, A.; Ahmad, W.; Aslam, F.; Joyklad, P. Compressive strength prediction of fly ash-based geopolymer concrete via advanced machine learning techniques. *Case Stud. Constr. Mater.* **2021**, *16*, e00840. [CrossRef]
96. Yuan, X.; Tian, Y.; Ahmad, W.; Ahmad, A.; Usanova, K.I.; Mohamed, A.M.; Khallaf, R. Machine Learning Prediction Models to Evaluate the Strength of Recycled Aggregate Concrete. *Materials* **2022**, *15*, 2823. [CrossRef]
97. Shang, M.; Li, H.; Ahmad, A.; Ahmad, W.; Ostrowski, K.A.; Aslam, F.; Joyklad, P.; Majka, T.M. Predicting the Mechanical Properties of RCA-Based Concrete Using Supervised Machine Learning Algorithms. *Materials* **2022**, *15*, 647. [CrossRef]
98. Ahmad, A.; Chaiyasarn, K.; Farooq, F.; Ahmad, W.; Suparp, S.; Aslam, F. Compressive strength prediction via gene expression programming (GEP) and artificial neural network (ANN) for concrete containing RCA. *Buildings* **2021**, *11*, 324. [CrossRef]

99. Farooq, F.; Ahmed, W.; Akbar, A.; Aslam, F.; Alyousef, R. Predictive modeling for sustainable high-performance concrete from industrial wastes: A comparison and optimization of models using ensemble learners. *J. Clean. Prod.* **2021**, *292*, 126032. [CrossRef]
100. Aslam, F.; Farooq, F.; Amin, M.N.; Khan, K.; Waheed, A.; Akbar, A.; Javed, M.F.; Alyousef, R.; Alabdulijabbar, H. Applications of gene expression programming for estimating compressive strength of high-strength concrete. *Adv. Civ. Eng.* **2020**, *2020*, 8850535. [CrossRef]
101. Singh, M.; Choudhary, K.; Srivastava, A.; Sangwan, K.S.; Bhunia, D. A study on environmental and economic impacts of using waste marble powder in concrete. *J. Build. Eng.* **2017**, *13*, 87–95. [CrossRef]

 *materials*

Article

Debonding of Thin Bonded Rubberised Fibre-Reinforced Cement-Based Repairs under Monotonic Loading: Experimental and Numerical Investigation

Syed Asad Ali Gillani [1,2], Shaban Shahzad [1], Wasim Abbass [2,*], Safeer Abbas [2], Ahmed Toumi [1], Anaclet Turatsinze [1], Abdeliazim Mustafa Mohamed [3,4] and Mohamed Mahmoud Sayed [5]

1. Laboratoire Matériaux et Durabilité des Constructions, Université de Toulouse, Institut National des Sciences Appliquées de Toulouse, UPS Génie Civil, 135 Avenue de Rangueil, CEDEX 04, 31077 Toulouse, France; asadgillani@uet.edu.pk (S.A.A.G.); shahzad@insa-toulouse.fr (S.S.); toumi@insa-toulouse.fr (A.T.); anaclet@insa-toulouse.fr (A.T.)
2. Civil Engineering Department, University of Engineering and Technology, Lahore 54890, Pakistan; safeer.abbas@uet.edu.pk
3. Department of Civil Engineering, College of Engineering, Prince Sattam Bin Abdulaziz University, Al-Kharj 16273, Saudi Arabia; a.bilal@psau.edu.sa
4. Building & Construction Technology Department, Bayan College of Science and Technology, Khartoum 210, Sudan
5. Architectural Engineering, Faculty of Engineering and Technology, Future University in Egypt, New Cairo 11745, Egypt; mohamed.mahmoud@fue.edu.eg
* Correspondence: wabbass@uet.edu.pk

Citation: Gillani, S.A.A.; Shahzad, S.; Abbass, W.; Abbas, S.; Toumi, A.; Turatsinze, A.; Mohamed, A.M.; Sayed, M.M. Debonding of Thin Bonded Rubberised Fibre-Reinforced Cement-Based Repairs under Monotonic Loading: Experimental and Numerical Investigation. *Materials* 2022, 15, 3886. https://doi.org/10.3390/ma15113886

Academic Editor: Jorge Otero

Received: 28 April 2022
Accepted: 19 May 2022
Published: 30 May 2022

Publisher's Note: MDPI stays neutral with regard to jurisdictional claims in published maps and institutional affiliations.

Abstract: In this study, the durability of cement-based repairs was observed, especially at the interface of debonding initiation and propagation between the substrate–overlay of thin-bonded cement-based material, using monotonic tests experimentally and numerically. Overlay or repair material (OM) is a cement-based mortar with the addition of metallic fibres (30 kg/m^3) and rubber particles (30% as a replacement for sand), while the substrate is a plain mortar without any addition, known as control. Direct tension tests were conducted on OM in order to obtain the relationship between residual stress-crack openings (σ-w law). Similarly, tensile tests were conducted on the substrate–overlay interface to draw the relationship between residual stress and opening of the substrate–overlay interface. Three-point monotonic bending tests were performed on the composite beam of the substrate–overlay in order to observe the structural response of the repaired beam. The digital image correlation (DIC) method was utilized to examine the debonding propagation along the interface. Based on the different parameters obtained through the above-mentioned experiments, a three-point bending monotonic test was modelled through finite elements using a software package developed in France called CAST3M. Structural behaviour of repaired beams observed by experimental results and that analysed by numerical simulation are in coherence. It is concluded from the results that the hybrid use of fibres and rubber particles in repaired material provides a synergetic effect by improving its strain capacity, restricting crack openings by the transfer of stress from the crack. This enhances the durability of repair by controlling propagation of the interface debonding.

Keywords: fibres; rubber particles; thin bonded overlay; debonding; DIC; CAST3M

Copyright: © 2022 by the authors. Licensee MDPI, Basel, Switzerland. This article is an open access article distributed under the terms and conditions of the Creative Commons Attribution (CC BY) license (https://creativecommons.org/licenses/by/4.0/).

1. Introduction

Concrete has been utilised abundantly in the construction sector over recent decades, and with the passage of time, a reduction in the load-retaining capacity of existing infrastructures has been observed. In order to rehabilitate damaged concrete structures, different techniques have been used. Among different rehabilitation techniques, a unique approach to reinstate the performance of a degraded structure is thin bonded cement-based overlay. This overlay technique is used to replace the decaying concrete, to provide smoothness to the damaged part of structure and to enhance the load carrying capacity through increased thickness, which also provides an extra margin for protecting it against corrosion [1]. Such a technique proves to be very efficient, specifically for larger surfaces of concrete such as pavements [2,3].

However, the durability of these overlays can be limited due to cracking of the repaired part, followed by the interface's delamination from the substrate [3,4]. This issue has already been well-reported in previous studies [1,5,6]. According to some previous research [1,3,4], mechanical loadings and differential shrinkage are the major causes for the delamination between overlay and substrate. The delamination normally begins from edges, cracks and joints in all mechanisms.

On the basis of previous literature, one can say that the long-term sustainability of the materials used for repair, and bonding between two layers, are merely influenced by the durability characteristics of thin bonded overlays. As for sustainable materials, reliable option to improve the durability properties of the repair system is to use rubber aggregates and steel fibres collectively [7–10]. The inclusion of rubber particles obtained by grinding scrape tyres in repair composites enhances their strain capacities [9,11–13], and fibres limit the crack opening, which assists in delaying debonding initiation and restricting the interfacial delamination to greater extent. Moreover, positive synergetic effects (enhancement in the strain capacity of material and in post-cracking residual tensile strength) were found by the collective use of rubber particles and fibres in mortar [7,8,10,12]. Due to these positive synergetic effects, use of rubber particles and fibre in cement-based overlay is most often adopted.

Several studies have been performed to analyse the crack and propagation of delamination in cement-based overlays under different kinds of mechanical or thermal loadings. Gillani et al. [14] studied the generation and movement of crack and delamination of the overlays under fatigue loading. They found that the addition of metallic fibres and rubberised particles help to control the debonding by restraining the crack (with addition of fibres), as well as by improving the strain capacity (with addition of rubber aggregates). Mateos et al. [15] reported the mechanical behaviour of the asphalt–concrete interface in a bonded concrete overlay of asphalt pavements (BCOA). Cylindrical specimens were used under various conditions such as wet and dry, and temperature ranges between 5 and 40 °C. The results indicate that the strength of the concrete–asphalt interface is strongly linked with the asphalt. Moreover, interface significantly softened under wet conditions, indicating that water is the decisive factor responsible for the failure of BCOA. However, one can conclude that concrete has not developed a good bond with asphalt. A. Toumi et al. [12] conducted experimental and analytical study on delamination of a thin rubberised and fibre-reinforced mortar repair. In this study, substrates of cement-based material (100 mm-thick) and an overlay with cement-based composites containing fibres and/or rubber particles (40 mm thick) were used. The study was conducted under a three-point bending test using a monotonic sequence of loading. They found that the addition of fibres in repair material helps to delay the debonding phenomena, and the inclusion of the rubber particles improves the strain capacity of the material, resulting in controlling of the debonding compared to the repair material without rubber aggregates. Studies were conducted to analyse the debonding of substrate and fibre-reinforced mortar (FRM) overlay by Q.T. Tran et al. [4]. A substrate in the form of a hollow metal beam was used in this study. The test was conducted under static three-point bending conditions. Experimental results were compared with the results obtained through the model. The finite element model (FEM) was based on the discrete crack model, which helps to model the crack and debonding propagation efficiently. The numerical results show that the developed model is an effective system to forecast the crack opening and debonding propagation.

A study on the behaviour of fibre-reinforced concrete (overlay) over the asphalt (substrate) was conducted by Isla et al. [16]. Bending tests were carried out on specimens of size 100 × 100 × 400 mm with centre-to-centre distance of supports of 350 mm, and a thickness of 50 mm per layer for concrete overlay and substrate of asphalt. Isla et al. [16]

reported that inclusion of fibres significantly improved the residual capacity of flexural member and composite beams as well. Hasani et al. [17] has also reported that the overlay of fibre-reinforced concrete also improves the mechanical and durability-related properties. Moreover, it was found that the compressive strengths, flexural strength, residual strength, and ductility of the FRC overlay material was improved. However, the modulus of elasticity was reduced.

On the basis of previous studies, it can be concluded that the bond between the overlay and substrate and characteristics of the repair material significantly affect the durability-related properties of thin overlay systems. Moreover, the debonding mechanism between the overlay and substrate initiates when the crack reaches at interface. In this regard, the hybrid use of fibre and rubber aggregates appears to be a viable option to improve the durability characteristics of repair system. The current research was planned to investigate the flexural behaviour of composite beams under monotonic load. The evolution of crack opening, deflection and debonding length was evaluated to study the potential of rubberized fibre-reinforced composite material for possible utilization as a repair material in cementitious overlays. To ensure a good bond between the repair material and substrate, the sandblasted substrate surfaces were used as per previous research findings [18]. To analyse the crack's evolution and delamination at the interface, the DIC method was used. The flexural tests were modelled using the finite element approach based on a discrete crack model to forecast the crack propagation and debonding mechanism under monotonic loading.

2. Materials

Cementitious mortar including rubber particles with the addition of fibres was used as a repair material in this study. Portland cement (CEM I 52.5R) in conformity with EN197-1:2011 [19] and natural sand (0–4 mm) were used. The chemical composition and physical properties of the Portland cement are shown in Table 1. Similar findings have also been reported in other research study [20]. Master Glenium 27, a modified polycarboxylic ether polymer-based superplasticizer, and Rheomac were used as superplasticizer and viscosity-modifying agent (VMA), respectively. The rubber particles were used as a partial replacement of sand in the same volumetric unit. The rubber aggregates were produced through grinding of scrap tyres. Rubber particles' specific gravity was 1.2, which is much less than sand, i.e., 2.7. Gradation curves of rubber aggregates and sand show slightly different particle size distribution for both materials, but in both cases, the maximum particle size is limited to 4 mm, as can be visualized in Figure 1. Fibraflex Saint-Gobain [21] provided the fibres, shown in Figure 2. The length of these fibres is 30 mm, which also meets the criteria for productive bridging properties, i.e., the maximum aggregate particle size should be equal or less than half of the length of fibre [22]. These fibres develop an excellent bond with cementitious composites because of the rough and large surface area. Properties of these amorphous metallic fibres are provided in Table 2 [3]. Similar mixture proportions from another study were adopted for the current work [4], [11], and are presented in Table 3. In view of previously conducted research [11], the maximum dosage of fibre of 30 kg/m^3 was investigated and the partial replacement of sand aggregate with rubber particles was 30%.

The mixtures were designated with R and F for rubber aggregates and fibres, respectively, for referencing of different mixtures. For instance, M0R30F is designated mixture refer as follows:

M represents mortar, 0R shows 0% rubber particle, and 30F denotes mixture with 30 kg/m^3 of fibres.

By keeping the same water-to-cement ratio (w/c), the quantity of super-plasticizer was changed to keep the same workability with a slump of 10 $\pm$ 2 mm. In fibre-reinforced and/or rubberized mortars, the quantity of super-plasticizer is required to be increased because of the decrease in the workability of mortar with the addition of fibres [17]. Additionally, it was observed that air content increases to 65% by the inclusion of rubber aggregates in mortar, as projected in the literature [23]. Rubber aggregates are lightweight and water-repellent which makes them very susceptible to segregation. The role of the viscosity agent is to avoid this detrimental phenomenon.

Table 1. Physical characteristics and chemical properties of Portland cement (CEM I 52.5R).

Physical Characteristics		
Properties	**Unit**	**Value**
Specific gravity	g/cm^3	3.13
Water demand	%	28.1
Fineness	cm^2/g	4067
28-day compressive strength	MPa	>50

Chemical properties (%)							
C$_3$S + C$_2$S	CaO/SiO$_2$	MgO	C$_3$S	C$_2$S	C$_3$A	C$_4$AF	Gypsum
78.1	2.9	0.6	68	12	7	9	4.2

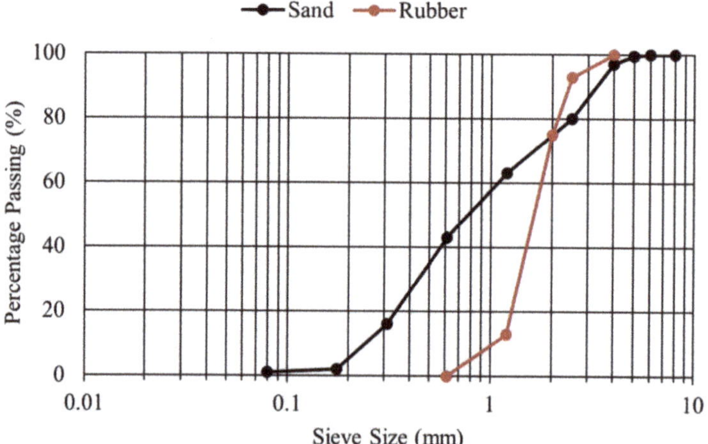

Figure 1. Gradation curve for rubber particles and sand.

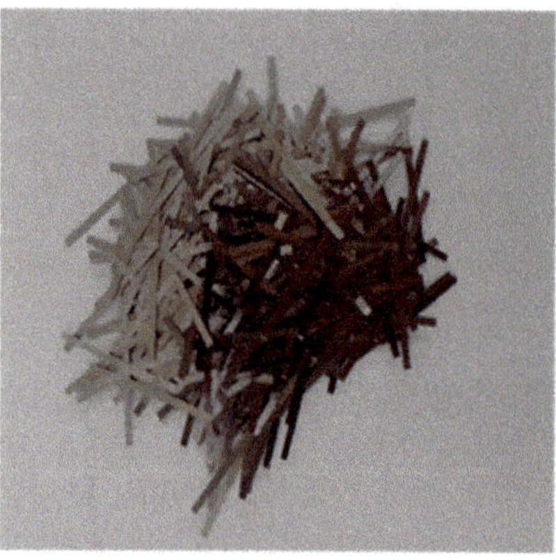

Figure 2. 30 mm-long metallic fibres.

Table 2. Properties of metallic fibres (Fibraexsaint-Gobain [21]).

Properties of Metallic Fibres	
Length, L	30 mm
Thickness	29 μm
Density	7200 kg/m^3
Tensile strength	More than 1400 MPa
Elastic modulus	140 GPa
Raw material	Amorphous metal (Fe, Cr)80, (P, C, Si)20

Table 3. Mixture design (kg/m^3).

Sr. No.	Mix Designation	Cement	Rubber Aggregates	Sand	Water	Fibres	Super-Plasticizer	Viscosity Modifying Agent
1	M0R0F	500	0	1600	250	0	1.2	0
2	M0R30F					30	5	
3	M30R0F		215	1120		0	4.5	2.5
4	M30R30F					30	10	

3. Mechanical Characterization

3.1. Compressive Tests

Tests for compressive strength were carried out in accordance with European standard, NF EN 12390-3 [24]. Specimens used for compressive tests were in accordance with EN 12390-1 [25]. Cylindrical specimens with a 110 mm diameter and height of 220 mm were prepared.

3.2. Modulus of Elasticity Tests

Elastic modulus tests were carried out on studied mixture composites, using the same specimen size as in Section 3.1 for each material. These tests were conducted by following the standard NF EN 12390-13 [26]. A cage with three attached extensometers at an equal angle from each other was used, as shown in Figure 3. The stress–strain relationship was plotted using average deformation in a longitudinal direction with mounted extensometers.

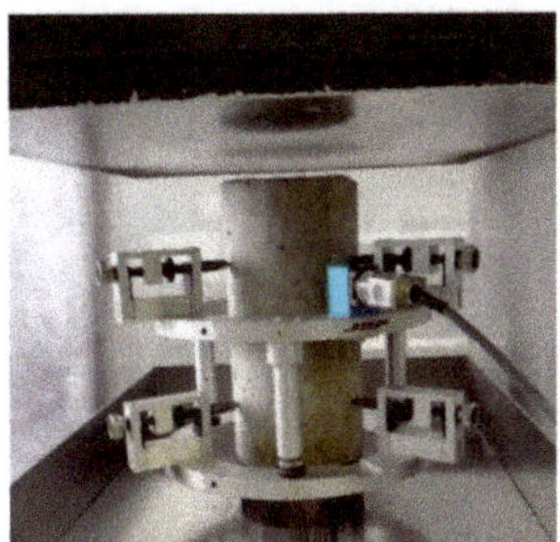

Figure 3. Testing arrangement for modulus of elasticity with cage.

3.3. Direct Tensile Tests

Prismatic notched specimens with a size of 100 × 100 × 200 mm and reduced cross sections of 50 × 50 mm, as shown in Figure 4, were prepared for direct tensile testing. These tests were conducted to assess the tensile properties and stress–crack relationships for various composites. These will be the input factors for the finite element model. These tests were conducted as per the RILEM recommendation [27]. An MTS press was used for conducting the test and CMOD was recorded by using a COD clip, as shown in Figure 5. One can analyse the capacity of the deformation linked with peak load and residual tensile strength beyond the peak through these tensile tests. A loading speed of 5 µm/min was adopted in the start of the test till the CMOD reaches 0.1 mm, and then the loading rate was increased to 100 µm/min until failure of the specimen.

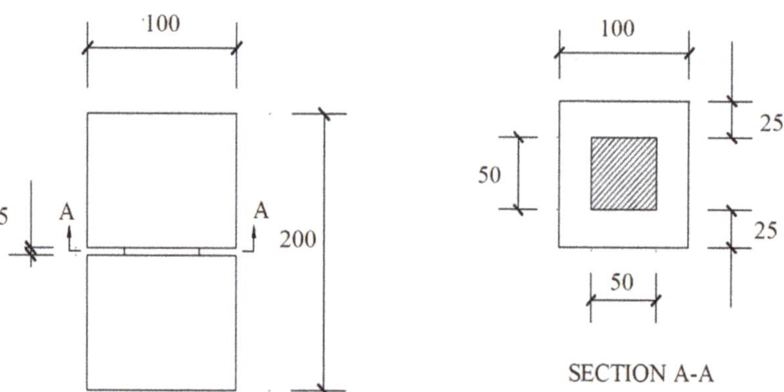

Figure 4. Prismatic notched specimens for direct tensile test (mm).

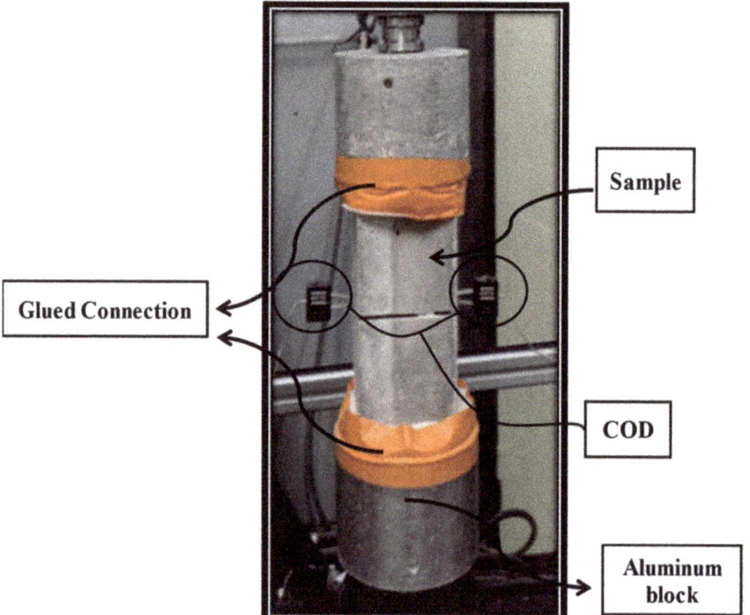

Figure 5. Experimental setup for direct tensile test.

4. Bending Monotonic Test

4.1. Specimen for the Monotonic Test

The composite samples were made of a thin repair layer applied on top of the substrate, which mimics the repaired beam. Cementitious substrates without rubber aggregates and fibre-reinforcement (M0R0F) were prepared to have a real application. The size of prismatic substrate was 100 mm × 100 mm × 500 mm. These substrate bases were placed in a control environment of 20 °C and relative humidity (RH) of 98% for curing purposes. Based on the results from previous studies [18,28–34], it is observed that the surface preparation of the substrate has an influential role on the performance of the repair as far as durability is concerned. So, substrates prepared by the sandblasting techniques were used in experiments. In reference to previous studies [34,35], a 40 mm-thick layer was used as the repair. So, a repair layer was cast on top of the 100 mm sandblasted substrates. A schematic diagram of the beam with a repair layer under three-point bending monotonic testing is shown in Figure 6. To predetermine the location of the crack, the repair layer was notched during the specimen's casting at the mid-span. These beams were placed under ambient conditions (20 °C and RH of 98%) for 28 days.

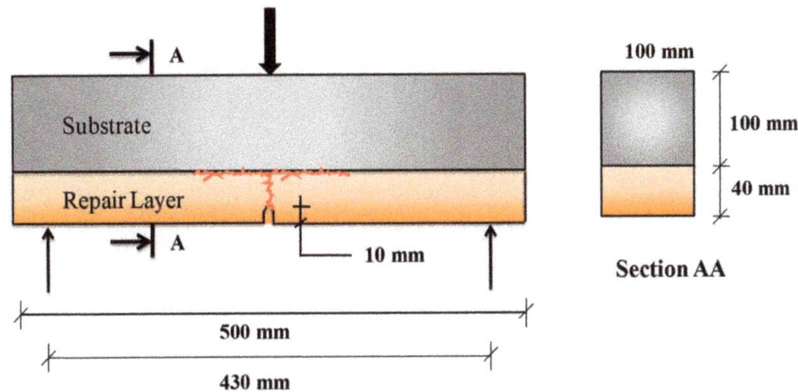

Figure 6. Schematic representation of tested specimen for use in bending test.

4.2. Testing Procedure

A three-point bending monotonic test was performed on the specimen to analyse the behaviour of the overlay–substrate under flexure. The schematic testing setup can be seen in Figure 6. CMOD was measured by using a COD sensor. A loading speed of 0.05 mm/min was adopted at the start of the test until the CMOD reached 0.1 mm, and then the loading speed was increased to 0.2 mm/min until failure of the specimen (when resisting load is equal to around zero). The LVDT sensor was used to monitor the vertical deflection of the composite specimens at the middle. For monitoring the interface delamination and crack propagation, a digital DIC technique was used. Under mechanical loading, crack initiated from the tip of the notch in the overlay, which eventually caused the delamination when it approached the interface. The main objective of the monotonic tests is to monitor the following parameters:

- The opening of the notch (CMOD) with the application of force.
- The deflection with force.
- The load at which the crack approaches the interface location with DIC.

4.3. Digital Image Correlation Technique

The DIC method was developed by researchers from University of California in the late 19th century [36–39]. DIC is a visual and non-contact measurement technique that is used for monitoring of surface displacements of an object under investigation by image registration techniques for accurate measurement of changes in images taken in series with test proceeding. Strain on the surface of the object is calculated using the displacements. Random speckles are made on the white painted surface of the object prior to the initiation of the test to obtain the most effective results [40].

Two images were taken from two cameras within the same period of time using the 3D DIC technique. The system must be calibrated prior to the test. After calibration, these results can be used correlate the images for the determination of the deflection and strain of the object under investigation [41]. For 3D image correlation, preparation of the specimen is necessary, as shown in Figure 7. The complete testing layout for DIC can be seen in Figure 8.

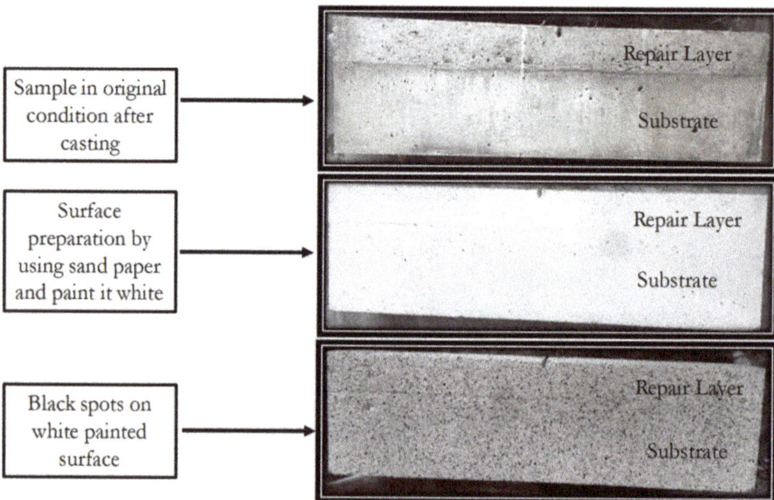

Figure 7. Surface preparation of specimen for the bending test along with DIC technique.

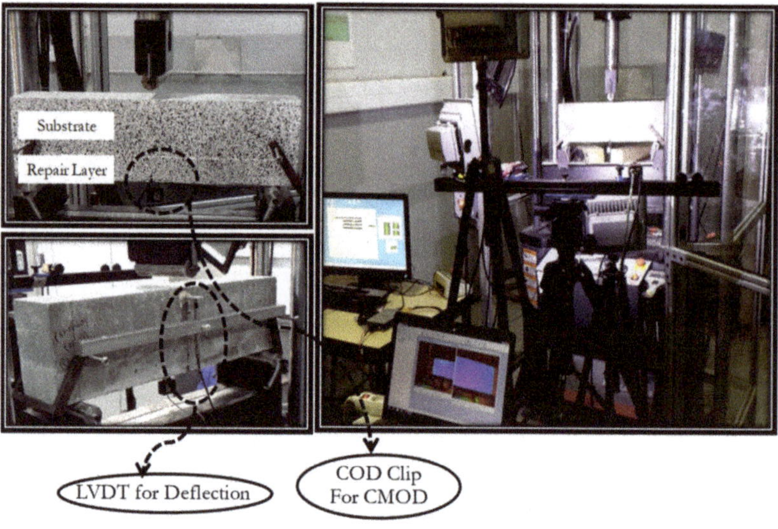

Figure 8. Complete testing layout for three-point bending monotonic test using DIC technique.

Three-point monotonic bending tests with DIC technique were conducted for all repairs to examine the pattern of the crack and to evaluate the load where the crack approaches the interface and debonding starts. The software Vic-3D, [42] was used for image processing. The generated displacement and strain on the surface of the object can be analysed through post-treatment of images taken during the bending test. The complete cracking pattern is shown by the post-treatment. An artificial extensometer is used to find the value of the load at which the crack propagates to the interface. The placement of an artificial extensometer on the surface is shown in Figure 9. The DIC method has the ability to carry out reverse analysis of obtained strains through post-processing. So, an artificial extensometer can be placed at an actual crack location after locating the crack path to detect the load where the crack approaches the interface. When the crack passes this extensometer, an abrupt variation in D1 value is observed (Figure 10). The D1 is an extension in the artificial extensometer that develops due to crack opening. At the location where there is an abrupt change in D1, the corresponding load value represents the propagation of the crack to the interface and initiation of the interface debonding.

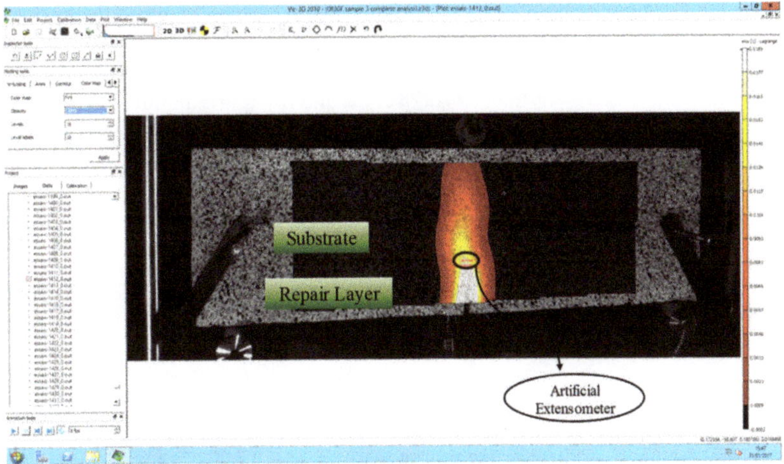

Figure 9. Placement of an artificial extensometer at the interface.

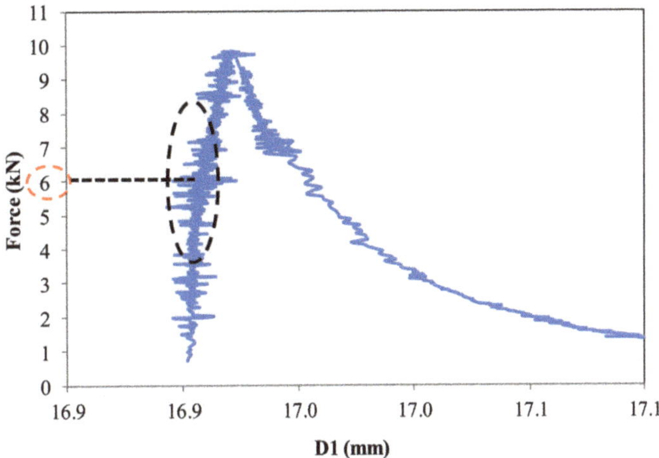

Figure 10. Force versus elongation in artificial extensometer (D1).

The loads required to start the delamination of various specimens with unique repair mixtures are provided in Table 4. For M0R0F and M30R0F repair mortars, the average force for debonding initiation is comparatively less for M0R30F and M30R30F. These fibres provide bridging properties throughout the crack path and reinstate the crack opening. For M30R0F repairs, the load needed to start the delamination is also improved in comparison to M0R0F mortar repair. The inclusion of rubber particles increases the capacity of strain in the material and helps in delaying crack initiation. Additionally, the delamination initiation load is notably enhanced for M30R30F. The increase in the load value is due to the synergetic effect induced by the collective use of rubber particles and fibres.

Table 4. Average interface debonding-initiation force.

Mix Composition	M0R0F	M0R30F	M30R0F	M30R30F
Average interface debonding-initiation force (kN)	6.5	8.0	7.0	9.0

5. Numerical Modelling

Various discrete crack models were developed for modelling and analysing the behaviour of normal concrete and fibre-reinforced concrete under monotonic loading. Petersson [43] used fictitious crack models based on fracture mechanics, enabling the prediction of the growth of crack and fracture zones in normal concrete or composite concrete. In the developed model, cracks in the overlay and interface debonding were propagated according to Mode I of fracture mechanics. The initiation as well as the propagation of the crack in the overlay followed the pre-damaged path, or along the zone of minimum strength (at the location of notch).

5.1. Mesh Size of Composite Beam

As per the symmetrical system, only one half of the beam was modelled for the optimization of the simulation by the FEM package. Figure 11 shows the model of half of the composite beam in the software. Triangular and rectangular elements were made with three and four nodes, respectively, to have an optimised mesh size. To obtain accurate and optimised results from model, Tran [34] analysed the behaviour of a composite beam by altering the size of the mesh to obtain stabilised results. He observed that while using inter-node distances less than 1mm, the effect of mesh size on the results became insignificant. In this study, a node-to-node distance of 1 mm was selected.

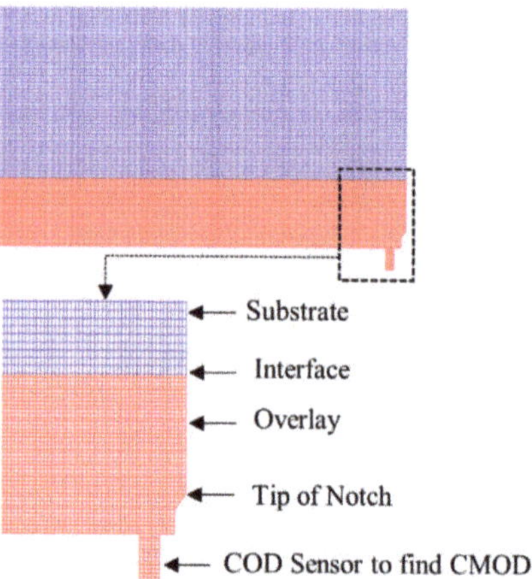

Figure 11. Substrate–overlay composite beam model in FEM software.

5.2. Cracking/Delamination Modelling Theory

Three-point bending tests under monotonic loading on the repaired beams as described in Section 3 were modelled using the discrete crack model mentioned earlier. To model mechanical response of fibre-rubberised mortar, material characteristics were described for the overlay and interface in Section 6. The stress–crack-opening relationships for overlay materials are given in Equations (1)–(4) as well as for interface in Equation (5).

CAST3M, developed in France by the Atomic Energy Commission, was used for calculation purposes in the finite element method (FEM). To control the propagation of delamination (debonding) or cracks, the stress premier node after the tip of crack of debonding, as proposed in previous studies [4,44], was considered. This type of technique prevents controlling the propagation by the condition or state of stress calculated at the interface or at the crack tip, where the stress singularity was predicted based on the strength theory analysis. So, the tip of the crack or delamination initiation node is advanced to the next node when the state of the stress at the commanding node does not satisfy a stability criterion (depending on the material tensile strength or on the one of the interfaces). The interlocking is considered by closing forces applied between the nodes in front of each other along the crack or along the de-bonded area. Debonding is started through tension perpendicular to the interface with the use of cementitious materials [45]. So, in the studied model, the loads were found through the residual stress–debonding opening/crack (σ-w) laws. By using a hypothesis of plain stress, trilateral and quadrilateral elements were modelled in 2D. At the time of calculation, the interlocking closure load of each particular pair of nodes facing each other were recalculated based on the crack or debonding openings each time a node was freed, to advance the crack or debonding. Residual closing forces were precisely fitted by an iterative approach until fracture widths or debonding widths were stable, according to the stated criterion. Then, for comparison with the propagation criterion, the same method is used on the next controlling node.

- Debonding or cracks moved to the next node, if $\sigma_t > R_t$ or $\sigma_{ti} > R_{ti}$, respectively, where σ_t is the tensile stress, R_t is the tensile strength of the repair materials and σ_{ti} and R_{ti} are the tensile stress and tensile strength of the substrate–repair interface, respectively.
- At the reach of stable state, and next step of increased loading was imposed to restart the propagation if the above-mentioned criterion is not met.

6. Results and Discussions

6.1. Modulus of Elasticity and Compressive Strength

Compressive strength and modulus of elasticity for all compositions are provided in Table 4. For mortars with rubber aggregates, notable depreciation in compressive strength is found. These results are in line with past studies on the effect of rubber particles partly replacing sand in cement-based materials [11,46]. The compressive strength of composites is not remarkably modified by the addition of fibre-reinforcement. Not only the low stiffness of rubberised aggregates, but also the high porosity and weak ITZ between cementitious and rubber particles had deleterious effects on the mechanical properties of mortar [47].

Similarly, a remarkable reduction in E values of the material was seen due to the inclusion of rubber particles, which is one of the same results found in past studies in this area [11,46]. Low stiffness and an increase in the porosity due to the addition of rubber particles are the main factors for the reduction in the E values. Like the results for compressive strength, the addition of metallic fibres has no effect or a very minute influence on E-values, as provided in the information in Table 5.

Table 5. Compressive strength and modulus of elasticity of various materials used for repair.

Repair Material	Compressive Strength (MPa)	Modulus of Elasticity (Mpa)
M0R0F	52.7	28,580
M0R30F	51.6	28,332
M30R0F	13.0	10,840
M30R30F	9.1	10,775

6.2. Tension Test for Material Used as Repair

Results of the direct tensile experiment for various repair materials are presented in Table 5. The reduction in tensile strength for M30R0F as well as the enhancement in the strain capacity is also observed (around 1.5 times increase in strain capacity vs. that of control mortar), as shown in Figures 12 and 13. Poor formation of ITZ and an increase in the porosity of the composite due to the addition of rubber particles are the main factors for the reduction in tensile strength. Even with the low strength in tension, the deformation at maximum load is higher for material containing rubber particles than the control one, because rubber particles have the capability to withstand failure deformation after peak [48]. From Figure 12, it can be analysed that for M0R30F, residual strength in tension is notably improved. A 3.5 times increase in the strain capacity is observed in M30R30F as well as enhancement in the residual strength in the post-cracking zone compared to the control. Figure 14 shows the experimental results obtained for various materials used for repair. By using best-fit curves of the results obtained through experiments, the following equations were finalised.

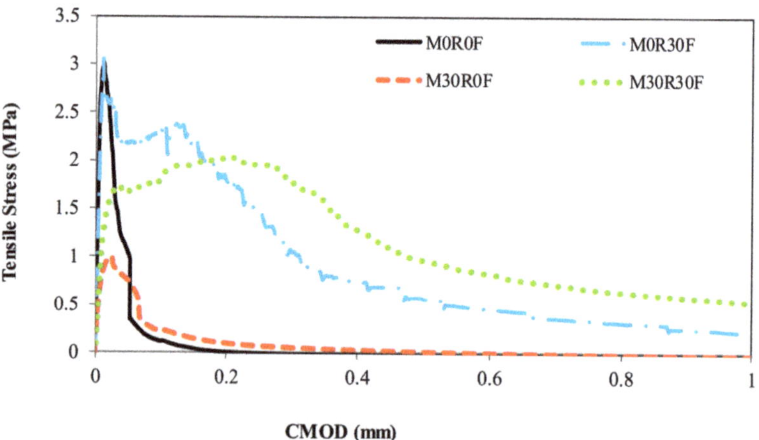

Figure 12. Impact of rubber aggregates and fibres on strain capacity and on residual post-peak strength.

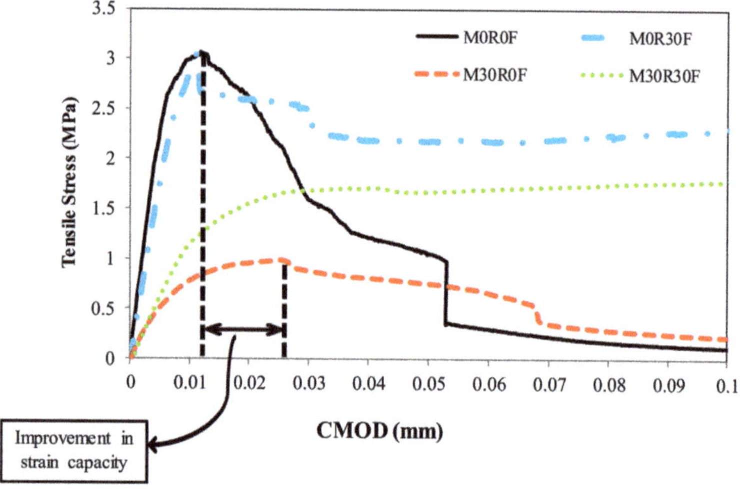

Figure 13. Impact of rubber aggregates and fibres on strain capacity and on residual post-peak strength (enlarged view).

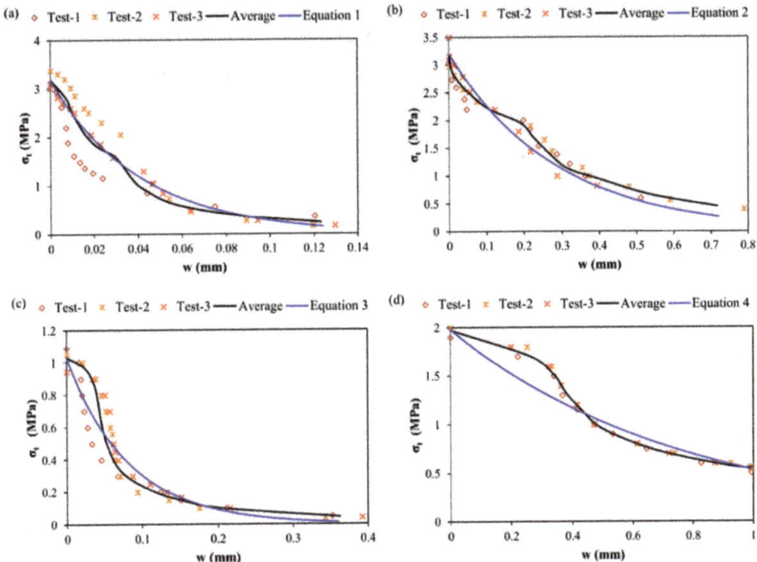

Figure 14. σ-w law for various composites (**a**) plain mortar, M0R0F (**b**) M0R30F (**c**) M30R0F (**d**) M30R30F.

For M0R0F

$$\sigma_t = R_t \times EXP\left(-3\frac{w}{w_l}\right) \tag{1}$$

For M0R30F

$$\sigma_t = R_t \times EXP\left(-2.5\,\frac{w}{w_l}\right) \tag{2}$$

For M30R0F

$$\sigma_t = R_t \times EXP\left(-1.5\,\frac{w}{w_l}\right) \tag{3}$$

For M30R30F

$$\sigma_t = R_t \times EXP\left(-1.3\frac{w}{w_l}\right) \tag{4}$$

where "σ_t" is the residual strength in tension, "R_t" is the strength in tension, "w" is evolution of the crack opening during loading and "w_l" is the controlling value of the crack opening, after which residual strength in tension becomes negligible or zero. "R_t" and "w_l" for the various materials are summarised in Table 6.

Table 6. Experimental data for calibrated model.

Repair Material	M0R0F	M0R30F	M30R0F	M30R30F
w_l (mm)	0.125	0.72	0.36	0.99
R_t (MPa)	3.17	3.2	1.02	1.98

6.3. Tension Test for Overlay/Repair–Substrate Interface

The principle of this test was as same as explained in Section 3.3. The objective was to have the analysis of the residual stress-delamination opening and tensile strength analytical relation for the substrate–overlay interface. The tested samples consist of old substrate and new overlay and were notched before testing at the interface on the ends facing each other. The model curve is also shown in Figure 15 based on the exponential model Equation (5).

$$\sigma_{ti} = R_{ti} \times EXP\left(-4\frac{w}{w_{li}}\right) \tag{5}$$

where, "σ_{ti}" is residual strength in tension, "R_{ti}" is the interface tensile strength (1.00 MPa), "w" is the opening of debonding and "w_{li}" is the controlling value of debonding opening beyond which strength in tension becomes negligible (0.1250 mm).

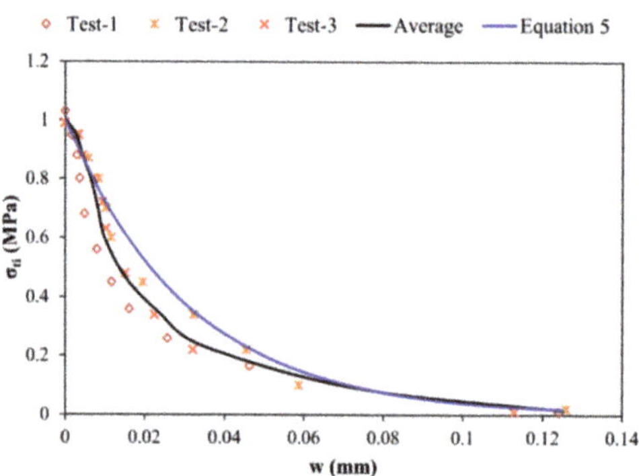

Figure 15. σ-w law for calibrated model and experimental results.

6.4. Relationship between Force and Opening of Notch

Figure 16 illustrates the relationship between notch opening and the force in the overlay. A comparison between numerically obtained results and experimental ones has been carried out and a good agreement have been observed.

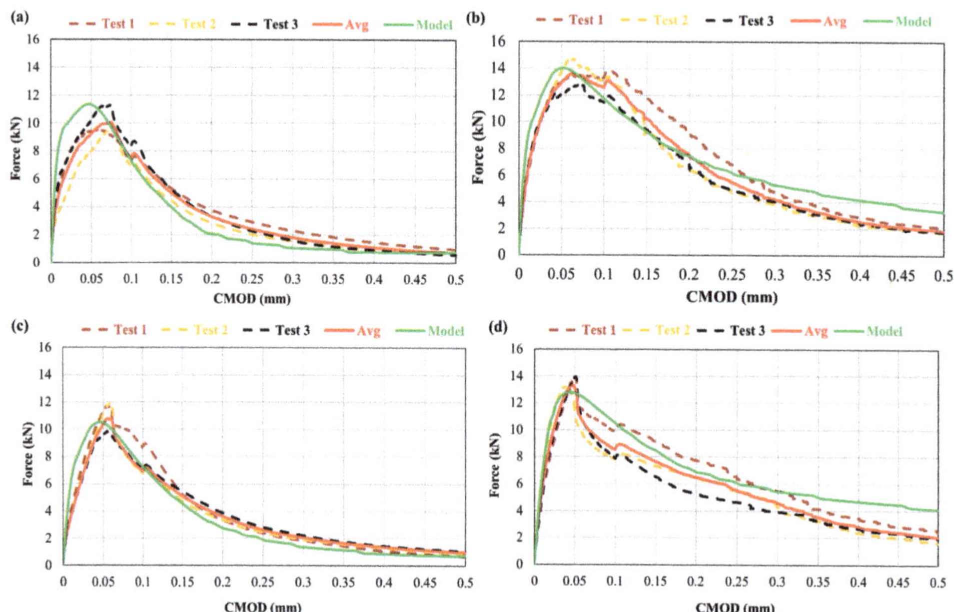

Figure 16. Force vs. opening of notch (CMOD) for (**a**) M0R0F-M0R0F, (**b**) M0R0F-M0R30F, (**c**) M0R0F-M30R0F, and (**d**) M0R0F-M30R30F.

The results obtained from model and the experimental campaign indicate that at any opening of the notch in the overlay, the corresponding load is higher in case of fibre-reinforced mortar compared to the control one. Repair material with fibre-reinforcement limits the notch opening during testing by controlling the opening of the crack.

The addition of rubber particles has no notable effect on the notch opening, as observed from the results. The M0R0F and M30R0F repair materials show an approximately similar response.

6.5. Relationship between Force and Deflection

Figure 17 show the relationship between force and deflection. A comparison between simulated and experimental results have been carried, out and an excellent coherence has been observed.

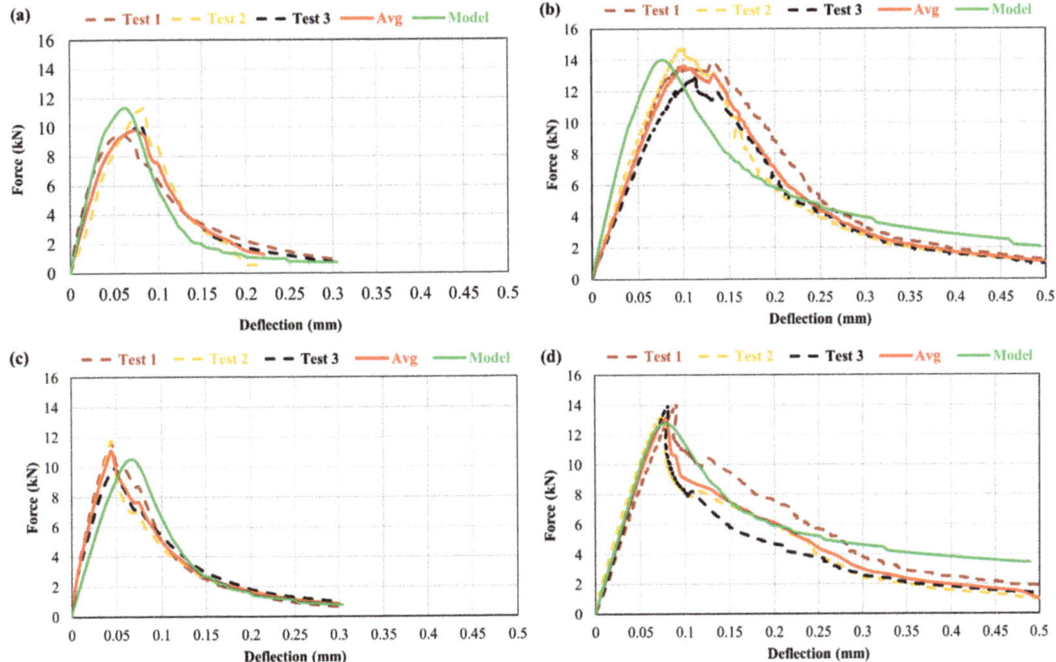

Figure 17. Force vs. deflection in composite beam for; (**a**) M0R0F-M0R0F, (**b**) M0R0F-M0R30F, (**c**) M0R0F-M30R0F, and (**d**) M0R0F-M30R30F.

From the results, it is observed that in beams repaired with fibre-reinforced composite, load carrying capacity of the repair material is increased at the corresponding deflection. This is because of the capacity of the fibres to transfer the stress across the crack, which restricts the deflection to a greater extent than in the other repair materials.

6.6. Relationship between Force and Debonding

Figure 18 show the relationship between force and debonding propagation at the interface of the composite beams repaired with different materials. In these figures, a comparison between simulated and experimental results has been carried out and an excellent coherence has been noticed.

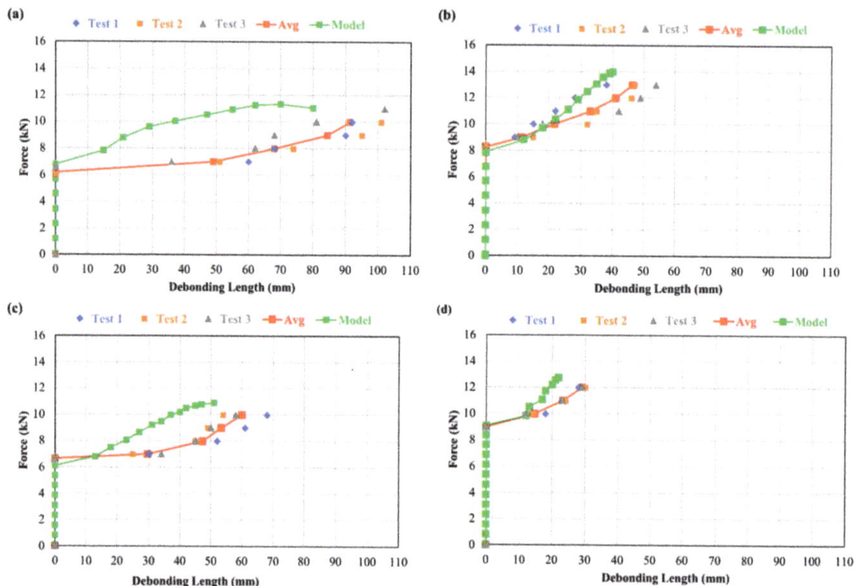

Figure 18. Force vs. debonding at interface in composite beam for; (**a**) M0R0F-M0R0F, (**b**) M0R0F-M0R30F, (**c**) M0R0F-M30R0F, and (**d**) M0R0F-M30R30F.

Figure 19 shows the results obtained from all repair materials in order to provide a good comparison. The test results show that propagation of debonding is more dominant in the repaired material without the fibres (M0R0F and M30R0F). With the addition of fibres in the repaired material (M0R30F and M30R30F), resistance in the debonding along the interface was noticed. This is closely linked with the crack opening in the repair layer. Therefore, the fibre-reinforced repair composites limit crack opening and thus delay the initiation interface debonding and limit propagation.

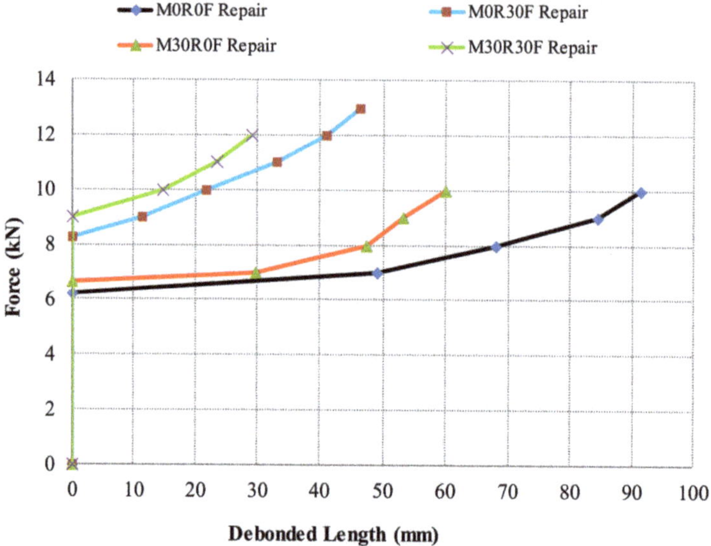

Figure 19. Force versus debonded length for various repair layers.

For M30R30F repair, it is also depicted that for the debonded length, the representing load is higher more than the other ones. For example, to obtain a 20 mm delamination at the interface, 6.5 kN of force is needed by M0R0F repair, 10 kN by M0R30F and 10.7 kN by M30R30F. Similarly, the force needed to initiate the delamination is also higher with M30R30F repair.

Notch opening plays a significant role in the transmission of debonding along the interface, as shown in Figure 20. The M0R0F as a repair material depicts the highest notch opening, and the corresponding debonded length is also the longest of the materials. On the contrary to the previous repair material, the M30R30F repair material restricts the opening of the notch and debonding propagation.

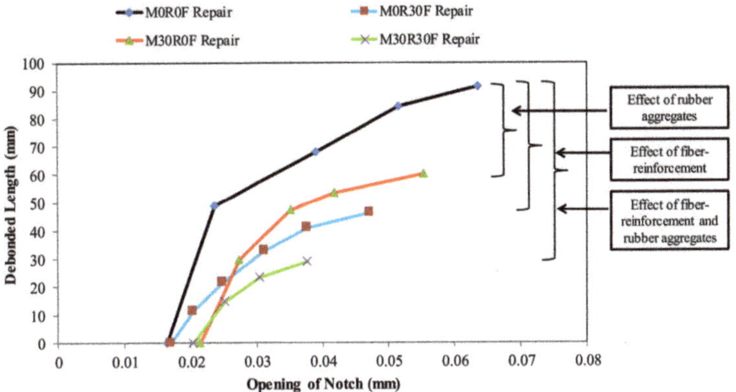

Figure 20. Debonded length vs. notch opening for different repair layers.

The notch openings of 15, 16, 21 and 22 µm were recorded, at which debonding is initiated in beams repaired with mixed compositions M0R0F, M0R30F, M30R0F and M30R30F, respectively. This indicates that the benefit of the inclusion of rubber aggregates in repair material is not only restricted to crack controlling, but it is also useful to retard the initiation of delamination.

Fibre-reinforcement has no notable effect on the start of delamination. The lengths of delamination are restricted with use of M0R30F and M30R30F as repair materials. Therefore, it is concluded that the repair material reinforced with fibres is not only limited to controlling the crack opening, but also to the debonding.

7. Conclusions

In this paper, a detailed experimental and numerical study has been conducted on the structural performance of beams with base and repair under three-point bending monotonic loading. The following conclusions are drawn from the experimental and numerical investigation:

- The maximum crack restraining was shown by repair material including fibres with or without the inclusion of rubber particles.
- To initiate the debonding, repair materials that are fibre-reinforced (either with or without rubber particles) required a greater value of load than the repair materials without the inclusion of fibres.
- M0R0F shows the minimum resistance to initiating interface debonding. However, for M30R0F repaired material, the transmission of delamination is restricted and retarded compared to the control one. The micro-cracking was controlled by rubber particles, which as a result increases and the load for debonding. Moreover, improved strain capacity of the repaired material with the addition of rubber particles also increases the notch opening at which the delamination starts.

- Debonding transmission at the interface is more significant in the materials without fibres (M0R0F or M30R0F) and is largely controlled in M30R30F repair. It shows the positive synergetic effect by the collective utilisation of rubber particles and fibres under static bending test.
- The FEM model results show that it accurately predicts the mechanical response of the material under monotonic flexure testing. In particular, it allows the kinetics of crack advancement and of delamination at the interface to be predicted.
- The developed FEM is an effective technique to predict and analyse the response of the repaired system under mechanical conditions of loading. In particular, it helps to highlight the benefit of incorporating rubber aggregates and fibre-reinforcement and the positive synergetic effect of both.
- The DIC technique is a suitable tool for the monitoring of debonding propagation along the interface.
- Finally, as added value to the contribution towards achieving more sustainable repair, the addition of rubber aggregates obtained through grinding of scrap tyres in cementitious materials can also be considered to maintain a clean environment by recycling used tyres and minimizing the use of landfill for residual waste.

Author Contributions: Conceptualization, S.A.A.G., A.T. (Ahmed Toumi) and A.T. (Anaclet Turatsinze); Data curation, S.A.A.G. and S.S.; Formal analysis, S.S. and W.A.; Investigation, S.A.A.G., S.S., W.A., S.A., A.T. (Ahmed Toumi) and A.T. (Anaclet Turatsinze); Methodology, S.A.A.G., S.A. and A.T. (Anaclet Turatsinze); Resources, W.A., A.M.M. and M.M.S.; Software, S.S. and A.T. (Ahmed Toumi); Supervision, A.T. (Ahmed Toumi) and A.T. (Anaclet Turatsinze); Validation, A.T. (Anaclet Turatsinze); Visualization, W.A., A.M.M. and M.M.S.; Writing—original draft, S.A.A.G.; Writing—review & editing, S.S., W.A., S.A., A.T. (Ahmed Toumi), A.T. (Anaclet Turatsinze), A.M.M. and M.M.S. All authors have read and agreed to the published version of the manuscript.

Funding: This research received no external funding.

Institutional Review Board Statement: Not applicable.

Data Availability Statement: Not applicable.

Acknowledgments: The authors are grateful to University of Engineering and Technology Lahore-Pakistan for the financial support of the thesis which is the source of this work.

Conflicts of Interest: The authors declare no conflict of interest.

References

1. Bissonnette, B.; Courard, L.; Fowler, D.W.; Granju, J.-L. *Bonded Cement-Based Material Overlays for the Repair, the Lining or the Strengthening of Slabs or Pavements*; State-of-the-Art Report of the RILEM Technical Committee 193-RLS; Springer Science & Business Media: Berlin/Heidelberg, Germany, 2011; Volume 3.
2. Chanvillard, G.; Aitcin, P.C.; Lupien, C. Field evaluation of steel-fiber reinforced concrete overlay with various bonding mechanisms. *Transp. Res. Rec.* **1989**, *1226*, 48–56.
3. Granju, J.L. Thin bonded overlays: About the role of fiber reinforcement on the limitation of their debonding. *Adv. Cem. Based Mater.* **1996**, *4*, 21–27. [CrossRef]
4. Tran, Q.T.; Toumi, A.; Granju, J.-L. Experimental and numerical investigation of the debonding interface between an old concrete and an overlay. *Mater. Struct.* **2006**, *39*, 379–389. [CrossRef]
5. Langlois, M.; Pigeon, M.; Bissonnette, B.; Allard, D. Durability of pavement repairs field experiment. *Concr. Int.* **1994**, *16*, 39–43.
6. Emmons, P.H.; Czarnecki, L.; Mcdonald, J.E.; Vaysburd, A.M. Durability of repair materials: Current practice and challenges. In Proceedings of the International Symposium on Brittle Matrix Composites, Warsaw, Poland, 9–11 October 2000; pp. 263–274.
7. Gillani, A.A.S.; Toumi, A.; Turatsinze, A. Effect of incorporating rubber aggregates and fiber reinforcement on the durability of thin bonded cement-based overlays. In Proceedings of the 8th RILEM International Conference on Mechanisms of Cracking and Debonding in Pavements, Nantes, France, 7–9 June 2016; Springer: Berlin/Heidelberg, Germany, 2016; pp. 619–625.
8. Gillani, S.A.A. Degradation of the residual strength of concrete: Effect of Fiber-Reinforcement and of Rubber Aggregates, Application to Thin Bonded Cement-Based Overlays. Ph.D. Thesis, Université de Toulouse, Université Toulouse III-Paul Sabatier, Toulouse, France, 2017.

9. Ali Gillani, S.A.; Rizwan Riaz, M.; Hameed, R.; Qamar, A.; Toumi, A.; Turatsinze, A. Fracture Energy of Fiber-Reinforced and Rubberized Cement-Based Composites: A Sustainable Approach Towards Recycling of Waste Scrap Tires. 2022. Available online: https://journals.sagepub.com/doi/abs/10.1177/0958305X221089223 (accessed on 11 April 2022).

10. Fiber-Reinforced and Rubberized Cement-Based Composite: A Sustainable Repair Material for Thin Bonded Overlays. Available online: https://scholar.google.com/citations?view_op=view_citation&hl=en&user=w8Ca5HIAAAAJ&sortby=pubdate&citation_for_view=w8Ca5HIAAAAJ:WF5omc3nYNoC (accessed on 24 April 2022).

11. Nguyen, T.H.; Toumi, A.; Turatsinze, A. Mechanical properties of steel fibre reinforced and rubberised cement-based mortars. *Mater. Des.* **2010**, *31*, 641–647. [CrossRef]

12. Toumi, A.; Nguyen, T.-H.; Turatsinze, A. Debonding of a thin rubberised and fibre-reinforced cement-based repairs: Analytical and experimental study. *Mater. Des.* **2013**, *49*, 90–95. [CrossRef]

13. Nguyen, T.-H.; Toumi, A.; Turatsinze, A.; Tazi, F. Restrained shrinkage cracking in steel fibre reinforced and rubberised cement-based mortars. *Mater. Struct.* **2012**, *45*, 899–904. [CrossRef]

14. Gillani, S.A.A.; Shahzad, S.; Toumi, A.; Turatsinze, A. Debonding of rubberised fibre-reinforced cement-based repairs under fatigue loading: Experimental study and numerical modelling. *Int. J. Pavement Eng.* **2021**, 1–17. [CrossRef]

15. Mateos, A.; Harvey, J.; Paniagua, J.; Paniagua, F.; Fan Liu, A. Mechanical characterisation of concrete-asphalt interface in bonded concrete overlays of asphalt pavements. *Eur. J. Environ. Civ. Eng.* **2017**, *21*, 43–53. [CrossRef]

16. Isla Calderón, F.A.; Luccioni, B.M.; Ruano Sandoval, G.J.; Torrijos, M.C.; Morea, F.; Giaccio, G.M.; Zerbino, R.L. Mechanical response of fiber reinforced concrete overlays over asphalt concrete substrate: Experimental results and numerical simulation. *Constr. Build. Mater.* **2015**, *93*, 1022–1033. [CrossRef]

17. Hasani, M.; Nejad, F.M.; Sobhani, J.; Chini, M. Mechanical and durability properties of fiber reinforced concrete overlay: Experimental results and numerical simulation. *Constr. Build. Mater.* **2021**, *268*, 121083. [CrossRef]

18. Gillani, S.A.A.; Toumi, A.; Turatsinze, A. Effect of surface preparation of substrate on bond tensile strength of thin bonded cement-based overlays. *Int. J. Pavement Res. Technol.* **2020**, *13*, 197–204. [CrossRef]

19. *BS EN 197-1: 2011*; Cement, Composition, Specifications and Conformity Criteria for Common Cements. British Standard Institution (BSI): London, UK, 2011.

20. Shahzad, S.; Toumi, A.; Balayssac, J.-P.; Turatsinze, A.; Mazars, V. Cementitious composites incorporating Multi-Walled Carbon Nanotubes (MWCNTs): Effects of annealing and other dispersion methods on the electrical and mechanical properties. *Matériaux Tech.* **2022**, *110*, 104. [CrossRef]

21. Fibraflex a New Generation of Metallic Fibers'. Available online: http://www.fibraflex.fr/sites/fibraflex.com/files/pdf/fibraflex_brochure_en_0.pdf (accessed on 11 April 2022).

22. Balouch, S.U.; Granju, J.-L. Corrosion of different types of steel fibres in SFRC and testing of corrosion inhibitors. In Proceedings of the Infrastructure Regeneration and Rehabilitation Improving the Quality of Life through Better Construction: A vision for the Next Millennium, Sheffield, UK, 28 June–2 July 1999; pp. 735–747.

23. Fiore, A.; Marano, G.C.; Marti, C.; Molfetta, M. On the fresh/hardened properties of cement composites incorporating rubber particles from recycled tires. *Adv. Civ. Eng.* **2014**, *2014*, 876158. [CrossRef]

24. *BS EN 12390-3*; Essais pour Béton Durci–Partie 3: Résistance à la Compression des Éprouvettes [Tests for Hardened Concrete-Part 3: Compressive Strength of the Specimens]. British Standards Institution (BSI): London, UK, 2012.

25. *BS EN 12390-1*; Testing hardened concrete–Part 1: Shape, dimensions and other requirements for specimens and moulds. British Standards Institution (BSI): London, UK, 2000.

26. *BS EN 12390-13*; Testing Hardened Concrete–Part 13: Determination of Secant Modulus of Elasticity in Compression. British Standards Institution: London, UK, 2013.

27. Vandewalle, L.; Nemegeer, D.; Balazs, G.L.; Barr, B.; Bartos, P.; Banthia, N.; Brandt, A.M.; Criswell, M.; Denarie, F.; Di Prisco, M. Rilem TC 162-TDF: Test and design methods for steel fibre reinforced concrete: Uni-axial tension test for steel fibre reinforced concrete. *Mater. Struct. Mater. Et Constr.* **2001**, *34*, 3–6.

28. Chausson, H. La Durabilité des Rechargements Minces Adhérents en Béton Renforcé de Fibres Métalliques. Ph.D. Thesis, Université Paul Sabatier, Toulouse, France, 1997.

29. Courard, L.; Piotrowski, T.; Garbacz, A. Near-to-surface properties affecting bond strength in concrete repair. *Cem. Concr. Compos.* **2014**, *46*, 73–80. [CrossRef]

30. Farhat, H. Durabilité des Réparations en Béton de Fibres: Effets du Retrait et de la Fatigue. Ph.D. Thesis, Université Paul Sabatier, Toulouse, France, 1999.

31. Garbacz, A.; Courard, L.; Kostana, K. Characterization of concrete surface roughness and its relation to adhesion in repair systems. *Mater. Charact.* **2006**, *56*, 281–289. [CrossRef]

32. Grandhaie, F. Le Béton de Fibres Métalliques Amorphes Comme Nouveau Matériau de Réparation. Ph.D. Thesis, Université Paul Sabatier, Toulouse, France, 1993.

33. Silfwerbrand, J. Improving concrete bond in repaired bridge decks. *Concr. Int.* **1990**, *12*, 61–66.

34. Tran, Q.T. Interface ancien-nouveau béton: Caractérisation du Comportement Adoucissant de L'interface au Cours de Décollement et son Évolution dans le cas de Sollicitation de Fatigue. Ph.D. Thesis, Université Paul Sabatier, Toulouse, France, 2006.

35. Nguyen, T.-H. Durabilité des Réparations à Base Cimentaire: Analyse Comparée de L'influence des Propriétés Mécaniques du Matériau de Réparation. Ph.D. Thesis, Université de Toulouse, Université Toulouse III-Paul Sabatier, Toulouse, France, 2010.

36. Chu, T.C.; Ranson, W.F.; Sutton, M.A. Applications of digital-image-correlation techniques to experimental mechanics. *Exp. Mech.* **1985**, *25*, 232–244. [CrossRef]
37. Peters, W.H.; Ranson, W.F.; Sutton, M.A.; Chu, T.C.; Anderson, J. Application of digital correlation methods to rigid body mechanics. *Opt. Eng.* **1983**, *22*, 226738. [CrossRef]
38. Peters, W.H.; Ranson, W.F. Digital imaging techniques in experimental stress analysis. *Opt. Eng.* **1982**, *21*, 213427. [CrossRef]
39. Sutton, M.A.; Mingqi, C.; Peters, W.H.; Chao, Y.J.; McNeill, S.R. Application of an optimized digital correlation method to planar deformation analysis. *Image Vis. Comput.* **1986**, *4*, 143–150. [CrossRef]
40. Gencturk, B.; Hossain, K.; Kapadia, A.; Labib, E.; Mo, Y.-L. Use of digital image correlation technique in full-scale testing of prestressed concrete structures. *Measurement* **2014**, *47*, 505–515. [CrossRef]
41. Pickerd, V. *Optimisation and Validation of the ARAMIS Digital Image Correlation System for Use in Large-Scale High-Strain-Rate Events*; Defense Science and Technology Organization Victoria: Maritime, Australia, 2013.
42. Kilonewton. Logiciel VIC-3D de Corrélation D'images. Available online: https://www.kilonewton.fr/correlation_images/vic_3d.html (accessed on 18 May 2022).
43. Petersson, P.-E. Crack Growth and Development of Fracture Zones in Plain Concrete and Similar Materials. Ph.D. Thesis, Lund University, Lund, Sweden, 1981.
44. Toumi, A.; Nguyen, T.-H.; Turatsinze, A. Modelling of the debonding of steel fibre reinforced and rubberised cement-based overlays under fatigue loading. *Eur. J. Environ. Civ. Eng.* **2015**, *19*, 672–686. [CrossRef]
45. Granju, J.-L. Debonding of thin cement-based overlays. *J. Mater. Civ. Eng.* **2001**, *13*, 114–120. [CrossRef]
46. Turatsinze, A.; Bonnet, S.; Granju, J.-L. Mechanical characterisation of cement-based mortar incorporating rubber aggregates from recycled worn tyres. *Build. Environ.* **2005**, *40*, 221–226. [CrossRef]
47. Turatsinze, A.; Granju, J.-L.; Bonnet, S. Positive synergy between steel-fibres and rubber aggregates: Effect on the resistance of cement-based mortars to shrinkage cracking. *Cem. Concr. Res.* **2006**, *36*, 1692–1697. [CrossRef]
48. Ho, A.C.; Turatsinze, A.; Hameed, R.; Vu, D.C. Effects of rubber aggregates from grinded used tyres on the concrete resistance to cracking. *J. Clean. Prod.* **2012**, *23*, 209–215. [CrossRef]

 *materials*

Article

The Influence of Cement Type on the Properties of Plastering Mortars Modified with Cellulose Ether Admixture

Edyta Spychał * and Przemysław Czapik

Faculty of Civil Engineering and Architecture, Kielce University of Technology, 25-314 Kielce, Poland; p.czapik@tu.kielce.pl
* Correspondence: espychal@tu.kielce.pl

Abstract: In this article, the effect of cement type on selected properties of plastering mortars containing a cellulose ether admixture was studied. In the research, commercial CEM I Portland cement, CEM II and CEM III, differing in the type and amount of mineral additives, and cement class, were used as binders. Tests of consistency, bulk density, water retention value (WRV), mechanical properties and calorimetric tests were performed. It was proved that the type of cement had no effect on water retention, which is regulated by the cellulose ether. All mortars modified with the admixture were characterized by WRV of about 99%. High water retention is closely related to the action of the cellulose ether admixture. As a result of the research, the possibility of using cement with additives as components of plasters was confirmed. However, attention should be paid to the consistency, mechanical properties of the tested mortars and changes in the pastes during the hydration process. Different effects of additives resulted from increasing or decreasing the consistency of mortars; the flow was in the range from 155 mm to 169 mm. Considering the compressive strength, all plasters can be classified as category III or IV, because the mortars attained the strength required by the standard, of at least 3.5 MPa. The processes of hydration of pastes were carried out with different intensity. In conclusion, the obtained results indicate the possibility of using CEM II and CEM III cements to produce plastering mortars, without changing the effect of water retention.

Keywords: cellulose ether; cement plastering mortar; mineral additives; consistency; bulk density; water retention; cement paste hydration; flexural and compressive strength

Citation: Spychał, E.; Czapik, P. The Influence of Cement Type on the Properties of Plastering Mortars Modified with Cellulose Ether Admixture. *Materials* **2021**, *14*, 7634. https://doi.org/10.3390/ma14247634

Academic Editor: Jorge Otero

Received: 17 November 2021
Accepted: 8 December 2021
Published: 11 December 2021

Publisher's Note: MDPI stays neutral with regard to jurisdictional claims in published maps and institutional affiliations.

Copyright: © 2021 by the authors. Licensee MDPI, Basel, Switzerland. This article is an open access article distributed under the terms and conditions of the Creative Commons Attribution (CC BY) license (https://creativecommons.org/licenses/by/4.0/).

1. Introduction

Human activities have an increasing impact on the surrounding natural environment. Reduction of CO_2 production has long been the main problem of the global economy, presenting challenges in areas such as engineering, environmental protection and the construction industry [1–12]. CO_2 emissions from fossil fuels and industry account for approximately 90% of all CO_2 emissions into the atmosphere from human activity. Cement production alone accounts for about 5% of global CO_2 emissions [1]. CO_2 emissions computed for the finished cement depend mainly on the clinker content, especially for CEM I cement [5]. The reduction of carbon dioxide emissions from cement production is, therefore, an important and urgent task for the cement industry. One of the possible ways to limit the use of clinker is the use of cement with a high content of mineral additives [4,6,10–13]. Siliceous fly ash, calcareous fly ash and granulated blast-furnace slag are traditionally used in the production of cement. These additives have pozzolanic and hydraulic properties, respectively, which advantageously influence cement properties. The use of these raw materials in the production of cement thus reduces carbon dioxide emissions [7–10,12].

Cement is one of the most popular binders used in dry-mix mortars, such as plastering mortars, masonry mortars and adhesive mortars [14–17]. Cement in these materials acts as a binder ensuring obtaining the appropriate strength class and the durability of the finished product. It is also largely responsible for the adhesion of the mortar to the substrate [15,17].

Portland cement CEM I is the basic binder in mortars, but more often this cement is replaced by CEM II multi-component and CEM III [17].

Modern plastering mortars are complex multi-component systems. Among the mortar components (besides binder, fine aggregate and water), cellulose ether admixture plays an important role in dry-mix mortars [18–27]. Cellulose ethers as polymer admixtures are being applied to a growing extent in the production of dry-mix mortars. This leads on the one hand to a great variety of areas of application and on the other hand to an increasing diversity of mortars. First of all, these polymers improve water retention [18,19,21–23,25]. Their function is to prevent water loss into porous, absorbent substrates [23]. High water retention provides proper conditions for the binding and hardening processes of a binder [26], and this ability has a positive effect on reducing mortar shrinkage [27–29]. Cellulose ethers have a significant impact on the rheology of fresh mortars [19,21,30,31].

In article [14], Chłądzyński assessed the suitability of cements with additives as a binder used in the production of adhesive mortars. The subject of the research was mortars prepared from CEM I Portland cements of various specific surface area and mortars with multi-component Portland cements CEM II (containing varying amounts of fly ash and granulated blast-furnace slag). Cements made in the laboratory were used for the tests, through joint grinding clinker, gypsum, silica fly ash or granulated blast-furnace slag. All samples contained a constant amount of cellulose ether and redispersible powder. Standard tests of physical and mechanical properties of cements were performed, as were calorimetric tests of the heat of hydration and standards tests of adhesive mortars. The results of tests of adhesive mortars with fly ash differed from the results obtained in the case of Portland cement mortars. The effect of fly ash addition was different for individual methods. On the one hand, the research showed slightly better results in terms of adhesion after thermal ageing, but on the other hand, the addition lowered the adhesion values under sample conditions (adhesion tests after immersion of samples in water, adhesion tests due to freeze-thaw cycles). Adhesive mortars made of cement with fly ash show smaller slip versus CEM I Portland cement mortars. The effect of granulated blast-furnace slag addition in adhesive mortars was similar to the effect of fly ash. The addition of granulated blast-furnace slag also improves the open time for tested mortars. As a result of the research, it was found that the tested cements with additives can be used as a binder in the composition of adhesive mortars. The influence of cement replacement by fly ash in brick masonry strength was experimentally verified by Seshu and Murthy in their article [32]. The research consisted of the casting and testing of brick masonry prisms, with two bricklayers. Cement and cement-fly ash mortars were prepared. In each mix the fly ash percentage replacing cement binder in the mortars was increased from 0% to 40%, in intervals of 10%. The results showed that replacement of cement with fly ash in cement mortars is possible up to 40%, without unfavorable effects on the properties of the masonry mortars. The tested additive replacement in leaner cement mortar mixes resulted in the loss of mechanical properties by more than 15%, so cement replacement with fly ash, in this case, may be not useful or profitable. Mortars containing cement and fly ash modified with chemical admixtures have been researched by Zhou and et.al. [33]. All samples contained a constant amount of cement and additive, but the amount of cellulose ether, starch ether, bentonite and redispersion emulsoid powder were variable. The research was an evaluation of the consistency, water retention, setting time, compressive strength, but the effect of the fly ash on the properties tested was not analyzed. The authors focused on the evaluation of the working and mechanical properties of ordinary dry-mixed mortars. It was found that cellulose ether admixtures had the biggest influence on the consistency, water retention and compressive strength of mortars, among all the analyzed chemical admixtures.

This paper describes how the type of cement affects the plastering mortars' selected properties, i.e., consistency, water retention, flexural and compressive strength and hydration process. The described experimental results constitute the first part of our research, concerning the assessment of the suitability of cements CEM II and CEM III as a binder in plastering mortar modified with cellulose ether admixture. The scope of further planned research is presented in the conclusions of this article. The research conducted so far has focused mainly on the use of cement CEM I, hydraulic lime or hydrated lime for the properties of the plasters. This article may be a supplement to the knowledge on the interaction of cellulose ether with cements containing additives. Nowadays, the use of additives in the production of cements is an important issue from the point of view of sustainable development, ecology and economic considerations. The goal of the investigation was the assessment of the suitability of the chosen cements CEM II and CEM III as binders in cement-based plastering mortars modified with cellulose ether—to determine the influence of these binders on the selected functional and mechanical properties of plastering mortars. Additionally, in order to complete the tests of flexural and compressive strength of mortars, calorimetric measurements of pastes were performed.

2. Materials and Methods

2.1. Materials and Sample Preparation

Commercial bag cement CEM I, CEM II and CEM III (from various cement plants), quartz sand 0.5–1.4 mm (Kreisel, Dąbrowa, Poland), cellulose ether admixture (WALOCEL, The Dow Chemical Company, Midland, MI, USA) and tap water were used. Cellulose ether used in tests is a hydroxyethyl methyl cellulose (HEMC) with the viscosity of 25,000 mPa·s. This admixture is in the form of white powder and it has a low level of chemical modification. Five main types of mortars were prepared for the tests. The first type of mortar (C1) was the reference one, which was prepared using an ordinary Portland cement, CEM I 42.5R, with cellulose ether admixture. The remaining mortars were prepared based on CEM III/A 32.5 N-LH, CEM II/B-V 42.5 R, CEM II/B-M (V-LL) 32.5 R, CEM II/B-V 32.5 R cements, marked sequentially as C2, C3, C4 and C5. All cements met the requirements of EN 197-1 standard. In addition, in the case of selected properties, the C0 mortar was prepared using CEM I 42.5 R cement and did not contain a cellulose ether admixture. The chemical composition and selected physical and mechanical properties of cements obtained from the cement plants are presented, respectively in Tables 1 and 2.

Table 1. Chemical composition of cements.

Components	CEM I 42.5 R	CEM III/A 32.5 N-LH	CEM II/B-V 42.5 R	CEM II/B-M (V-LL) 32.5 R	CEM II/B-V 32.5 R
SiO_2	20.14	23.96	27.3	20.0	28.75
Al_2O_3	4.88	6.74	10.9	6.9	11.32
Fe_2O_3	3.44	1.99	3.8	2.9	4.23
CaO	64.05	50.38	46.2	53.1	44.42
Na_2O	0.16	0.33	0.4	0.3	0.48
MgO	1.61	4.28	1.7	1.5	2.14
SO_3	2.71	1.48	2.6	2.8	2.38
Cl	0.025	0.037	0.1	0.1	0.065
K_2O	0.61	0.54	1.4	1.0	1.59

All samples were prepared and tested in an air-conditioned laboratory at the temperature of 20 ± 2 °C and at a relative humidity of 65 ± 5%.

The mortar mix proportion is detailed in Table 3. The samples were made with a binder to fine aggregate weight ratio 1:3. The water to binder ratio was 0.7 for all mortars. The amount of water was selected in such a way that the C1 mortar had a flow of 165 mm (consistency within borders 175 ± 10 mm). All samples from C1 to C5 contained a constant amount of cellulose ether admixture, in quantity 4 g. The amount of the admixture was selected experimentally and based on the analysis of the literature [18–21,26,31].

Table 2. Physical and mechanical properties of cements.

Properties	CEM I 42.5 R	CEM III/A 32.5 N-LH	CEM II/B-V 42.5 R	CEM II/B-M (V-LL) 32.5 R	CEM II/B-V 32.5 R
Water requirement of normal consistency (%)	27.1	29.3	30.3	26.7	28.0
Initial setting time (min)	215	277	254	215	315
Specific surface area (cm^2/g)	3942	4146	4433	4532	3424
28 days compressive strength (MPa)	58.8	51.3	55.2	41.6	40.8

Table 3. Mortar mix proportion of all samples.

Component (g)	C0 CEM I 42.5 R	C1 CEM I 42.5 R	C2 CEM III/A 32.5 N-LH	C3 CEM II/B-V 42.5 R	C4 CEM II/B-M (V-LL) 32.5 R	C5 CEM II/B-V 32.5 R
Cement	450	450	450	450	450	450
Fine aggregate	1350	1350	1350	1350	1350	1350
Admixture	-	4	4	4	4	4
Water	315	315	315	315	315	315
w/c Ratio	0.7	0.7	0.7	0.7	0.7	0.7

2.2. Methods

The measurements of standard consistency were done according to PN-EN 1015-3:2000 + A2:2007 [34] and PN-B-04500:1985 standards [35].

The bulk density of fresh mortars was determined in accordance with PN-EN 1015-6:2000 + A1:2007 standard [36], but the bulk density of hardened mortars was determined in accordance with PN-EN 1015-10:2001 + A1:2007 standard [37].

Water retention value was determined in the accordance with the defined guidelines [38]. These tests were performed after 10, 30 and 60 min and were defined as WRV10, WRV30 and WRV60. This parameter was determined by weighing absorbent materials (filter paper) placed on the fresh sample before and after the predetermined measurement time. Water retention was calculated according to the formula [38]:

$$WRV = 100 - W3 \; [\%] \tag{1}$$

$$W3 = \frac{W2}{W1} \cdot 100 \; [\%] \tag{2}$$

In Formula (1), W3 means the relative water loss in the mortar, expressed as a percentage. In Formula (2), W2 means water mass absorbed by the filter paper, but W1 means water content in the tested mortar in the plastic ring (expressed in grams) [38].

The flexural strength and compressive strength of cement mortars were determined in accordance with PN-EN 1015-11:2001 + A1:2007 standard [39]. For each mortar, three cuboid samples of mortar of 40 mm $\times$ 40 $\times$ mm $\times$ 160 mm dimensions were prepared. Mechanical properties measurements were performed after 2, 7 and 28 days.

Samples intended for testing properties of hardened mortars (bulk density and mechanical properties), after their disassembly (2 days after preparation), were stored for 5 days in polyethylene bags, and then for another 21 days in dry air conditions.

The hydration heat evolution of cement pastes was investigated using a differential conducting microcalorimeter at 20 °C for 72 h. The pastes were prepared as mixtures of 4.5 g of cement, 3.15 g of water and 0.04 g of admixture. The w/c ratio of all samples was 0.7. The research used the BT2.15CS low-temperature differential scanning microcalorimeter (Setaram, Plan-Ies-Ouates, Geneva, Switzerland) operating under non-isothermal and non-adiabatic conditions.

3. Results

3.1. Consistency Measurements

In Table 4, the results of the consistency for all samples are presented (measurements made with the flow table method in mm and measurements made with the drop cone in cm).

Table 4. Consistency results for all mortars.

Symbol of Mortar	Flow [1] (mm)	Cone Penetration [2] (cm)
C0	205	12.9
C1	165	8.5
C2	169	8.0
C3	165	7.7
C4	155	6.6
C5	164	7.9

[1] Consistency was determined in accordance with standard [34]. [2] Consistency was determined in accordance with standard [35].

The flow of C1 mortar was 165 mm. This value was established as the baseline. All mortars from C1 to C5 are characterized by plastic consistency, according to the standard PN-EN 1015-3:2000+A2:2007 [34] (flow diameter in the range from 140 mm to 200 mm) [36,40]. The lowest flow among mortars modified with admixture was observed with C4 mortar (155 mm), but the largest was observed with C2 mortar (169 mm). In both cases the type of additive influenced the consistency. Ground granulated blast-furnace slag increases the flow of the mortars, while the use of limestone increases the water demand of mortars, thus reducing their flow. A similar trend can be observed in the case of the cone penetration test (consistency test according to the standard PN-B-04500:1985). Taking into account the results of consistency of C1–C5 mortars in accordance with [35], it can be concluded that all tested materials achieve the consistency value characteristic of typical plasters used in practice [38,40]. In the case of plastering mortars intended for manually applied plasters, their consistency (according by PN-B-04500:1985 standard) should be 6–9 cm, while for mechanical (by machine) application it should be 8–11 cm [38]. All mortars modified with cellulose ether admixture can be applied manually. Only C1 and C2 mortars can be applied by a machine.

Table 4 shows the results of mortar consistency tests without admixture (sample marked with the symbol C0). It is clearly visible that the mortar without admixture has the greatest consistency in comparison to mortars modified with cellulose ether (these differences vary from 18% to even 35%). Cellulose ether significantly reduces the consistency. Mortars containing this admixture in their composition are characterized by good workability, no segregation of ingredients, which can be seen when comparing the appearance of the tested materials—flow test (Figure 1a,b).

Figure 1a,b shows the appearance of the C0 sample during the flow test. Even before the final measurements are taken, water is separating immediately after removing the mold. After measuring the flow diameter, one can also see water separating from the sample. This phenomenon is not observed in the case of other materials. Figure 2 shows the appearance of a mortar sample with CEM I cement and an admixture. The mortar is consistent, there are no visible signs of segregation of ingredients. The consistency measurements thus confirm the advantages of using cellulose ether admixtures, which improves the rheological and application properties of plastering mortars.

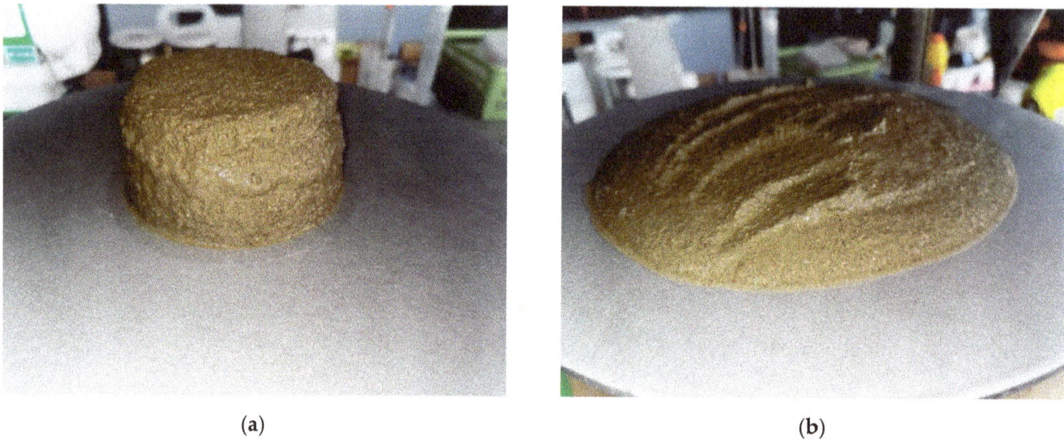

(a)   (b)

Figure 1. (**a**) View of the C0 sample after removing the form for flow research; (**b**) View of the C0 sample after flow test.

Figure 2. View of the C1 sample after flow test.

3.2. Water Retention

Table 5 and Figure 3 present the results of the water retention values WRV10, WRV30 and WRV60 (the tests were made after 10, 30 and 60 min measurements).

Table 5. Water retention results.

Symbol of Mortar	WRV10 (%)	WRV30 (%)	WRV60 (%)
C0	86.5	79.7	73.5
C1	99.6	99.3	98.9
C2	99.6	99.3	99.0
C3	99.6	99.3	98.9
C4	99.5	99.3	98.9
C5	99.5	99.4	99.0

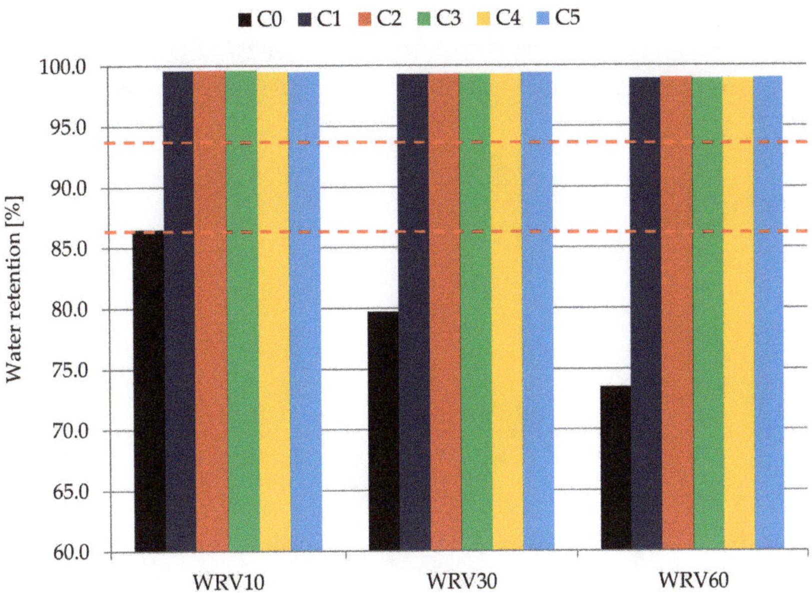

Figure 3. Change in water retention value of mortars C0–C5 over time.

Based on the research, it can be concluded that all mortars modified with cellulose ether admixture are characterized by a high water retention value throughout the whole test. Changes in water retention during the 60 min of the measurement are practically imperceptible (within 1%). Mortars from C1 to C5 can be classified according to the classification given by Brumaud et al. [22] as materials with high water retention (WRV > 94%), but mortar C0 has low water retention (WRV < 86%). A high water retention level is marked with a solid line in the Figure 3. A high water retention for plasters C1–C5 is related to the action of the admixture. Cellulose ether impacts the viscosity of mortar and causes greater water retention [18,19,21,41]. A part of the water is bonded in the first stage of cement hydration. At the same time, the remaining amount of water forms a gel with the admixture. In this gel, the water molecules are attracted by the functional groups from the polymer and agglomeration process takes place. As the hydration process occurs, this gel can release water into the system [26]. These conclusions also confirm the results obtained for mortar C0. The water retention value for this sample differed significantly from the others; moreover, it underwent changes over time. After 10 min, retention was 86.5% and after 60 min it was 73.5%. Water loss for mortar C0 was thu 26.5%, while it was a maximum of 1.1% for all the modified mortars. In conclusion, there is no apparent influence of type of cement on the water retention value and the change in water retention over time.

3.3. Bulk Density Measurements

Table 6 and Figure 4 present the results for the bulk density for mortars in the plastic and hardened state.

Table 6. Bulk density results.

Symbol of Mortar	Bulk Density in Plastic State (kg/m^3)	Bulk Density in Hardened State (kg/m^3)
C1	1434	1392
C2	1421	1375
C3	1513	1454
C4	1422	1377
C5	1455	1440

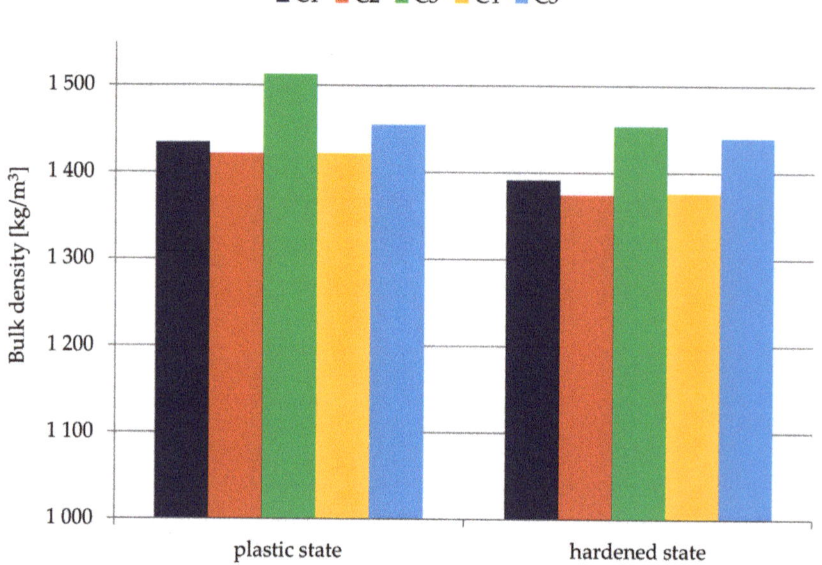

Figure 4. Bulk density results of mortars in plastic and hardened states.

The bulk density of fresh mortars is different. The parameter ranges from 1421 kg/m^3 (C2 mortar) to 1513 kg/m^3 (C3 mortar). The results for the bulk density of four of the tested samples are within the limits 1421–1455 kg/m^3, while the bulk density of the C3 sample differs from the others and amounts to 1513 kg/m^3. The plaster performance can be indirectly assessed on the basis of the parameters affecting the application properties of mortars (ease of application on the substrate, processing time) [19,21,26]. Taking into account the obtained results, mortars C2 and C4, are characterized by the biggest efficiency. Use of these plasters would be the most advantageous in terms of economy (bigger efficiency—lower costs related to material consumption) [21]. Due to the obtained results for bulk density ($\geq$1300 kg/m^3), the tested plasters are defined as ordinary mortars [40]. When it comes to the results for mortar bulk density in the hardened state, these range from 1375 kg/m^3 to 1454 kg/m^3. The lowest bulk density in the plastic and hardened state was achieved by C2 and C4 mortars. Mortar C3 with CEM II/B-V 42.5 R cement obtained the highest bulk density.

3.4. Results of Mechanical Properties

The strength measurements were done after 2, 7 and 28 days of curing. The values from three bars (flexural strength) or six bars (compressive strength) were calculated as an average. The results for flexural strength are shown in Table 7 and in Figure 5.

Table 7. Average strength of the mortars.

Sample Designation	Flexural Strength (MPa)			Compressive Strength (MPa)		
	2d	7d	28d	2d	7d	28d
C1	1.10	1.73	3.38	2.30	3.69	8.38
	(0.041) [1]	(0.042)	(0.155)	(0.084) [2]	(0.094)	(0.146)
C2	0.44	0.61	2.76	0.96	1.39	7.27
	(0.005)	(0.005)	(0.261)	(0.047)	(0.056)	(0.264)
C3	0.93	1.37	3.96	1.99	2.80	9.28
	(0.014)	(0.097)	(0.076)	(0.056)	(0.221)	(0.301)
C4	0.37	0.62	2.42	0.72	1.39	6.03
	(0.005)	(0.005)	(0.195)	(0.043)	(0.105)	(0.247)
C5	0.49	0.97	3.06	1.05	2.16	7.85
	(0.008)	(0.029)	(0.285)	(0.023)	(0.113)	(0.672)

[1] Standard deviation of flexural strength measurements. [2] Standard deviation of compressive strength measurements.

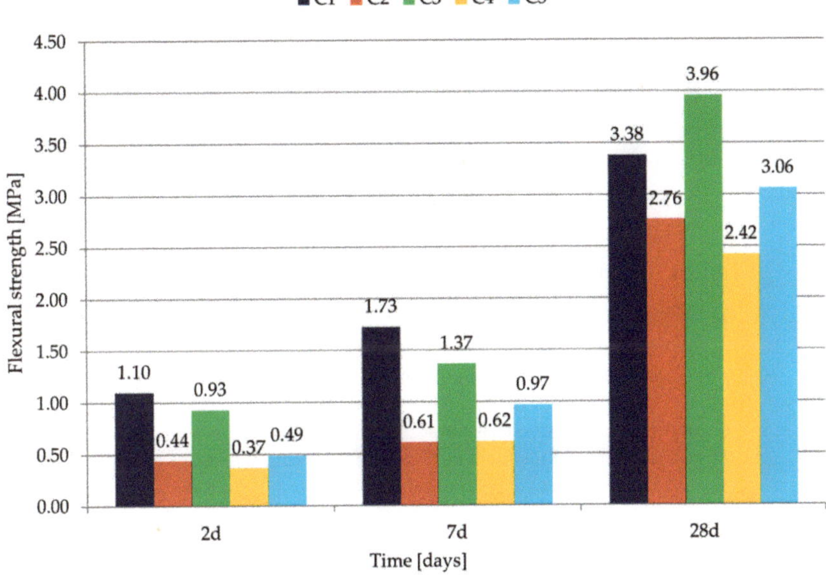

Figure 5. Change of flexural strength after different curing times.

Mortar C1 with cement CEM I 42.5 R (cement without addition) is characterized by the highest strength after 2 and 7 days of maturation. The early strength of the mortars C2–C5 was lower than that of the reference sample C1—the difference after 2 days of maturing was in the range of 15% to 66%. This was a result of the type of binder (class of cement and type of addition). The use of cement CEM II/B-V 42.5 R as cement CEM I 42.5 R replacement brings about strength increase at a later age. Mortar C3 (with cement CEM II/B-V 42.5 R) has the highest strength after 28 days.

The results of compressive strength are shown in Table 7 and in Figure 6.

The results of the compressive strength tests are similar to the results of the flexural strength tests. Mortar with cement CEM I 42.5 R is characterized by the highest strength after 2 and 7 days of maturation. This is due to the lower content of Portland cement clinker in CEM II and CEM III. Mortar with CEM II/B-V 42.5 R is characterized by the highest compressive strength after 28 days. Mortar with this cement has a bigger strength than the base mortar, made of cement without additives. Similar conclusions from the research were obtained in [32]. The authors concluded that fly ash as partial replacement of cement is very useful in mortar with high cement content. 40% replacement is possible without much affecting the strength of the mortars.

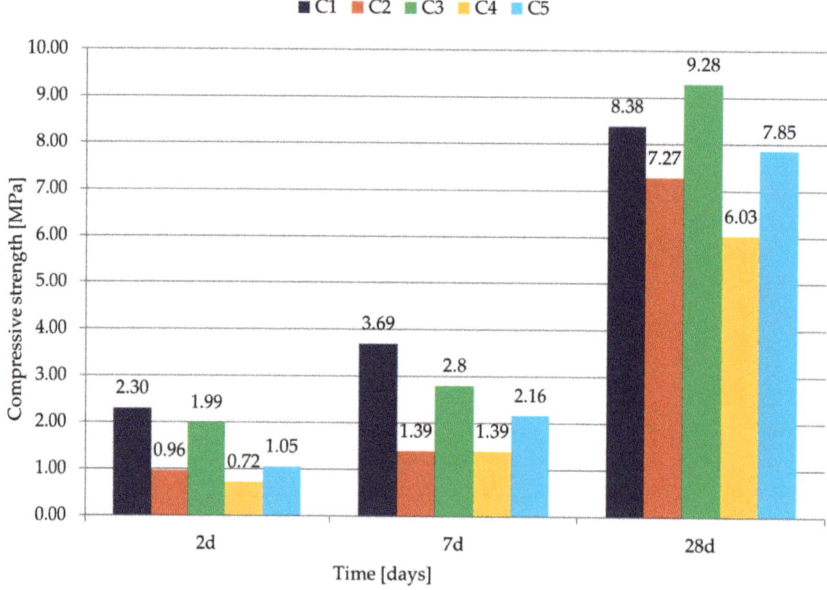

Figure 6. Change of the compressive strength after different curing times.

As one could expect, mortars with the cement of class 42.5 of high early strength (C1 and C3) are characterized by the highest flexural and compressive strength after 28 days, regardless of the type of cement, due to the additives.

Comparing C4 and C5 mortars with CEM II cement, differing in the type of additives, it can be concluded that the strength of mortars with fly ash is only greater than the results for mortars with fly ash and limestone (during the study period).

According to the classification of plastering mortars included in the PN-EN 998-1:2012 standard [42], all mortars can be classified as categories III and IV due to the compressive strength after 28 days.

Tables 8 and 9 show a comparison of the strength in relation to the reference mortar (C1); the results are given as a percentage. Changes in the increment of flexural and compressive strength in MPa were also determined, relating the strength results obtained after 7 and 28 days to the test results after 2 days of specimen maturation. Figure 7a,b shows the gain of flexural and compressive strength over time.

Table 8. Additional information obtained on the basis of flexural strength test.

Sample Designation	Flexural Strength (%)			Strength Gain (MPa)	
	2d	7d	28d	2d to 7d [1]	7d to 28d [2]
C1	100	100	100	0.63	2.30
C2	40.00	35.26	81.66	0.17	2.32
C3	84.55	79.19	117.16	0.44	3.03
C4	33.64	35.84	71.60	0.25	2.05
C5	44.55	56.07	90.53	0.48	2.57

[1] The difference in endurance between the 7th and the 2nd day of maturation. [2] The difference in endurance between the 28th and the 7th day of maturation.

Table 9. Additional information obtained on the basis of a compressive strength test.

Sample Designation	Compressive Strength (%)			Strength Gain (MPa)	
	2d	7d	28d	2d to 7d [1]	7d to 28d [2]
C1	100	100	100	1.39	6.08
C2	41.74	37.67	86.75	0.43	6.31
C3	86.52	75.88	110.74	0.81	7.29
C4	31.30	37.67	71.96	0.67	5.31
C5	45.65	58.54	93.68	1.11	6.80

[1] The difference in endurance between the 7th and the 2nd day of maturation. [2] The difference in endurance between the 28th and the 7th day of maturation.

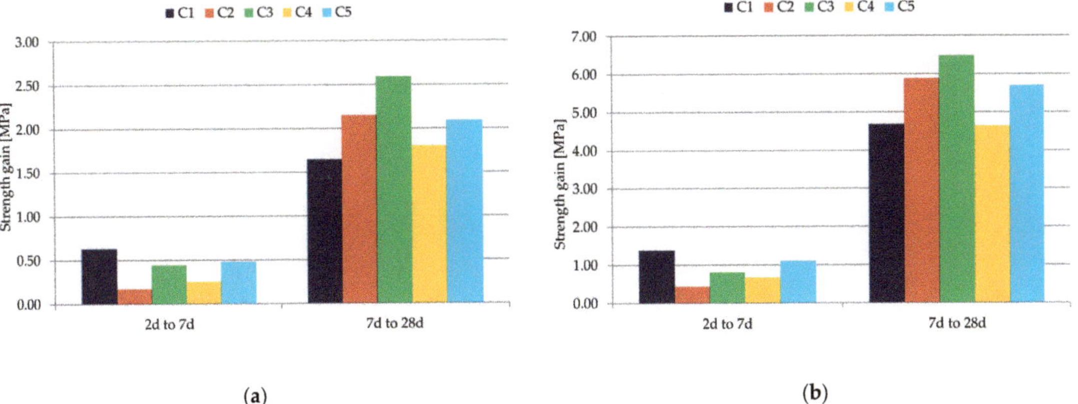

Figure 7. (**a**) Flexural strength gain over time. (**b**) Compressive strength gain over time.

The biggest increase in strength after 7 days was recorded for C1 mortar with CEM I 42.5 R cement, and the lowest for C2 mortar with CEM III/A 32.5 N-LH cement, in which the largest amount of Portland clinker is replaced by a mineral additive in the form of ground granulated blast-furnace slag. Other results can be seen when comparing the strength gains after 28 days. The biggest increase in strength after 28 days was recorded for C3 mortar with CEM II/B-V 42.5 R cement, and the lowest for C4 mortar with CEM II/B-M (V-LL) 32.5 R cement. Cements with chemically active mineral additives allow us to obtain significantly higher strength. However, in the case of the C4 sample, this effect is significantly reduced by the use of a chemically inactive additive—limestone.

3.5. Heat of Hydration for Pastes

The rate of heat evolution and the total heat released during the hydration of the tested pastes C1–C5 are shown in Figures 8 and 9. Induction time and the total heat released by cement pastes after 12, 24, 36, 41, 48, 72 h of hydration are given in Table 10. The results of calorimetric measurements of cements modified with cellulose ether were supplemented with the results for the hydration heat of C0 paste (cement paste with CEM I without admixture). The microcalorimetric curves for cement paste containing CEM II/B-V 42.5 R show that both the total amount of evolved heat and the rate of heat evolution over time do not differ significantly, as compared to a base paste with cement CEM I 42.5 R.

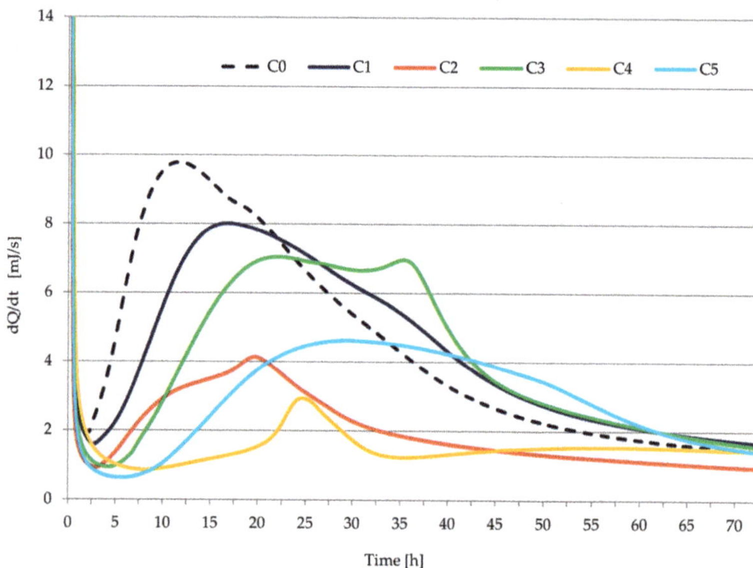

Figure 8. Rate of heat evolution as a function of time for all pastes used in the study.

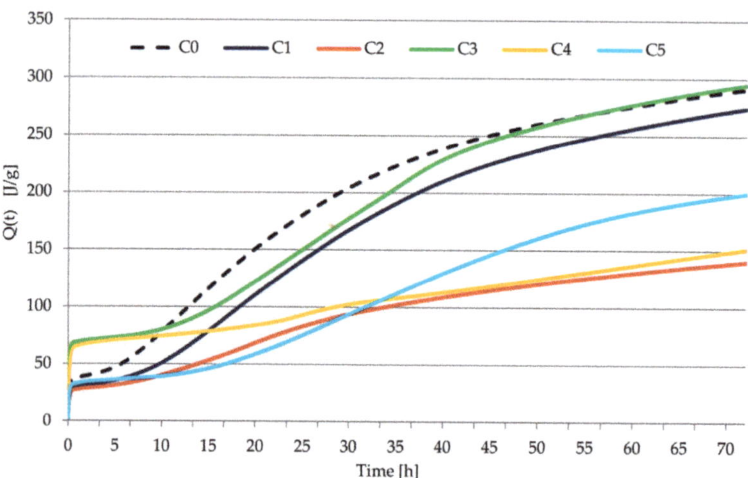

Figure 9. Total heat evolved as a function of time for all pastes used in the study.

Table 10. Heat of hydration of pastes.

Symbol of Paste	Induction Period (h)	Heat after Hours of Hydration (J/g)					
		12	24	36	41	48	72
C0	2 h 21 min	93.18	174.97	226.97	241.32	256.61	291.22
C1	2 h 51 min	60.95	134.97	195.06	213.67	233.07	274.41
C2	3 h 5 min	45.26	80.65	103.57	110.33	118.60	140.67
C3	5 h 3 min	85.58	144.88	210.24	233.25	253.48	295.02
C4	8 h 10 min	76.14	90.81	109.19	114.37	122.42	151.60
C5	6 h 15 min	41.55	72.13	115.95	133.27	155.17	199.95

In the case of the C0 paste, the typical course of the heat release curve during the hydration of Portland cement is visible. During this process, the highest indications for the deepened heat effect related to the hydration of alite and tricalcium aluminate [13,43] were achieved, and at the same time the shortest induction period. Cellulose ether admixture caused an extension of the induction period and a delay and suppression of the main heat release peak. As a result, the amount of heat release during hydration was reduced. The use of mineral additions in the cement (fly ash and ground granulated blast-furnace slag) caused longer shifts in time and reduced the occurrence of the main thermal effect as well as extending the induction period. This is usually related to the reduction of the cumulative amount of heat released. The amount of exhausted heat exceeded that determined for the Portland cement samples (C0 and C1) only in the case of sample C3. It is related to the occurrence of an additional, clear effect with a maximum recorded after about 37 h of hydration. It can be explained by the formation of calcium silicates rich in silicon, resulting from the initiation of the pozzolanic reaction [12,13,38,43–45].

There is a clear division between samples made of 32.5 and 42.5 class cements (about 30% to 50% compared to the cumulative amount of heat released after a certain period of hydration time). In the case of sample C5, in which, as in sample C3, CEM II/B-V cement was used, but of a lower class, no separate thermal effect was observed, which can be identified with a pozzolan reaction. Instead, the main heat effect was significantly extended over time and had a less pronounced maximum value than can be observed in other tests. This can be explained by the overlapping of two thermal effects. In the case of the C4 sample, where apart from silica fly ash, it is present in the form of a filler, the limestone had the longest induction period with a short-lived heat effect and the lowest intensity. Such a course of the thermal curve can be explained by the cement dilution caused by a non-reactive material that did not emit heat during the test.

It should be noted that the biggest heat emission from the hydration process was obtained with the C3 paste, and in the case of the C3 mortar, the biggest strength parameters were achieved after 28 days, as well as the biggest bulk density. In contrast, the smallest heat emission from hydration was obtained with samples of C2 and C3 paste, while the smallest mechanical properties and bulk density were noted in the case of the C2 and C4 mortars. However, C2 mortar from CEM III cement was characterized by a significant (second after C3 mortar) increase in strength in the period between 7 and 28 days. This proves that the processes essential for the strength of the sample took place in a later period, not covered by calorimetric measurements.

4. Conclusions

In the presented research, tests of consistency, bulk density, water retention value, mechanical properties of mortars and the heat of hydration pastes were performed using commercial cement CEM I, CEM II and CEM III. The possibility of using these binders as components of plaster mortars modified with a cellulose ether admixture was assessed.

Based on the experimental results presented in this paper, the following conclusions can be drawn:

- All mortars containing cellulose ether had a lower consistency (flow and cone penetration) than the cement mortar without admixture. The use of a polymer admixture was advisable due to the need to obtain a homogeneous, coherent material, with no visible signs of component segregation, characterized by high water retention. High water retention value is indicated for plastering mortars.
- The flow of plasters modified with cellulose ether admixture (C1, C2, C3, C5 samples) was in the range of 164 mm to 169 mm, and the cone penetration was in the range of 7.7 cm to 8.5 cm. The standard consistency of mortars modified with cellulose ether did not show significant differences with respect to mortars with CEM II and CEM III, except for mortar C4 with CEM II/B-M (V-LL) 32.5 R cement. Mortar with this binder showed the lowest flow (155 mm) and the lowest cone penetration (6.6 cm). In order to obtain the consistency as for the reference mortar, the water-cement ratio should be

increased or use can be made of appropriate admixtures, the compatibility of which with cellulose ether should be tested early.

- No effect of the type of cement on the water retention value was noted. The WRV value for all plastering mortars modified with cellulose ether was about 99% (during the study period). All plasters were characterized by high water retention during the test period (the water retention values were greater than 94%).

- The smallest bulk density of mortars in a plastic state (1421 kg/m^3, 1422 kg/m^3) and, at the same time, the highest efficiency were achieved by plasters made with the use of cements CEM III/A 32.5 N-LH (mortar C2) and CEM II/B-M (V-LL) 32.5 R (mortar C4).

- The type of cement, in particular the amount of clinker and additives, and the class of cement, have a key influence on the mechanical properties of mortars. However, regardless of the type of binder used, all plasters met the standard requirements for compressive strength and can be classified in categories III and IV [42]. The compressive strength (after 28 days) for mortars was in the range from 6.03 MPa to 9.28 MPa. Plasters qualify for category III if their compressive strength after 28 days is in the range 3.5–7.5 MPa, and for category IV when their compressive strength is above 6.0 MPa.

- The increase in flexural and compressive strength of the tested mortars was different, depending on the type and amount of the additive/additives. Comparing the results of compressive strength (after 28 days) of the mortar C1 with CEM I cement with the results of tests of mortars with cement CEM II and CEM III, the compressive strengths of mortars C2-C5 were from 72% to 111% of the compressive strength of mortar C1 (8.38 MPa). The smallest result was obtained for mortar with CEM II/B-M (V-LL) 32.5 N (6.03 MPa), but the largest result was obtained for mortar with CEM II B/V 42.5 R (9.28 MPa). Similar conclusions can be drawn on the basis of calorimetric tests.

- All mortars modified with cellulose ether, regardless of the type of cement, according to the applicable standard requirements, can be classified as ordinary mortars of plastic consistency [34,36,40]. Their bulk density in plastic state is greater than 1300 kg/m^3 and their flow is in the range 140–200 mm.

- These results indicate the possibility of using cement with mineral additives chosen for this research as a binder in plastering mortars. The selection of the appropriate cement gave similar and in some cases even better results than for plastering mortars in which Portland cement CEM I (without additives) was applied. The use of tested cements CEM II and CEM III did not lessen water retention, which is an important parameter in the case of plastering mortars.

- Considering the consistency determined in accordance with the standard [35] and the possible methods of application of the plasters, and taking into account the composition of mortars specified in the adopted composition, all materials were suitable for manual application, and additionally the mortar with CEM III/A 32.5 N-LH (besides cement CEM I 42.5 R) can be applied by a machine.

The next stage of this work will be the performance of other tests relating to the standards for plastering mortars, assessment of mechanical properties after more than 28 days and assessment of the microstructure of these plasters.

Author Contributions: Conceptualization, E.S.; methodology, E.S. and P.C.; validation, E.S. and P.C.; formal analysis, E.S.; investigation, E.S.; resources, E.S.; data curation, E.S.; writing—original draft preparation, E.S.; writing—review and editing, E.S. and P.C.; visualization, E.S. and P.C.; supervision, E.S.; project administration, E.S.; funding acquisition, E.S. All authors have read and agreed to the published version of the manuscript.

Funding: The APC was funded by the program of the Minister of Science and Higher Education under the name: Regional Initiative of Excellence in 2019–2022 project number 025/RID/2018/19 financing amount PLN 12.000.000.

Institutional Review Board Statement: Not applicable.

Informed Consent Statement: Not applicable.

Data Availability Statement: No new data were created or analyzed in this study. Data sharing is not applicable to this article.

Conflicts of Interest: The authors declare no conflict of interest.

References

1. Siemienuk, J.; Szatyłowicz, E. Zmniejszenie emisiji CO_2 w procesie produkcji cementu. *Civ. Environm. Eng.* **2018**, *9*, 81–87. (In Polish)
2. Gawlicki, M. Belite in cements with low emission of CO_2 during clinker formation. *Cement Wapno Beton* **2020**, *5*, 348–357. [CrossRef]
3. Baran, T. The use of waste and industrial by-products and possibilities of reducing CO_2 emission in the cement industry-industrial trials. *Cement Wapno Beton* **2021**, *3*, 169–184. [CrossRef]
4. Giergiczny, Z.; Garbacik, A. Properties of cements with calcareous fly ash addition. *Cement Wapno Beton* **2012**, *4*, 217–224.
5. Chatterjee, A.K. Alternative cements and concretes in development with recycled plant-emitted carbon dioxide. *Cement Wapno Beton* **2017**, *2*, 120–137.
6. Środa, B. Przemysł cementowy na drodze do Zielonego Ładu. *Bud. Technol. Archit.* **2020**, *7–9*, 68–74.
7. Gartner, E. Industrially interesting approaches to "low-CO_2" cements. *Cem. Concr. Res.* **2004**, *34*, 1489–1498. [CrossRef]
8. Dziuk, D.; Giergiczny, Z.; Garbacik, A. Calcareous fly ash as a main constituent of common cements. *Roads Bridges–Drogi i Mosty* **2013**, *12*, 57–69.
9. Król, A.; Giergiczny, Z.; Kuterasińska-Warwas, J. Properties of concrete made with low-emission cements CEM II/C-M and CEM VI. *Materials* **2020**, *13*, 2257. [CrossRef] [PubMed]
10. Chłądzyński, S.; Garbacik, A. *Multicomponents Cements in Construction Industry*; Association of Cement Producers: Cracow, Poland, 2008. (In Polish)
11. Anastasiou, E.K. Effect of high calcium fly ash, ladle furnace slag, and limestone filler on packing density, consistency and strength of cement pastes. *Materials* **2021**, *14*, 301. [CrossRef]
12. Czapik, P.; Czechowicz, M. Effects of natural zeolite particle size on the cement paste properties. *Struct. Environ.* **2017**, *9*, 180–190.
13. Czapik, P.; Cebulski, M. The properties of cement mortar with natural zeolite and silica fume additions. *Struct. Environ.* **2018**, *10*, 106–113. [CrossRef]
14. Chłądzyński, S. Influence of cement on properties of adhesive for tiles. *Res. ISCMIiB* **2008**, *1*, 141–152. (In Polish)
15. Chłądzyński, S.; Bąk, Ł. The role of cement in shaping the properties of dry mixtures of construction chemicals. *Izolacje* **2011**, *11–12*, 32–35. (In Polish)
16. Chłądzyński, S.; Walusiak, K. Influence of cement strength on the properties of cement-based adhesives for thermal insulation systems. *Izolacje* **2018**, *4*, 38–42. (In Polish)
17. Chłądzyński, S.; Wójcik, A.; Bąk, Ł. The chosen aspects of influence of cement on dry construction mixes properties. In Proceedings of the Conference "8th DNI BETONU", Wisła, Poland, 13–15 October 2014. (In Polish).
18. Kotwa, A.; Spychał, E. The influence of cellulose ethers on the chosen properties of cement mortar in the plastic state. *Struct. Environ.* **2016**, *3*, 153–160.
19. Spychał, E. The rheology of cement pastes with the addition of hydrated lime and cellulose ether in comparison with selected properties of plastering mortars. *Cement Wapno Beton* **2020**, *25*, 21–30. [CrossRef]
20. Spychał, E.; Czapik, P. The influence of HEMC on cement and cement-lime composites setting processes. *Materials* **2020**, *13*, 5814. [CrossRef]
21. Spychał, E.; Dachowski, R. The influence of hydrated lime and cellulose ether admixture on water retention, rheology and application properties of cement plastering mortars. *Materials* **2021**, *14*, 5487. [CrossRef] [PubMed]
22. Brumaud, C.; Bessaies-Bey, H.; Mohler, C.; Baumann, R.; Schmitz, M.; Radlre, M.; Roussel, N. Cellulose ether and water retention. *Cem. Concr. Res.* **2013**, *53*, 176–184. [CrossRef]
23. Bülichen, D.; Kainz, J.; Plank, J. Working mechanism of methyl hydroxyethyl cellulose (MHEC) as water retention agent. *Cem. Concr. Res.* **2012**, *42*, 953–959. [CrossRef]
24. Skripkiunas, G.; Karpova, E.; Bendoraitiene, J.; Barauskas, I. Rheological properties and flow behaviour of cement-based materials modified by carbon nanotubes and plasticizing admixtures. *Fluids* **2020**, *5*, 169. [CrossRef]
25. Žižlavský, T.; Bayer, P.; Vyšvařil, M. Bond properties of NHL-based mortars with viscosity-modifying water-retentive admixtures. *Minerals* **2021**, *11*, 685. [CrossRef]
26. Pichniarczyk, P.; Niziurska, M. Properties of ceramic tile adhesives modified by different viscosity hydroxypropyl methylcellulose. *Constr. Build. Mat.* **2015**, *77*, 227–232. [CrossRef]
27. Spychał, E. The influence of viscosity modifying admixture on the shrinkage of plastering mortars. *Struct. Environ.* **2017**, *9*, 248–255.
28. Gołaszewski, J.; Cygan, G. Wpływ domieszek zwiększających lepkość na skurcz wczesny zapraw. *Civ. Environm. Eng.* **2011**, *2*, 263–266.

29. Messan, A.; Ienny, P.; Nectoux, D. Free and restrained early-age shrinkage of mortar: Influence of glass fiber, cellulose ether and EVA (ethylene-vinyl acetate). *Cem. Concr. Compos.* **2011**, *33*, 402–410. [CrossRef]

30. Liu, C.; Gao, J.; Chen, X.; Zhao, Y. Effect of polysaccharides on setting and rheological behaviour of gypsum-based materials. *Const. Build. Mat.* **2021**, *267*, 1–10. [CrossRef]

31. Žižlavský, T.; Vyšvařil, M.; Rovnanikova, P. Rheological study on influence of hydroxypropyl derivatives of guar gum, cellulose, and chitosan on the properties of natural hydraulic lime pastes. *IOP Conf. Ser. Mater. Sci. Eng.* **2019**, *583*. [CrossRef]

32. Seshu, D.R.; Murthy, N.R.D. A study on fly ash cement mortar as brick masonry. *Cement Wapno Beton* **2020**, *25*, 45–51. [CrossRef]

33. Zhou, S.-C.; Lv, H.-L.; Li, N.; Zhang, J. Effects of chemical admixtures on the working and mechanical properties of Ordinary Dry-Mixed Mortar. *Adv. Mat. Sci. Eng.* **2019**, 1–10. [CrossRef]

34. *PN-EN 1015-3:2000+A2:2007 Test Methods of Mortars for Walls. Determining the Consistency of Fresh Mortar (Using a Flow Table)*; Polish Committee for Standardization: Warszawa, Poland, 2007. (In Polish)

35. *PN-B-04500:1985 Mortars–Testing Physical and Strength Characteristics*; Polish Committee for Standarization: Warszawa, Poland, 1985. (In Polish)

36. *PN-EN 1015-6:2000+A1:2007 Test Methods of Mortars for Walls. Determination of the Bulk Density of the Fresh Mortar*; Polish Committee for Standardization: Warszawa, Poland, 2007. (In Polish)

37. *PN-EN 1015-10:2001+A1:2007 Methods of Test for Mortar for Masonry–Part 10: Determination of Dry Bulk Density of HARDENED Mortar*; Polish Committee for Standarization: Warszawa, Poland, 2007. (In Polish)

38. Małolepszy, J.; Gawlicki, M.; Pichór, W.; Brylska, E.; Brylicki, W.; Łagosz, A.; Nocuń-Wczelik, W.; Petri, M.; Pytel, Z.; Roszczynialski, W.; et al. *Podstawy Technologii Materiałów Budowlanych i Metody Badań Pod Redakcją Jana Małolepszego*; Wydawnictwo AGH: Cracow, Poland, 2013. (In Polish)

39. *PN-EN 1015-11:2001+A1:2007 Methods of Test for Mortar for Masonry–Part 11: Determination of Flexural and Compressive Strength of Hardened Mortar*; Polish Committee for Standarization: Warszawa, Poland, 2007. (In Polish)

40. Gantner, E.; Chojczak, W. *Materiały Budowlane. Ćwiczenia Laboratoryjne. Spoiwa, Kruszywa, Zaprawy, Betony*; Oficyna Wydawnicza Politechniki Warszawskiej: Warsaw, Poland, 2013. (In Polish)

41. Patural, L.; Marchal, P.; Govin, A.; Grosseau, P.; Ruot, B.; Devès, O. Cellulose ethers influence on water retention and consistency in cement-based mortars. *Cem. Concr. Res.* **2011**, *41*, 46–55. [CrossRef]

42. *PN-EN 998-1:2012. Specification for Mortar for Masonry–Part 1: Rendering and Plastering Mortar*; Polish Committee for Standarization: Warszawa, Poland, 2012. (In Polish)

43. Zhang, M.-H.; Sisomphon, K.; Ng, T.S.; Sun, D.J. Effect of superplasticizers on workability retention and initial setting time of cement pastes. *Constr. Build. Mater.* **2010**, *24*, 1700–1707. [CrossRef]

44. Usherov-Marshak, A.V.V.; Ciak, M.J. Isothermal calorimetry in the standard ASTM C1679-08. *Cement Wapno Beton* **2010**, *15*, 108–110.

45. Nocuń-Wczelik, W.; Bobrowski, A.; Gawlicki, M.; Łagosz, A.; Łój, G. *Cement. Metody Badań. Wybrane Kierunki Stosowania*; Wydawnictwo AGH: Cracow, Poland, 2010. (In Polish)

Article

Forecasting Compressive Strength of RHA Based Concrete Using Multi-Expression Programming

Muhammad Nasir Amin [1,*], Kaffayatullah Khan [1], Muhammad Faisal Javed [2], Dina Yehia Zakaria Ewais [3], Muhammad Ghulam Qadir [4], Muhammad Iftikhar Faraz [5], Mir Waqas Alam [6], Anas Abdulalim Alabdullah [1] and Muhammad Imran [7]

1   Department of Civil and Environmental Engineering, College of Engineering, King Faisal University, P.O. Box 380, Al-Hofuf, Al-Ahsa 31982, Saudi Arabia; kkhan@kfu.edu.sa (K.K.); 218038024@student.kfu.edu.sa (A.A.A.)
2   Department of Civil Engineering, Abbottabad Campus, COMSATS University Islamabad, Abbottabad 22060, Pakistan; arbabfaisal@cuiatd.edu.pk
3   Structural Engineering, Faculty of Engineering and Technology, Future University in Egypt, New Cairo 11835, Egypt; dina.yehya@fue.edu.eg
4   Department of Environmental Sciences, Abbottabad Campus, COMSATS University Islamabad, Abbottabad 22060, Pakistan; hashir785@gmail.com
5   Department of Mechanical Engineering, College of Engineering, King Faisal University, P.O. Box 380, Al-Hofuf, Al-Ahsa 31982, Saudi Arabia; mfaraz@kfu.edu.sa
6   Department of Physics, College of Science, King Faisal University, P.O. Box 380, Al-Hofuf, Al-Ahsa 31982, Saudi Arabia; wmir@kfu.edu.sa
7   School of Civil and Environmental Engineering (SCEE), National University of Sciences & Technology (NUST), Islamabad 44000, Pakistan; imran.nice@nust.edu.pk
*   Correspondence: mgadir@kfu.edu.sa; Tel.: +966-13-589-5431; Fax: +966-13-581-7068

Citation: Amin, M.N.; Khan, K.; Javed, M.F.; Ewais, D.Y.Z.; Qadir, M.G.; Faraz, M.I.; Alam, M.W.; Alabdullah, A.A.; Imran, M. Forecasting Compressive Strength of RHA Based Concrete Using Multi-Expression Programming. *Materials* **2022**, *15*, 3808. https://doi.org/10.3390/ma15113808

Academic Editor: Jorge Otero

Received: 23 April 2022
Accepted: 24 May 2022
Published: 26 May 2022

Publisher's Note: MDPI stays neutral with regard to jurisdictional claims in published maps and institutional affiliations.

Copyright: © 2022 by the authors. Licensee MDPI, Basel, Switzerland. This article is an open access article distributed under the terms and conditions of the Creative Commons Attribution (CC BY) license (https://creativecommons.org/licenses/by/4.0/).

Abstract: Rice husk ash (RHA) is a significant pollutant produced by agricultural sectors that cause a malignant outcome to the environment. To encourage the re-use of RHA, this work used multi expression programming (MEP) to construct an empirical model for forecasting the compressive nature of concrete made with RHA (CRHA) as a cement substitute. Thus, the compressive strength of CRHA was developed comprising of 192 findings from the broad and trustworthy database obtained from literature review. The most significant characteristics, namely the specimen's age, the percentage of RHA, the amount of cement, superplasticizer, aggregates, and the amount of water, were used as input for the modeling of CRHA. External validation, sensitivity analysis, statistical checks, and Shapley Additive Explanations (SHAP) analysis were used to evaluate the models' performance. It was discovered that the most significant factors impacting the compressive strength of CRHA are the age of the concrete sample (AS), the amount of cement (C) and the amount of aggregate (A). The findings of this study have the potential to increase the re-use of RHA in the production of green concrete, hence promoting environmental protection and financial gain.

Keywords: rice husk ash; machine learning; waste material; external validation; compressive strength

1. Introduction

Different researchers have suggested different methods to lessen the malignant impacts of the construction industry on the atmosphere. Some researchers suggested replacing the natural coarse aggregate in concrete with recycled concrete aggregate, oil palm shell aggregate, lightweight aggregate, rubber, and so on, while others suggested replacing natural sand with sugarcane bagasse ash, rice husk ash (RHA), eggshell ash, and other different types of industrial and agricultural wastes [1–4]. However, it is observed to be more beneficial if cement is replaced with concrete, as cement is the main culprit in concrete which affects the environment. The partial replacement of cement with natural pozzolanic materials, industrial wastes, and agricultural wastes has been a point of interest for different researchers for the last couple of decades [5,6]. One of the common agricultural wastes is RHA, which is highly pozzolanic and contains a high amount of silica content. RHA is a byproduct of the cultivation of rice. RHA is formed as a result of heating husks in processing industries in order to process rice paddy. Rice is one of the world's most important food crops and is consumed in vast amounts by the global population. As of 2020/2021, it is estimated that 497.7 million tons of rice are produced globally. Therefore, RHA is prevalent in agricultural nations that produce millions of metric tons of rice annually. As it includes roughly 85–90% amorphous silica, RHA may be effectively recycled as a pozzolanic material as opposed to being discarded publicly. The use of RHA in concrete has been researched by different scientists [7–9]. The research on RHA is mostly conducted in Agricultural countries as shown in Figure 1. The gathered data is up to April 2022 as illustrated in Figure 1. The number of publications from India is more than twice that of any other country on RHA. Most of the research performed on RHA is published in high-impact Journals as shown in Figure 2. RHA is mainly utilized as a partial replacement of cement (as Supplementary Cementitious Material) and provides better properties than normal concrete (concrete without RHA). RHA can be used for many other purposes as shown in Figure 3, but they are out of the scope of this study. Concrete made with RHA (CRHA) is reported to be more durable and posseses higher mechanical properties when compared with normal concrete [9–11]. In addition, the use of RHA in concrete provides sustainability to the construction industry in two ways. First, it reduces the amount of cement (C) used, and second, it helps in the disposal of waste RHA. Furthermore, concrete made with RHA is more economical as some percentage of cement (the most expensive material in concrete) is being replaced with waste material. The behavior of RHA concrete is anomalous due to numerous factors, i.e., concrete mix design, amount of RHA used, and physical properties of concrete ingredients [12–15]. Therefore, the use of RHA requires prior experimental testing to be used in mega projects. However, the presence of reliable, trustworthy models and formulas to relate the compressive strength of RHA concrete with its ingredients may provide ease to construction engineers to use RHA concrete in their projects. The wide use of RHA concrete may help in reducing the carbon footprint of the construction industry. The use of modern computing techniques like artificial intelligence algorithms (AIA) can be used to achieve this objective.

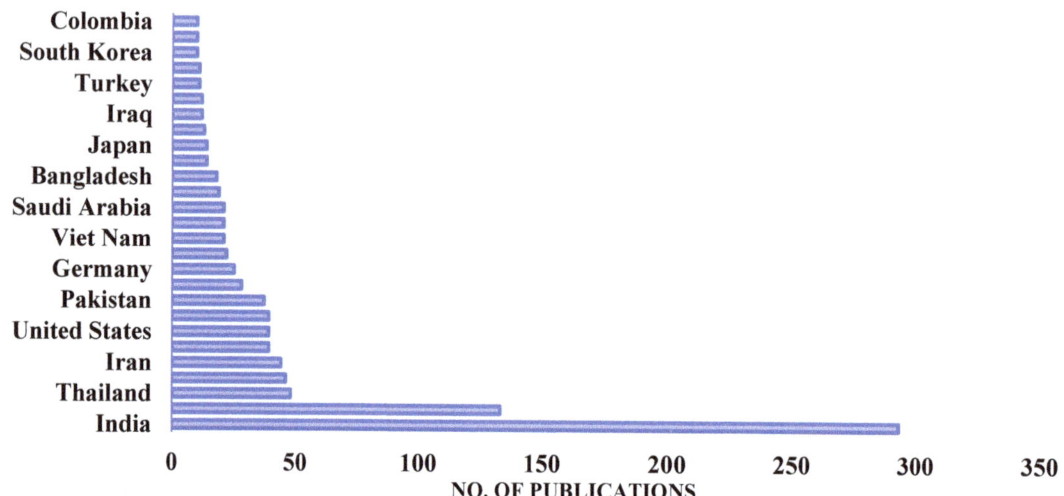

Figure 1. Number of publications on RHA.

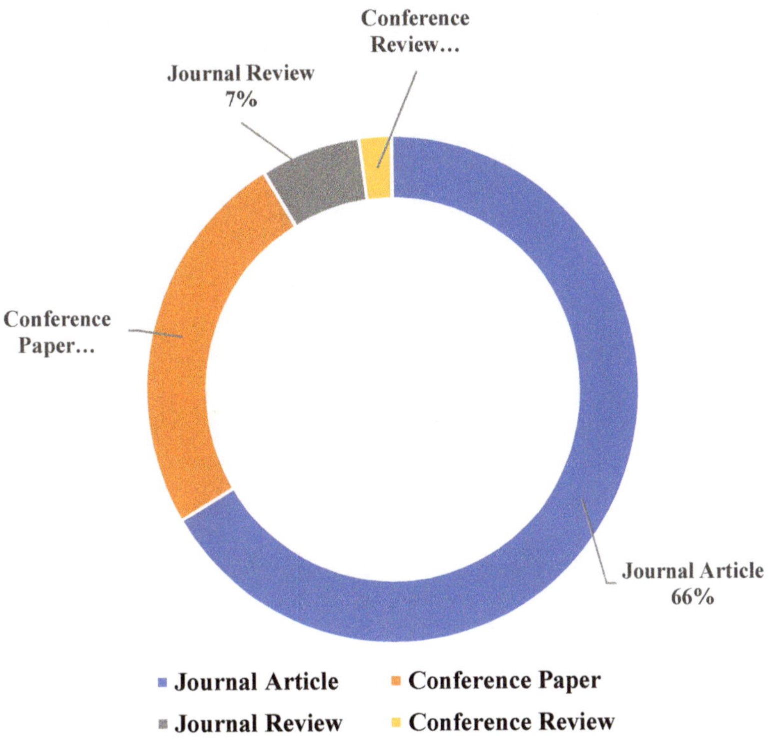

Figure 2. Publications of RHA in high-impact factor journals.

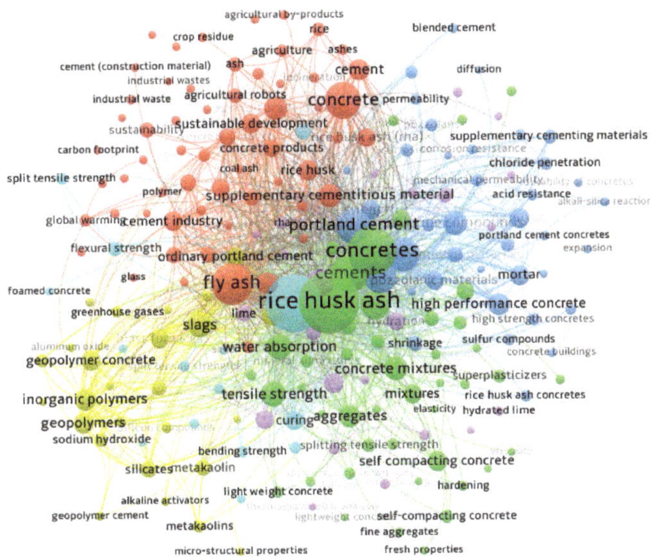

Figure 3. Importance of RHA in the construction industry.

The use of AIA is rising in every field [16–23]. AIA has distinctive features like pattern recognition and object recognition, which can be used to solve various engineering problems [24–30]. However, AIA is generally termed as black-box algorithms (BBA), because it does not give an insight into the problem being solved [31,32]. AIA ignores any knowledge or physical occurrences related to the subject at hand. The majority of ANN approaches lag in the development of an advanced mathematical formulation for estimating output based on input factors [33–35]. A correlation between input and output is referred to as an ANN-based model, and the relationship seems to be either nonlinear or based on a pre-defined structure [36–38]. To address these challenges, numerous evolutionary algorithms (EA) are being used to simulate concrete features, including genetic programming, convolutional neural networks (NN), and the model tree algorithm [19,39–41]. The advantage of EA is that they enable the production of realistic algebraic expressions, as well as a high degree of generality and prediction capabilities [1,3–6]. A few recent research have attempted to simulate the characteristics of waste foundry sand concrete using AIA. Different EA was used to create decision tree structures for the purpose of estimating the mechanical characteristics of waste foundry sand concrete [42]. Numerous influencing factors, a robust correlation coefficient, and minor arithmetical errors were obtained for the constructed models. Nevertheless, parametric research was not possible due to the linear character of decision trees, which reduces their effectiveness when applied to unknown data. Similarly, in a recent study, a genetic programming approach was used to estimate the compressive strength of waste foundry sand concrete [43]. To assess the suggested models' dependability, parametric, and error, sensitivity analyses were conducted. However, the gene expression programming (GEP) approach has drawbacks in that it was powerless to contain a few differing datasets into the model construction process, hence limiting its application range [44]. To improve the performance of the models, the differing datapoints required to be eliminated from the set processes. Additionally, genetic algorithms (GA) program uses a solitary chromosome, and are useful when the relationship between the targeted and predicted is reasonably basic.

To overcome the drawbacks of AIA, an enhanced modeling approach known as multi-expression programming (MEP) was utilized to predict the mechanical characteristics, i.e., compressive strength of CRHA based on the most influential factors. MEP is unique in that it can encode many expressions in one computer program [45,46]. To guarantee that the models are effectively trained, a big database was compiled from the literature and subdivided into three sets: training, validation, and testing. The effectiveness of the models is assessed by using statistical error analysis, external validation, and various statistical analyses to ensure that the models are generalizable and reliable. The article is arranged as follows: a description of the MEP algorithm, a database of experimental findings, a modeling approach, results and discussion, external validation, sensitivity analysis, and lastly, a brief discussion of the conclusion and significant discoveries.

2. Multi-Expression Programming (MEP)

The goal of a machine learning model is to produce a mathematical expression for output prediction that is accurate and practicable based on a collection of independent parameters. Koza (1992) suggested the GEP as an evolution of the GA based on Darwin's selection concept [47]. It is important to note that the main difference between the two techniques is that in GEP, fixed-length binary strings are replaced with non-linear parse trees. Several other types of EAs have been proposed in recent years, with linearity being a key one. Individuals (chromosomes) can be modeled as variable-length entities in the case of MEP [48]. MEP simulation output may be characterized as a linear string of instructions consisting of variables or mathematical operations (functions). Figure 4 illustrates the procedures involved in implementing MEP [48]. The process of MEP starts with the initialization of functions and expressions. After that, the chromosomes population is increased randomly based on the binary selection of the connection functions as shown in Figure 4b. When the chromosomes population reach a certain point, off-springs are produced and evaluated with the help of the evaluation function. The process is terminated when the required fitness value is achieved. The MEP method evolves by creating a random chromosomal population, selecting two parents via a binary tournament, and recombination with a set cross-over frequency, the generation of two offspring through recombination of the selected parents, mutation of the offspring, and replacement of the population's worst individuals with the best are some of the steps followed in MEP. The process is cyclical and repeats itself until convergence is attained.

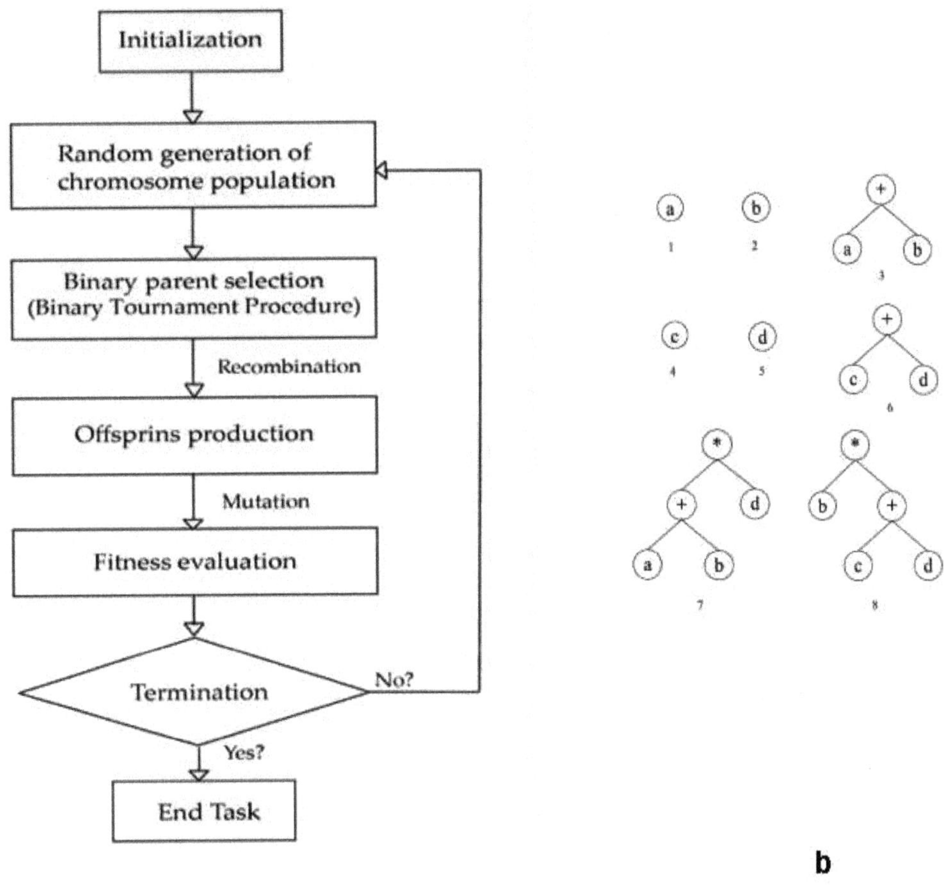

a

b

Figure 4. (**a**) Procedures involved in implementing MEP, (**b**) Flowchart for expressions encoded by an MEP chromosome.

Most of the research over the last decade has been on the application of artificial neural network (ANN) and GEP approaches to model the characteristics of green concrete. However, MEP has several benefits over comparable algorithms. Typically, a large database is used to represent concrete characteristics. In GEP, just a cross-over genetic operator is used, resulting in the generation of a large population of parse trees, increasing simulation time and requiring a considerable amount of memory [47,49,50]. Additionally, because GEP's non-linear structure functions like gene expression patterns, the algorithm has a hard time proposing a simple mathematical representation for the required attribute. The integration of linear variants enables MEP to discriminate between an individual's genotype and phenotype. Moreover, up to a certain point, the amount of genes on chromosomes improves the likelihood of GEP success. The model's usefulness in the construction industry is limited by overfitting, which manifests itself in the predicted strength qualities in the construction industry. In fact, MEP is particularly useful when the objective expression is uncertain, as in material engineering problems where a small change in a concrete mix parameter might have a huge impact on the strength [48]. Due to the linearity of chromosomes and the encoding of numerous solutions in one chromosome in MEP, the software may search for a larger space for the output prediction. Due to the evident benefits of MEP over other EAs, accurate models in the field of civil engineering may be developed. It has been used in several research to forecast different soil properties using physical properties of soil as input parameter [34], to predict the elasticity of concrete by using mix design ratios, and to create predicting modeling for concrete columns confined with thermoplastic fiber reinforced polymer [51]. The present work used the MEP approach to develop models to predict the parameters of CRHA. Further validation of the model is made by applying various statistical checks to the model. The availability of trustworthy models will encourage the use of CRHA in the building sector since it circumvents the time-consuming testing process necessary for such an unconventional construction material. This would help to waste reduction, sustainable building, and natural resource conservation. Additionally, the provided modeling technique will pave the way for correctly modeling comparable complicated engineering processes.

3. Data Collection

To build a computational equation that properly predicted the compressive strength of CRHA, a database of 192 data points from the published research was employed (Table S1) [52–58]. The CRHA is composed of the same components: OPC, RHA, aggregates (A), water (W), and superplasticizer (SP). All mixtures obtained from the literature utilized the same type of cement with identical age of concrete (AS). The correlation matrix for the inputs and compressive strength (CS) of CRHA is shown in Table 1.

Table 1. Coefficient of correlation (R) for explanatory variables.

	AS * (Day)	C * (kg/m^3)	RHA * (kg/m^3)	W * (kg/m^3)	SP * (kg/m^3)	A * (kg/m^3)	CS (MPa)
AS (day)	1.00						
C30 (kg/m^3)	−0.11	1.00					
RHA (kg/m^3)	−0.03	−0.22	1.00				
W (kg/m^3)	0.01	0.08	0.14	1.00			
SP (kg/m^3)	0.00	0.25	−0.02	0.27	1.00		
A (kg/m^3)	−0.06	−0.24	−0.14	−0.55	−0.21	1.00	
CS (MPa)	0.49	0.37	−0.02	−0.24	0.30	0.15	1.00

* AS = age of concrete sample, C30 = cement with 30% replacement, W = water, SP = superplasticizer, A = aggregate.

The compressive strength of cubic specimens was converted to the compressive strength of cylinders using a conversion ratio of 0.8 [59]. The purpose of this research was to determine the compressive strength of various CRHA mixtures using MEP. As input parameters, variables such as the amount of cement (C), the amount of water (W), amount of RHA, age of concrete (AS), amount of aggregate (A), and dosage of SP were collected from the literature. Figure 5 depicts histograms for all variables utilized in this investigation. Additionally, Table 2 has a statistical description of the gathered data. The mean and median for all AS values were obtained to be 34.57 and 28, respectively. While the value of skewness is positive for all the variables except for water and fine aggregate.

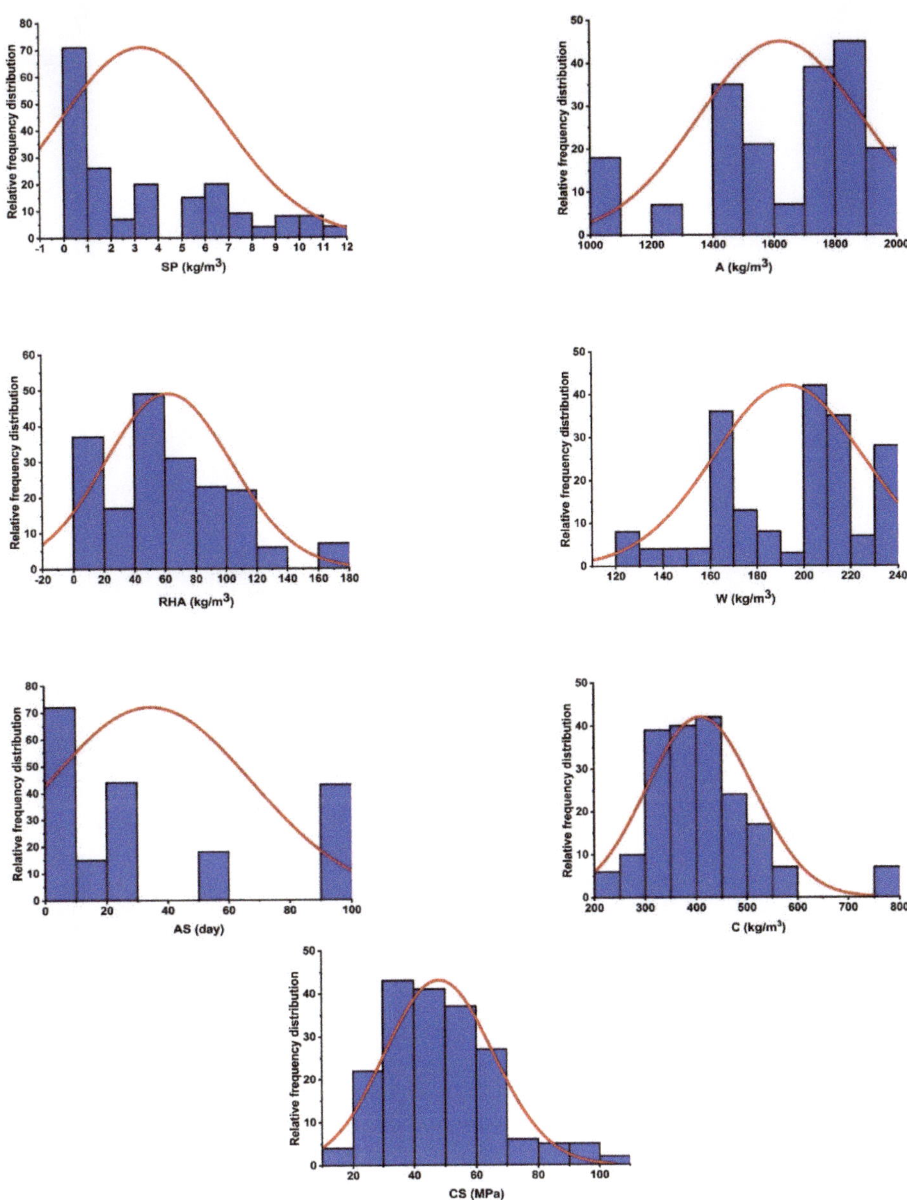

Figure 5. Histogram of variables used in making model.

Table 2. Statistical description of variables.

Description of Variables	AS (Day)	C (kg/m^3)	RHA (kg/m^3)	W (kg/m^3)	SP (kg/m^3)	A (kg/m^3)	CS (MPa)
Mean	34.57	409.02	62.33	193.54	3.34	1621.51	48.14
Median	28.00	400.00	57.00	203.00	1.85	1725.00	45.95
Mode	28.00	400.00	0.00	203.00	0.00	1725.00	47.00
Standard Deviation	33.52	105.47	41.55	31.93	3.52	267.77	17.54
Sample Variance	1123.61	11,124.88	1726.77	1019.71	12.37	71,702.44	307.70
Skewness	0.75	1.55	0.44	−0.42	0.69	−0.74	0.83
Range	89.00	534.00	171.00	118.00	11.25	930.00	88.10
Minimum	1.00	249.00	0.00	120.00	0.00	1040.00	16.00
Maximum	90.00	783.00	171.00	238.00	11.25	1970.00	104.10
Sum	6638.00	78,531.00	11,967.10	37,158.91	640.35	311,330.00	9243.10
Count	192.00	192.00	192.00	192.00	192.00	192.00	192.00

4. Model Development

One of the objectives of this study is to develop a new formulation for the compressive strength of CRHA using the MEP model. The essential parameters recommended in the literature were used as input variables. Therefore, formulation of the compressive strength (CS) of CRHA was assumed using Equation (1) as follows:

$$CS = f(AS, OPC, A, SP, W, RHA) \tag{1}$$

In order to develop a strong and generic model, a large number of MEP fitting parameters must be defined before modeling begins. The relevant variables are chosen in accordance with prior suggestions and a trial-and-error method. The number of programs that will develop in a population is determined by the size of the population. It would be more complex and precise to run a model with a huge population size, and it may take a long time for the model to converge. The method was begun by assuming a total of ten subpopulations. Table 3 summarizes the parameters used in the study. All these values are calculated after running several trials on different combinations as shown in Table 4. It should be noted that several parameters (like code length, connecting functions) can further increase the accuracy of the developed model, but they increase the computation time as well as the complexity of the model. Hence, they were kept at an optimum level. For simplicity in the final formulations, the function set includes the fundamental mathematical operations of multiplication, square root, natural log, subtraction, division, and addition. The number of generations indicates the amount of accuracy that the algorithm should reach before being terminated. Similarly, the rate of mutation and cross-over indicates the likelihood that the progeny will experience similar genetic processes. The incidence of cross-over varies between 50% and 95%. Numerous combinations of these parameters were tested on the data, and the optimal combination was chosen as shown in Table 4. The final parameters selected are shown in Table 3. One of the challenges with AI-based modeling is data overfitting. A model works admirably on the original data, but drastically degrades on the unseen data. To circumvent this issue, it has been proposed to test the trained model on an unknown or testing dataset. As a result, the whole database was randomly partitioned into training, validation, and testing sets. While modeling, the training and validation data were processed. The validated model is next evaluated on a third dataset, i.e., one that was not utilized to construct the model. It was assured that the distribution was uniform across all datasets. The resulting models outperformed the baseline models on all three datasets. MPX v1.0, a commercially available computer tool, was used to implement the MEP algorithm [44–46].

Table 3. MEP parameter used in making a model.

Parameters	MEP
Num of subpopulation	20
Subpopulation size	1000
Code length	50
Crossover probability	0.9
Crossover type	Uniform
Mutation probability	0.001
Tournament size	2
Operators	0.5
Variables	0.5
Number of generations	1000
Function set	$+, -, \times, /$
Terminal set	Problem input
Replication number	10
Error measure	Mean squared error
Problem type	Regression
Simplified	Yes
Random seed	0
Number of runs	10
Number of threads	1

Table 4. MEP optimal combination.

Trial No.	No. of Subpopulation	Subpopulation Size	Code Length	No. of Generation	Functions Used	R^2	RMSE	MAE	RRSE	Time (Min)
MP1	10	200	20	200	$+, -, \times, /$	0.9275	71.1	48.03	0.2693	0–2
MP2	20		20		$+, -, \times, /$	0.9448	62.17	41.82	0.2355	
MP3	50		25		$+, -, \times, /$	0.9454	61.94	45.67	0.2346	
MP4	70		25		$+, -, \times, /$	0.9233	74.09	47.03	0.2806	
MP5	100		35		$+, -, \times, /$	0.9221	74.33	46.89	0.2815	
MP6	20	400	35		$+, -, \times, /$	0.9156	88.17	60.35	0.334	
MP7		600	35		$+, -, \times, /$	0.9496	59.68	41.9	0.226	
MP10			40	400	$+, -, \times, /$	0.9614	53.41	38.12	0.2023	15
MP11			40	600	$+, -, \times, /$	0.9376	66.01	42.78	0.25	25
MP12		1000	50		$+, -, \times, /$	0.9298	70.13	43.56	0.2656	
MP13			50	1000	$+, -, \times, /$	0.9362	66.97	45.06	0.2536	45

4.1. Shapley Additive Explanations (SHAP)

Even though numerous ML research on concrete structures have attained great accuracy in their predictions, the applicability of the ML models receives little consideration. Numerous research assesses the feature relevance for tree-based models single decision process, heuristic techniques, or model-agnostic methods [47,48]. However, these approaches are frequently impractical and skewed for EML models, particularly those with a significant bias. In this study, SHAP is utilized to demonstrate the interpretation of every input parameter. SHAP is expressed as the mean marginal contribution to a feature value over all conceivable coalitions, in accordance with the collaborative game theory. In particular, the SHAP value of a data is the mean prediction rate of samples having the characteristic minus the mean prediction value of samples lacking the feature. To enhance the interpretability of a machine learning (ML) model, its output is stated as the linear sum of its input data multiplied by their respective SHAP values.

To check the performance criteria, Root mean square error (RMSE), coefficient of correlation (R), mean absolute error (MAE), coefficient of regression (R^2), relative root mean square error (RRMSE), relative squared error (RE), and performance index ρ (Equations (2)–(8), respectively) have been used in this study.

$$RMSE = \sqrt{\frac{\sum_{i=1}^{n}(x_i - y_i)^2}{n}} \tag{2}$$

$$R = \frac{\sum_{i=1}^{n}(x_i - \overline{x}_i)(y_i - \overline{y}_i)}{\sqrt{\sum_{i=1}^{n}(x_i - \overline{x}_i)^2 \sum_{i=1}^{n}(y_i - \overline{y}_i)^2}} \tag{3}$$

$$\text{MAE} = \frac{\sum_{i=1}^{n}|x_i - y_i|}{n} \tag{4}$$

$$R^2 = 1 - \frac{\sum_{i=1}^{n}(x_j - y_j)^2}{\sum_{i=1}^{n}(x_j - \overline{y})} \tag{5}$$

$$\text{RRMSE} = \frac{1}{|\overline{e}|}\sqrt{\frac{\sum_{i=1}^{n}(x_i - y_i)^2}{n}} \tag{6}$$

$$\text{RE} = \frac{\sum_{i=1}^{n}(x_i - y_i)^2}{\sum_{i=1}^{n}(\overline{x} - x_i)^2} \tag{7}$$

$$\rho = \frac{\text{RRMSE}}{1 + \text{R}} \tag{8}$$

$$\text{OBJ} = \left(\frac{n_L - n_T}{n}\right)\rho_L + 2\left(\frac{n_T}{n}\right)\rho_T \tag{9}$$

where, x_i and y_i are the ith experimental and predicted output values, respectively; and denote the experimental and expected output values, respectively; and n denotes the complete number of observations. Lower values of RMSE, MAE, and higher values of R, and R^2, as well as the pre-selected significance value, i.e., alpha (usually 0.05) from F and *t*-tests, indicate that the predictive model performs well and has a better accuracy. Additionally, it implies that the experimental and anticipated values are highly connected. Additionally, it is worth noting that a R value larger than 0.8, an R^2 value nearer to 1, an RMSE value nearer to or equal to zero, and ρ value (0 to infinity) approaching zero all contribute to improved model calibration. Unlike the RMSE, MAE is a positive evolution metric when the original data is relatively smooth [60]. On the other hand, the normalized mean square error (NSE) runs between 0 and 1.0 (1 inclusive), with 1 regarded as the best number. Additionally, a significant issue linked with AI systems is overfitting, which occurs because of extensive training and results in higher mistakes in the testing set. As demonstrated in Equation (9), the objective function (OBF) is assessed and decreased prior to selecting the best predictive mode [61]. The OBF is used to evaluate the trained model's performance by including changes in the error function (RRMSE) and correlation coefficient (R). A low OBF value aids in overcoming the issue of overfitting.

4.2. Cross-Validation Using 10 K-Fold Method

Generally, cross-validation procedure is applied using 10 k-fold to decrease the random sampling-related distortion of training and residual set of inputs. According to the findings of Kohavi, the ten-fold validation test yields a dependable variance and the ideal computing time (Kohavi, 1995). This study employed a stratified 10 k-fold cross-validation method to evaluate the performance of a model that categorizes a given number of data samples into 10 subgroups. In each of 10 rounds of model development and validation, a separate data subset is used for testing while the remaining data subsets are used to train the model. As seen in Figure 6, the test subset is used to validate model precision. The algorithm's precision is then reported as the average precision gained by the 10 models during ten rounds of validation.

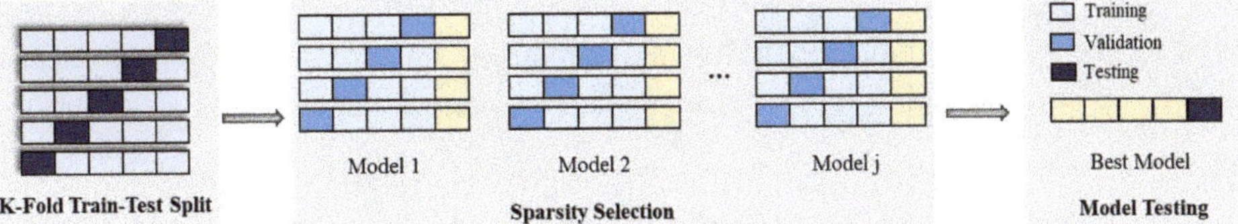

Figure 6. K-fold cross-validation algorithm [61].

5. Results and Discussion

5.1. MEP Analysis of CRHA

Appendix A contains the optimized MEP code for compressive strength prediction of CRHA utilizing specified input variables. The compressive strength of CRHA for the training dataset is displayed in Figure 7 along with the slope. The optimal location of the regression line is 45°, with a slope equal to 1, but it must be closer to 1 for good association. As shown in Figure 7, the proposed model accurately predicts the compressive strength of CRHA (R for the entire dataset is 0.97). Additionally, the RMSE, MAE, and the NSE for estimating the training dataset of compressive strength of CRHA are 3.98, 0.6, and 0.77, respectively. The near proximity of the points to the ideal fit and the inclusion of most points within the acceptable confidence interval demonstrates the suggested MEP model's validity. As previously stated, R values greater than 0.8 [45] and NSE values near unity indicate that the suggested models for the compaction parameters function effectively. Figure 8 shows the compressive strength of CRHA for validation and testing set. For simplicity, both sets are combined in the Figure 8.

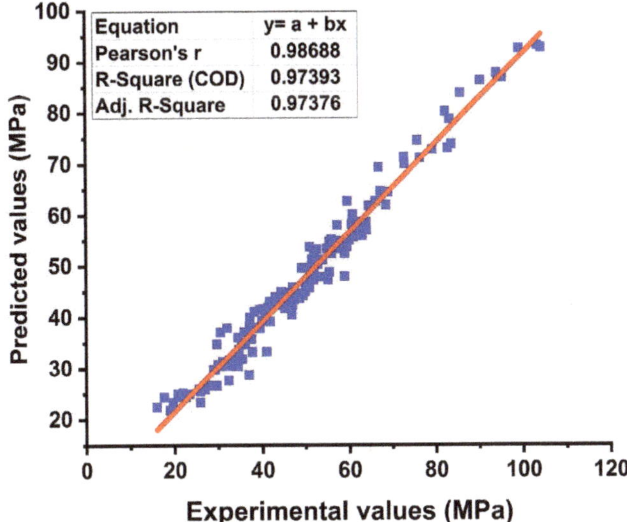

Figure 7. Regression analysis of training set of MEP.

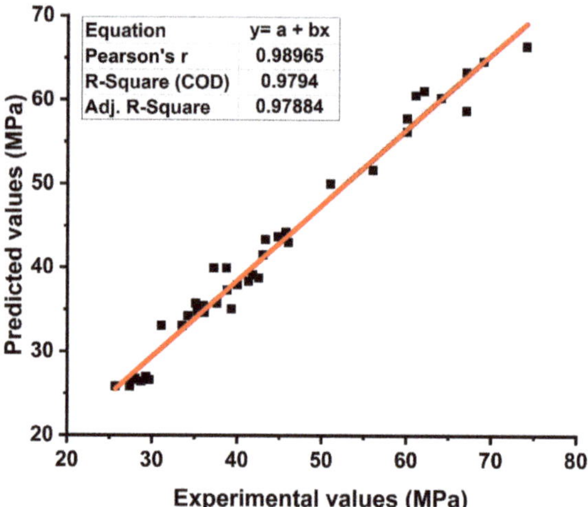

Figure 8. Regression analysis of testing set of MEP.

The created MEP model's adaptability was further measured by calculating the error distribution between the experimental and predicted values in both datasets (training and validation sets). The error pattern for the training and validation sets is depicted in Figures 9 and 10 for both sets. The deeper red color indicates the greater error levels. The model's error value is small, indicating that it successfully simulates the compressive strength of CRHA. The whole database is displayed with the absolute error in each data point to see the model's maximum error percentage, as shown in Figure 10. As can be observed, the model and predicted outputs are quite near, with an average error of 2 MPa and a peak error of less than 6 MPa for the compressive strength of CRHA. Additionally, the frequency of occurrence of maximal error is rather low. It has been discovered that around 80% of CRHA results estimated compressive strengths have an inaccuracy of less than 4 MPa.

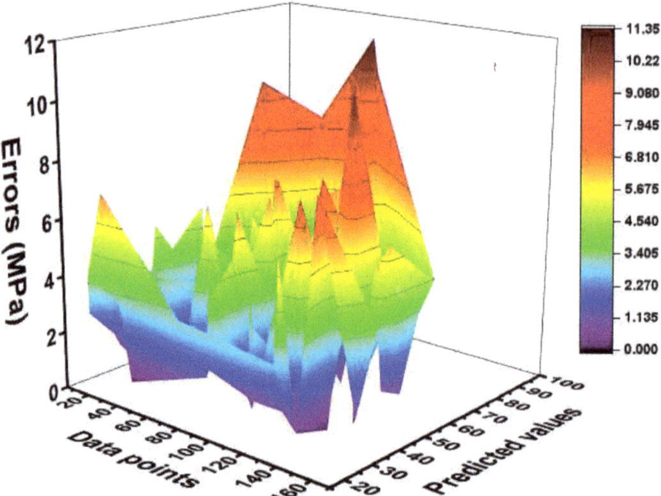

Figure 9. Error graphs of training set of MEP model.

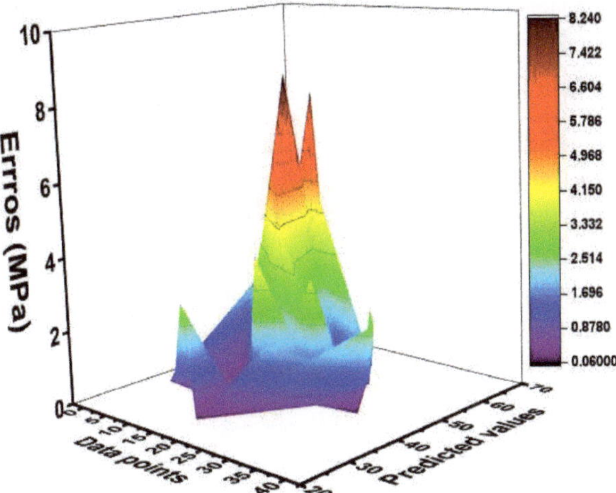

Figure 10. Error graphs of validation set of MEP model.

5.2. Performance Evaluation of MEP Model

According to Iqbal et al. [43], the database-to-input ratio should be at least three for good models and preferably greater than five for perfect models. The ratio is substantially greater in this research, at 32. Table 5 exhibit the statistical parameters for the validation and training sets for the MEP model. These results demonstrate that the models have been trained efficiently and that there is a strong correlation between expected and experimental output with low error levels. The MAE, RMSE, and RE values for the training set of the MEP model are 3.067, 3.843, and 0.047, respectively, while the values for the validation phase are 2.317, 3.406, and 0.048. The statistical measurements are nearly the same for the validation and training sets, demonstrating a greater capability for generalization and the ability to predict trustworthy outcomes for previously unknown data. As seen in Table 5, the ρ of the MEP projected model approaches zero (zero for ideal model). The OBF values of 0.04 adequately solved the issue of data overfitting.

Table 5. Statistical indictors for training and validation set.

Indicators	Training	Validation
R^2	0.976419	0.971378
R	0.988139	0.985585
RMSE	3.843116	3.406354
MAE	3.067433	2.317413
RRMSE	0.079188	0.072075
RE	0.047253	0.048581
ρ	0.03983	0.0363
OBF	0.04	

5.3. External Validation

External validation of the MEP model was also examined, owing to its substantially improved efficiency, which is shown in Table 6. As per literature, at least one regression slope line (k or k′) going through the origin must approach one [62]. The performance indices must have values less than 0.1. For the situation of optimal moisture content, the requirement of additional external validation, namely, $R_m > 0.5$, is met [63–65]. Additionally, the squared correlation coefficient (R'^2_0) between the estimated and experimental datasets, as well as the correlation coefficient (R^2_0) between the experimental and estimated values, must approach one [66–68]. As seen in Table 6, the suggested MEP model meets nearly all the stated requirements, which is consistent with the findings of existing literature and recommendations [69–72].

Table 6. External validation of data.

S. No.	Equation	Condition	MP	Suggested by
1	$R = \dfrac{\sum_{i=1}^{n}(x_i - \bar{x}_i)(y_i - \bar{y}_i)}{\sqrt{\sum_{i=1}^{n}(x_i - \bar{x}_i)^2 \sum_{i=1}^{n}(y_i - \bar{y}_i)^2}}$	$R > 0.8$	0.98	[63–65]
2	$k = \dfrac{\sum_{i=1}^{n}(x_i \times y_i)}{x_i^2}$	$0.85 < k < 1.15$	0.975	[62]
3	$k' = \dfrac{\sum_{i=1}^{n}(x_i \times y_i)}{y_i^2}$	$0.85 < k' < 1.15$	0.976	
4	$R_m = R^2 \times \left(1 - \sqrt{\left\| R^2 - R_0^2 \right\|}\right)$ where $R_0^2 = 1 - \dfrac{\sum_{i=1}^{n}\left(y_i - x_i^o\right)^2}{\sum_{i=1}^{n}\left(y_i - \bar{y_i^o}\right)^2},\ x_i^o = k \times y_i$ $R'^2_o = 1 - \dfrac{\sum_{i=1}^{n}\left(x_i - y_i^o\right)^2}{\sum_{i=1}^{n}\left(x_i - \bar{x_i^o}\right)^2},\ y_i^o = k' \times x_i$	$R_m > 0.5$ $R_0^2 \cong 1$ $R'^2_o \cong 1$	0.856 0.989 1.000	[66–68] [69–72]

5.4. 10-K Fold Cross Validation

A desired level of accuracy is required for the validity of prediction models. The 10 K-fold cross-validation method is used to ensure the accuracy of the model by shuffling the available data. By using this technique, the bias associated with a random sampling of training data set is minimized. This technique divides the experimental data samples into ten equal subsets and utilizes the nine subsets for developing and shaping the strong learner. Meanwhile, the last subset is utilized to gauge the validity of the developed model. The validation process repeats for ten times, and at the end, the average accuracy is obtained from the ten times repetition. The generalization performance and the reliability of the model are well represented by 10 K-fold cross-validations [65]. The cross-validation tests for individual MEP model are represented in Figure 11. The results of 10 K-fold cross-validations are assessed by using the coefficient of determinant, R^2 (regression tool) along with MAE and RMSE (statistical error tools) as shown in Table 7. In Figure 11, fluctuation in the value R^2 is observed for the 10 K-fold validation of different ML techniques, but still, a high level of accuracy is maintained in each fold. The accuracy of the cross-validation was also assessed in terms of MAE and RMSE and is given in Figure 11, respectively. The average value of MAE for is 4.2 MPa, respectively, as shown in Figure 11.

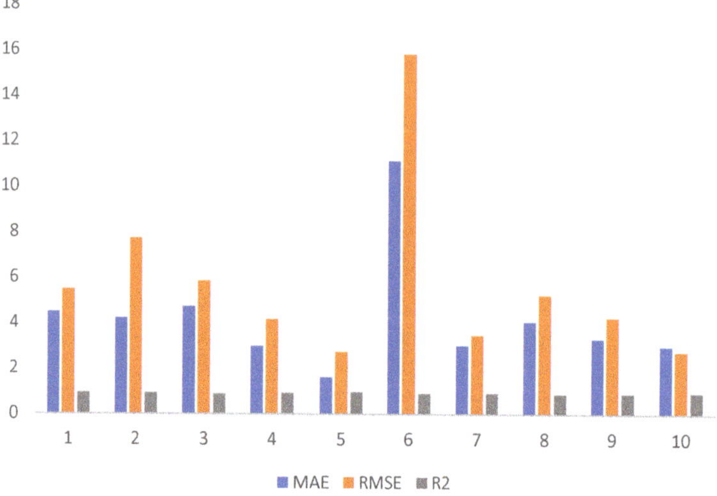

Figure 11. Results of K-fold validation.

Table 7. Statistics for K-fold Validation.

MAE	RMSE	R²
4.47	5.4	0.919
4.209	7.68	0.91
4.71	5.82	0.86
2.97	4.14	0.91
1.60	2.71	0.95
11.1	15.	0.89
2.99	3.45	0.90
4.04	5.21	0.87
3.30	4.22	0.89
2.97	2.73	0.93

Figure 11 shows the RMSE values of 10 K-fold validation and gives an average value of 5.7 MPa, respectively. The results of the 10 K-fold cross-validation method reflect the accuracy and reliability of the concerned developed models.

5.5. Explanation Using MEP Model

A detailed explanation of the machine learning model, as well as the feature correlations and interactions, is performed. To begin, better global depictions of feature impacts are created by aggregating local descriptions from the SHAP tree integrator over the whole dataset. Figure 12 illustrates a SHAP summary graphic in which each mark corresponds to a single data point in the dataset. The dots along the x-axis represent the effect of each feature values on the compressive strength of CRHA prediction. The marks are heaped together to demonstrate the density of several dots landing at the same x-axis point. According to Figure 12, the top three characteristics that have the most effect on compressive strength of CRHA prediction, in order of importance, are the age of concrete (AS), the amount of cement (C), and the amount of aggregate (A).

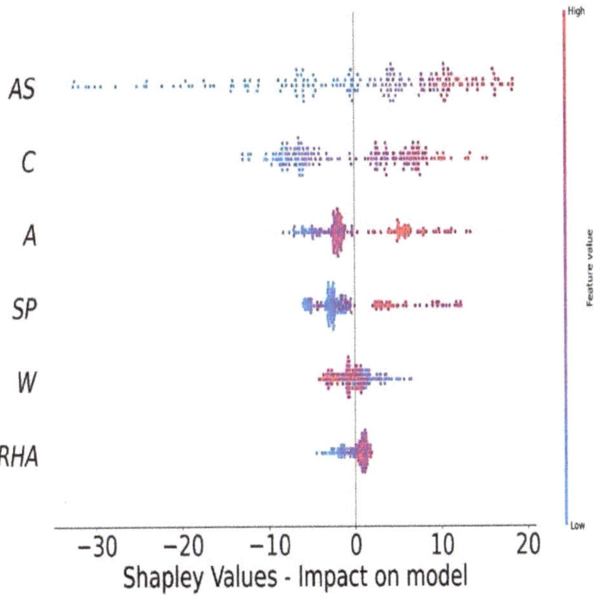

Figure 12. Shapley values of MEP model.

Figure 13 illustrates the feature reliance on the machine learning model in further depth by evaluating every single value in the dataset independently. On the *x*- and *y*-axes, the feature values and their related SHAP values are shown. The plots are additionally enhanced by feature interactions (shown by color bars) that indicate the combined influence of many features. One must keep in mind that SHAP values do not indicate causal linkages but rather characterize the model's behavior. A greater SHAP value implies that the model is attempting to forecast higher compressive strengths from the associated feature values. Similarly, a SHAP value less than zero indicates that the model is seeking to reduce the predicted compressive strength. These microscopic representations demonstrate interactions between various feature pairs impact the related SHAP values, which correlate to the comparable compressive strength values.

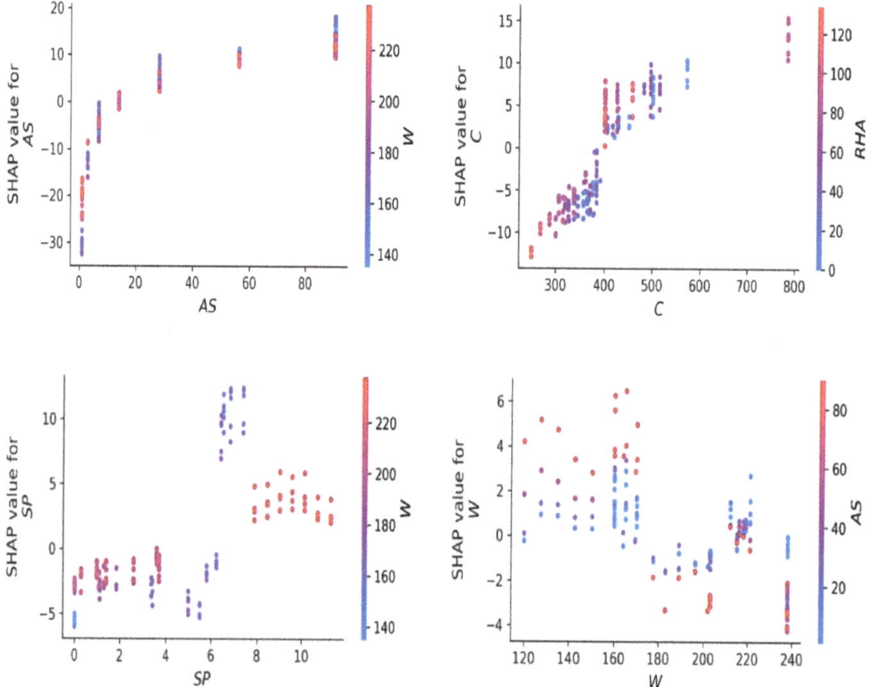

Figure 13. Feature reliance of the model.

Historically, AIA were mostly viewed as black boxes that served as a significant barrier between research and practice [73–75]. Because of AIA's lack of explainability and credibility, practitioners avoid it [75]. However, due to the improved predictability and explainability of the MEP model described in this study, it may be used by a broader range of experts to make some real-world judgments. This amount of data regarding the composition versus strength connection of concrete enhances one's comprehension of the concrete's nature and the optimization of the concrete mixture.

5.6. Sensitivity Analysis

Figure 14 demonstrates that each parameter is crucial for predicting the compressive strength of CRHA. According to sensitivity analysis, cement and age have a significant part in the total contribution to compressive strength, which is greater than fifty percent. Age of concrete (AS) provides around 29.47 percent, whereas cement quantity (C) contributes approximately 27.93 percent. The remaining four factors, namely RHA, water (W), SP, and aggregate (A), contribute about 8.26%, 12.85%, 13.49%, and 7.99%, respectively.

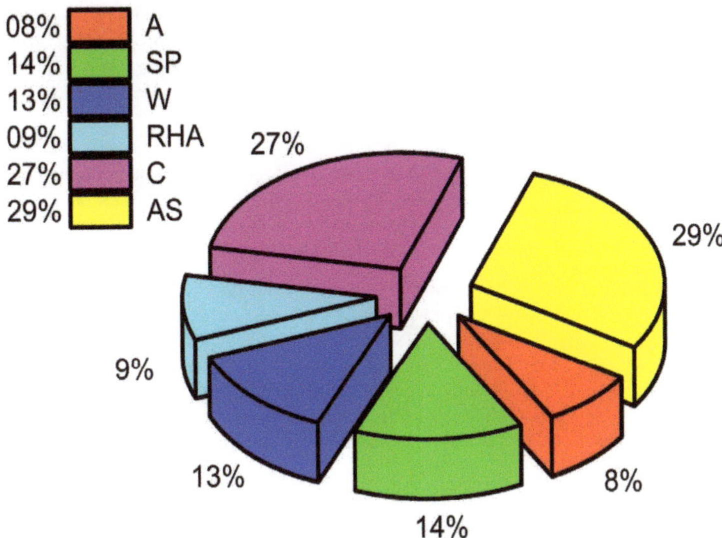

Figure 14. Sensitivity analysis of CRHA concrete.

6. Conclusions

Experts have been examining several AIA techniques for predicting the compressive strength of CRHA as feasible alternatives to the highly time-consuming and costly experimental compression testing. However, little effort has been made to improve the predictive powers and explainability of these commercial AIAs, which function as a significant barrier between research and practice, since practitioners avoid adopting AIA owing to their lack of understandability and reliability. To address this, an MEP model is employed to increase the predictability of the compressive strength of CRHA's. Advanced AIA principles such as model pipelining, model optimization, and feature selection via cross-validation are employed to help in the generation of more accurate models to forecast the compressive strength of CRHA. A comparison of the findings demonstrates that the created model generates the most precise prediction when compared to previously published models over the last two decades.

It is proved that the created MEP model generates verifying data (not available in the current literature) regarding the feature impacts, dependencies, and interactions with the compressive strength of CRHA. The core concept of this study was to explain a prediction model by calculating the contribution of each feature to the prediction of CRHA's compressive strength. In addition, the relationship between different variables affecting the strength of CRHA is calculated using SHAP analysis. It was discovered that the most significant factors impacting the compressive strength of CRHA are the age of concrete (AS), amount of cement (C), and the amount of aggregate (A). Furthermore, the dependency factors and relationship between different variables may help in future research to make a novel CRHA mix design as per the requirement of the site without compromising on cost, mechanical properties, available time, and availability of the mix ingredients.

Future Recommendation

The CRHA can effectively replace OPC concrete. Recommendation: comprehensive research of CRHA that includes more parameters. Including more input parameters and expanding the database can yield more trustworthy results for more generic expressions. These parameters should include resistance to acid attack and high temperature, sulphate and chloride resistance, and corrosion. For additional predictions, sophisticated techniques such as particle swarm programming and ensemble methods can be utilized.

ML approaches can be used with heuristic methods, such as the whale optimization, ant colony optimization, and PSO, for improved outcomes. These procedures may then be compared to those utilized in this investigation. In addition, MEP is an expanded and enhanced version of GEP. It is necessary to apply and analyze Honeybee algorithm to overcome the limits of ensemble algorithms.

Supplementary Materials: The following supporting information can be downloaded at: https://www.mdpi.com/article/10.3390/ma15113808/s1, Table S1: A database of 192 data points based on the literature review and the published data to build computational equation for predicting the compressive strength of CRHA [52–58].

Author Contributions: Conceptualization, M.N.A., K.K. and M.F.J.; Data curation, M.F.J. and D.Y.Z.E.; Formal analysis, M.F.J. and D.Y.Z.E.; Funding acquisition, M.N.A., K.K., M.G.Q., M.I.F. and M.I.; Investigation, M.N.A.; Methodology, K.K., M.F.J. and D.Y.Z.E.; Project administration, M.N.A.; Resources, M.N.A., K.K. and M.W.A.; Software, M.F.J. and D.Y.Z.E.; Supervision, M.N.A.; Validation, M.F.J., A.A.A. and M.W.A.; Visualization, M.G.Q., M.I., M.W.A., A.A.A. and M.I.F.; Writing—original draft, M.N.A., K.K. and M.F.J.; Writing—review & editing, M.N.A. All authors have read and agreed to the published version of the manuscript.

Funding: This work was supported by the Deanship of Scientific Research, Vice Presidency for Graduate Studies and Scientific Research, King Faisal University, Saudi Arabia [Project No. AN000163]. The APC was funded by the same "Project No. AN000163".

Institutional Review Board Statement: Not applicable.

Informed Consent Statement: Not applicable.

Data Availability Statement: The data used in this research has been properly cited and reported in the main text.

Acknowledgments: The authors acknowledge the support from the Deanship of Scientific Research, Vice Presidency for Graduate Studies and Scientific Research, King Faisal University, Saudi Arabia [Project No. AN000163]. The authors extend their appreciation for the financial support that has made this study possible.

Conflicts of Interest: The authors declare no conflict of interest.

Abbreviations

RHA	Rice husk ash
MEP	Multi-expression programming
CRHA	Concrete made with rice husk ash
SHAP	SHapley Additive exPlanations
OPC	Ordinary Portland cement
AIA	Artificial intelligence algorithms
BBA	Black-box algorithms
EA	Evolutionary algorithms
GA	Genetic algorithm
NN	Neural network
GEP	Gene expression programming
SP	Superplasticizer
C	Amount of cement
W	Amount of water
A	Amount of aggregate
AC	Age of concrete
CS	Compressive strength
RMSE	Root mean square error
R	Coefficient of correlation
MAE	Mean absolute error
R^2	Coefficient of regression
RRMSE	Relative root mean square error

RE   Relative squared error
ρ   Performance index
OBF   Objective function
NSE   Normalized mean square error

Appendix A

```
pg[0] = x[0];
pg[1] = sqrt(pg[0]);
pg[2] = x[1];
pg[3] = x[3];
pg[4] = exp(pg[1]);
pg[5] = pg[1] × pg[2];
pg[6] = pg[4] / pg[3];
pg[7] = pg[0] + pg[3];
pg[8] = pg[2] − pg[7];
pg[9] = x[4];
pg[10] = pg[8] × pg[9];
pg[11] = pg[10] + pg[5];
pg[12] = pg[1] × pg[0];
pg[13] = pg[8] + pg[11];
pg[14] = pow(pg[1], pg[9]);
pg[15] = pg[13] − pg[12];
pg[16] = sqrt(pg[15]);
pg[17] = x[2];
pg[18] = pg[6] + pg[8];
pg[19] = pg[9] × pg[17];
pg[20] = pg[12] / pg[18];
pg[21] = pg[19] + pg[18];
pg[22] = sqrt(pg[11]);
pg[23] = pg[5] / pg[21];
pg[24] = sqrt(pg[14]);
pg[25] = pg[24] × pg[4];
pg[26] = pg[6] − pg[20];
pg[27] = pg[22] / pg[25];
pg[28] = pg[26] / pg[23];
pg[29] = pg[16] − pg[27];
pg[30] = pg[29] + pg[28];
```

References

1. Imtiaz, L.; Rehman, S.K.U.; Ali Memon, S.; Khizar Khan, M.; Faisal Javed, M. A review of recent developments and advances in eco-friendly geopolymer concrete. *Appl. Sci.* **2020**, *10*, 7838. [CrossRef]
2. Yang, H.; Liu, L.; Yang, W.; Liu, H.; Ahmad, W.; Ahmad, A.; Aslam, F.; Joyklad, P. A comprehensive overview of geopolymer composites: A bibliometric analysis and literature review. *Case Stud. Constr. Mater.* **2022**, *16*, e00830. [CrossRef]
3. Akbar, A.; Farooq, F.; Shafique, M.; Aslam, F.; Alyousef, R.; Alabduljabbar, H. Sugarcane bagasse ash-based engineered geopolymer mortar incorporating propylene fibers. *J. Build. Eng.* **2021**, *33*, 101492. [CrossRef]
4. Alyousef, R.; Ahmad, W.; Ahmad, A.; Aslam, F.; Joyklad, P.; Alabduljabbar, H. Potential use of recycled plastic and rubber aggregate in cementitious materials for sustainable construction: A review. *J. Clean. Prod.* **2021**, *329*, 129736. [CrossRef]
5. Ahmad, W.; Ahmad, A.; Ostrowski, K.A.; Aslam, F.; Joyklad, P.; Zajdel, P. Sustainable approach of using sugarcane bagasse ash in cement-based composites: A systematic review. *Case Stud. Constr. Mater.* **2021**, *15*, e00698. [CrossRef]
6. Ahmad, W.; Ahmad, A.; Ostrowski, K.A.; Aslam, F.; Joyklad, P. A scientometric review of waste material utilization in concrete for sustainable construction. *Case Stud. Constr. Mater.* **2021**, *15*, e00683. [CrossRef]
7. Santhosh, K.G.; Subhani, S.M.; Bahurudeen, A. Recycling of palm oil fuel ash and rice husk ash in the cleaner production of concrete—A review. *J. Clean. Prod.* **2022**, *354*, 131736. [CrossRef]
8. Yao, W.; Bai, M.; Pang, J.; Liu, T. Performance degradation and damage model of rice husk ash concrete under dry–wet cycles of sulfate environment. *Environ. Sci. Pollut. Res.* **2022**, 1–17. [CrossRef]
9. Siddika, A.; Al Mamun, M.A.; Alyousef, R.; Mohammadhosseini, H. State-of-the-art-review on rice husk ash: A supplementary cementitious material in concrete. *J. King Saud Univ.-Eng. Sci.* **2021**, *33*, 294–307. [CrossRef]
10. Thomas, B.S. Green concrete partially comprised of rice husk ash as a supplementary cementitious material–A comprehensive review. *Renew. Sustain. Energy Rev.* **2018**, *82*, 3913–3923. [CrossRef]
11. Amran, M.; Fediuk, R.; Murali, G.; Vatin, N.; Karelina, M.; Ozbakkaloglu, T.; Krishna, R.; Kumar, S.A.; Kumar, D.S.; Mishra, J. Rice husk ash-based concrete composites: A critical review of their properties and applications. *Crystals* **2021**, *11*, 168. [CrossRef]

12. Iftikhar, B.; Alih, S.C.; Vafaei, M.; Elkotb, M.A.; Shutaywi, M.; Javed, M.F.; Deebani, W.; Khan, M.I.; Aslam, F. Predictive modeling of compressive strength of sustainable rice husk ash concrete: Ensemble learner optimization and comparison. *J. Clean. Prod.* **2022**, *348*, 131285. [CrossRef]

13. Aslam, F.; Elkotb, M.A.; Iqtidar, A.; Khan, M.A.; Javed, M.F.; Usanova, K.I.; Khan, M.I.; Alamri, S.; Musarat, M.A. Compressive strength prediction of rice husk ash using multiphysics genetic expression programming. *Ain Shams Eng. J.* **2022**, *13*, 101593. [CrossRef]

14. Das, S.K.; Mishra, J.; Mustakim, S.M. Rice husk ash as a potential source material for geopolymer concrete: A. *Int. J. Appl. Eng. Res.* **2018**, *13*, 81–84.

15. Das, S.K.; Mishra, J.; Singh, S.K.; Mustakim, S.M.; Patel, A.; Das, S.K.; Behera, U. Characterization and utilization of rice husk ash (RHA) in fly ash–Blast furnace slag based geopolymer concrete for sustainable future. *Mater. Today Proc.* **2020**, *33*, 5162–5167. [CrossRef]

16. Ahmadi, M.H.; Baghban, A.; Sadeghzadeh, M.; Zamen, M.; Mosavi, A.; Shamshirband, S.; Kumar, R.; Mohammadi-Khanaposhtani, M. Evaluation of electrical efficiency of photovoltaic thermal solar collector. *Eng. Appl. Comput. Fluid Mech.* **2020**, *14*, 545–565. [CrossRef]

17. Asadi, E.; Isazadeh, M.; Samadianfard, S.; Ramli, M.F.; Mosavi, A.; Nabipour, N.; Shamshirband, S.; Hajnal, E.; Chau, K.W. Groundwater quality assessment for sustainable drinking and irrigation. *Sustainability* **2019**, *12*, 177. [CrossRef]

18. Farooq, F.; Jin, X.; Javed, M.F.; Akbar, A.; Shah, M.I.; Aslam, F.; Alyousef, R. Geopolymer concrete as sustainable material: A state of the art review. *Constr. Build. Mater.* **2021**, *306*, 124762. [CrossRef]

19. Ghalandari, M.; Shamshirband, S.; Mosavi, A.; Chau, K.W. Flutter speed estimation using presented differential quadrature method formulation. *Eng. Appl. Comput. Fluid Mech.* **2019**, *13*, 804–810. [CrossRef]

20. Nabipour, M.; Nayyeri, P.; Jabani, H.; Shahab, S.; Mosavi, A. Predicting stock market trends using machine learning and deep learning algorithms via continuous and binary data; a comparative analysis. *IEEE Access* **2020**, *8*, 150199–150212. [CrossRef]

21. Aldrees, A.; Khan, M.A.; Tariq, M.A.U.R.; Mustafa Mohamed, A.; Ng, A.W.M.; Bakheit Taha, A.T. Multi-Expression Programming (MEP): Water Quality Assessment Using Water Quality Indices. *Water* **2022**, *14*, 947. [CrossRef]

22. Khan, M.A.; Farooq, F.; Javed, M.F.; Zafar, A.; Ostrowski, K.A.; Aslam, F.; Malazdrewicz, S.; Maślak, M. Simulation of Depth of Wear of Eco-Friendly Concrete Using Machine Learning Based Computational Approaches. *Materials* **2022**, *15*, 58. [CrossRef]

23. Farooq, F.; Czarnecki, S.; Niewiadomski, P.; Aslam, F.; Alabduljabbar, H.; Ostrowski, K.A.; Śliwa-Wieczorek, K.; Nowobilski, T.; Malazdrewicz, S. A comparative study for the prediction of the compressive strength of self-compacting concrete modified with fly ash. *Materials* **2021**, *14*, 4934. [CrossRef] [PubMed]

24. Band, S.S.; Janizadeh, S.; Chandra Pal, S.; Saha, A.; Chakrabortty, R.; Melesse, A.M.; Mosavi, A. Flash flood susceptibility modeling using new approaches of hybrid and ensemble tree-based machine learning algorithms. *Remote Sens.* **2020**, *12*, 3568. [CrossRef]

25. Band, S.S.; Janizadeh, S.; Chandra Pal, S.; Saha, A.; Chakrabortty, R.; Shokri, M.; Mosavi, A. Novel ensemble approach of deep learning neural network (DLNN) model and particle swarm optimization (PSO) algorithm for prediction of gully erosion susceptibility. *Sensors* **2020**, *20*, 5609. [CrossRef] [PubMed]

26. Shabani, S.; Samadianfard, S.; Sattari, M.T.; Mosavi, A.; Shamshirband, S.; Kmet, T.; Várkonyi-Kóczy, A.R. Modeling pan evaporation using Gaussian process regression K-nearest neighbors random forest and support vector machines; comparative analysis. *Atmosphere* **2020**, *11*, 66. [CrossRef]

27. Shamshirband, S.; Mosavi, A.; Rabczuk, T.; Nabipour, N.; Chau, K.W. Prediction of significant wave height; comparison between nested grid numerical model, and machine learning models of artificial neural networks, extreme learning and support vector machines. *Eng. Appl. Comput. Fluid Mech.* **2020**, *14*, 805–817. [CrossRef]

28. Taherei Ghazvinei, P.; Hassanpour Darvishi, H.; Mosavi, A.; Yusof, K.B.W.; Alizamir, M.; Shamshirband, S.; Chau, K.W. Sugarcane growth prediction based on meteorological parameters using extreme learning machine and artificial neural network. *Eng. Appl. Comput. Fluid Mech.* **2018**, *12*, 738–749. [CrossRef]

29. Torabi, M.; Hashemi, S.; Saybani, M.R.; Shamshirband, S.; Mosavi, A. A Hybrid clustering and classification technique for forecasting short-term energy consumption. *Environ. Prog. Sustain. Energy* **2019**, *38*, 66–76. [CrossRef]

30. Farooq, F.; Ahmed, W.; Akbar, A.; Aslam, F.; Alyousef, R. Predictive modeling for sustainable high-performance concrete from industrial wastes: A comparison and optimization of models using ensemble learners. *J. Clean. Prod.* **2021**, *292*, 126032. [CrossRef]

31. Azim, I.; Yang, J.; Javed, M.F.; Iqbal, M.F.; Mahmood, Z.; Wang, F.; Liu, Q. feng Prediction model for compressive arch action capacity of RC frame structures under column removal scenario using gene expression programming. *Structures* **2020**, *25*, 212–228. [CrossRef]

32. Iqbal, M.F.; Javed, M.F.; Rauf, M.; Azim, I.; Ashraf, M.; Yang, J.; Liu, Q.F. Sustainable utilization of foundry waste: Forecasting mechanical properties of foundry sand based concrete using multi-expression programming. *Sci. Total Environ.* **2021**, *780*, 146524. [CrossRef] [PubMed]

33. Iqbal, M.; Zhang, D.; Jalal, F.E.; Faisal Javed, M. Computational AI prediction models for residual tensile strength of GFRP bars aged in the alkaline concrete environment. *Ocean. Eng.* **2021**, *232*, 109134. [CrossRef]

34. Jalal, F.E.; Xu, Y.; Iqbal, M.; Jamhiri, B.; Javed, M.F. Predicting the compaction characteristics of expansive soils using two genetic programming-based algorithms. *Transp. Geotech.* **2021**, *30*, 100608. [CrossRef]

35. Jalal, F.E.; Xu, Y.; Iqbal, M.; Javed, M.F.; Jamhiri, B. Predictive modeling of swell-strength of expansive soils using artificial intelligence approaches: ANN, ANFIS and GEP. *J. Environ. Manag.* **2021**, *289*, 112420. [CrossRef] [PubMed]

36. Zha, T.H.; Castillo, O.; Jahanshahi, H.; Yusuf, A.; Alassafi, M.O.; Alsaadi, F.E.; Chu, Y.M. A fuzzy-based strategy to suppress the novel coronavirus (2019-NCOV) massive outbreak. *Appl. Comput. Math.* **2021**, *20*, 160–176.

37. Zhao, T.H.; Khan, M.I.; Chu, Y.M. Artificial neural networking (ANN) analysis for heat and entropy generation in flow of non-Newtonian fluid between two rotating disks. *Math. Methods Appl. Sci.* **2021**. [CrossRef]

38. Zhao, T.H.; Qian, W.M.; Chu, Y.M. Sharp power mean bounds for the tangent and hyperbolic sine means. *J. Math. Inequalities* **2021**, *15*, 1459–1472. [CrossRef]

39. Ghalandari, M.; Ziamolki, A.; Mosavi, A.; Shamshirband, S.; Chau, K.W.; Bornassi, S. Aeromechanical optimization of first row compressor test stand blades using a hybrid machine learning model of genetic algorithm, artificial neural networks and design of experiments. *Eng. Appl. Comput. Fluid Mech.* **2019**, *13*, 892–904. [CrossRef]

40. Joloudari, J.H.; Hassannataj Joloudari, E.; Saadatfar, H.; Ghasemigol, M.; Razavi, S.M.; Mosavi, A.; Nabipour, N.; Shamshirband, S.; Nadai, L. Coronary artery disease diagnosis; ranking the significant features using a random trees model. *Int. J. Environ. Res. Public Health* **2020**, *17*, 731. [CrossRef]

41. Mahmoudi, M.R.; Heydari, M.H.; Qasem, S.N.; Mosavi, A.; Band, S.S. Principal component analysis to study the relations between the spread rates of COVID-19 in high risks countries. *Alex. Eng. J.* **2021**, *60*, 457–464. [CrossRef]

42. Behnood, A.; Golafshani, E.M. Machine learning study of the mechanical properties of concretes containing waste foundry sand. *Constr. Build. Mater.* **2020**, *243*, 118152. [CrossRef]

43. Iqbal, M.F.; Liu, Q.F.; Azim, I.; Zhu, X.; Yang, J.; Javed, M.F.; Rauf, M. Prediction of mechanical properties of green concrete incorporating waste foundry sand based on gene expression programming. *J. Hazard. Mater.* **2020**, *384*, 121322. [CrossRef] [PubMed]

44. Farooq, F.; Amin, M.N.; Khan, K.; Sadiq, M.R.; Javed, M.F.; Aslam, F.; Alyousef, R. A comparative study of random forest and genetic engineering programming for the prediction of compressive strength of high strength concrete (HSC). *Appl. Sci.* **2020**, *10*, 7330. [CrossRef]

45. Chu, H.H.; Khan, M.A.; Javed, M.; Zafar, A.; Khan, M.I.; Alabduljabbar, H.; Qayyum, S. Sustainable use of fly-ash: Use of gene-expression programming (GEP) and multi-expression programming (MEP) for forecasting the compressive strength geopolymer concrete. *Ain Shams Eng. J.* **2021**, *12*, 3603–3617. [CrossRef]

46. Fallahpour, A.; Wong, K.Y.; Rajoo, S.; Tian, G. An evolutionary-based predictive soft computing model for the prediction of electricity consumption using multi expression programming. *J. Clean. Prod.* **2021**, *283*, 125287. [CrossRef]

47. Zou, Y.; Zheng, C.; Alzahrani, A.M.; Ahmad, W.; Ahmad, A.; Mohamed, A.M.; Elattar, S. Evaluation of Artificial Intelligence Methods to Estimate the Compressive Strength of Geopolymers. *Gels* **2022**, *8*, 271. [CrossRef]

48. Oltean, M.; Groşan, C. Evolving evolutionary algorithms using multi expression programming. In *European Conference on Artificial Life*; Springer: Berlin/Heidelberg, Germany, 2003; pp. 651–658.

49. Koza, J.R.; Koza, J.R. *Genetic Programming: On the Programming of Computers by Means of Natural Selection*; MIT Press: Cambridge, MA, USA, 1992; Volume 1.

50. Khan, S.; Ali Khan, M.; Zafar, A.; Javed, M.F.; Aslam, F.; Musarat, M.A.; Vatin, N.I. Predicting the Ultimate Axial Capacity of Uniaxially Loaded CFST Columns Using Multiphysics Artificial Intelligence. *Materials* **2022**, *15*, 39. [CrossRef]

51. Ilyas, I.; Zafar, A.; Javed, M.F.; Farooq, F.; Aslam, F.; Musarat, M.A.; Vatin, N.I. Forecasting Strength of CFRP Confined Concrete Using Multi Expression Programming. *Materials* **2021**, *14*, 7134. [CrossRef]

52. Ameri, F.; Shoaei, P.; Bahrami, N.; Vaezi, M.; Ozbakkaloglu, T. Optimum rice husk ash content and bacterial concentration in self-compacting concrete. *Constr. Build. Mater.* **2019**, *222*, 796–813. [CrossRef]

53. Chao-Lung, H.; Le Anh-Tuan, B.; Chun-Tsun, C. Effect of rice husk ash on the strength and durability characteristics of concrete. *Constr. Build. Mater.* **2011**, *25*, 3768–3772. [CrossRef]

54. Bui, D.; Hu, J.; Stroeven, P. Particle size effect on the strength of rice husk ash blended gap-graded Portland cement concrete. *Cem. Concr. Compos.* **2005**, *27*, 357–366. [CrossRef]

55. Ganesan, K.; Rajagopal, K.; Thangavel, K. Rice husk ash blended cement: Assessment of optimal level of replacement for strength and permeability properties of concrete. *Constr. Build. Mater.* **2008**, *22*, 1675–1683. [CrossRef]

56. Ramezanianpour, A.; Mahdikhani, M.; Ahmadibeni, G. The effect of rice husk ash on mechanical properties and durability of sustainable concretes. *Int. J. Civ. Eng.* **2009**, *7*, 83–91.

57. Sakr, K. Effects of silica fume and rice husk ash on the properties of heavy weight concrete. *J. Mater. Civ. Eng.* **2006**, *18*, 367–376. [CrossRef]

58. de Sensale, G.R. Strength development of concrete with rice-husk ash. *Cem. Concr. Compos.* **2006**, *28*, 158–160. [CrossRef]

59. Elwell, D.J.; Fu, G. *Compression Testing of Concrete: Cylinders vs. Cubes*; New York State Department of Transportation: Albany, NY, USA, 1995.

60. Lopes, H.S.; Weinert, W.R. A Gene Expression Programming System for Time Series Modeling. In Proceedings of the XXV Iberian Latin American Congress on Computational Methods in Engineering (CILAMCE), Recife, Brazil, 21–25 November 2004; pp. 10–12.

61. Ahmad, A.; Ahmad, W.; Aslam, F.; Joyklad, P. Compressive strength prediction of fly ash-based geopolymer concrete via advanced machine learning techniques. *Case Stud. Constr. Mater.* **2022**, *16*, e00840. [CrossRef]

62. Golbraikh, A.; Tropsha, A. Beware of q2! *J. Mol. Graph. Model.* **2002**, *20*, 269–276. [CrossRef]

63. Hajiseyedazizi, S.N.; Samei, M.E.; Alzabut, J.; Chu, Y.M. On multi-step methods for singular fractional q-integro-differential equations. *Open Math.* **2021**, *19*, 1378–1405. [CrossRef]

64. Nazeer, M.; Hussain, F.; Khan, M.I.; El-Zahar, E.R.; Chu, Y.M.; Malik, M. Theoretical study of MHD electro-osmotically flow of third-grade fluid in micro channel. *Appl. Math. Comput.* **2022**, *420*, 126868. [CrossRef]

65. Rashid, S.; Sultana, S.; Karaca, Y.; Khalid, A.; Chu, Y.M. Some further extensions considering discrete proportional fractional operators. *Fractals* **2022**, *30*, 2240026. [CrossRef]

66. Chu, Y.M.; Nazir, U.; Sohail, M.; Selim, M.M.; Lee, J.R. Enhancement in thermal energy and solute particles using hybrid nanoparticles by engaging activation energy and chemical reaction over a parabolic surface via finite element approach. *Fractal Fract.* **2021**, *5*, 119. [CrossRef]

67. Zhao, T.; Wang, M.; Chu, Y. On the bounds of the perimeter of an ellipse. *Acta Math. Sci.* **2022**, *42*, 491–501. [CrossRef]

68. Zhao, T.H.; Wang, M.K.; Hai, G.J.; Chu, Y.M. Landen inequalities for Gaussian hypergeometric function. *Rev. Real Acad. Cienc. Exactas Físicas Naturales. Ser. A. Matemáticas* **2022**, *116*, 53. [CrossRef]

69. Darban, S.; Ghasemzadeh Tehrani, H.; Karballaeezadeh, N.; Mosavi, A. Application of Analytical Hierarchy Process for Structural Health Monitoring and Prioritizing Concrete Bridges in Iran. *Appl. Sci.* **2021**, *11*, 8060. [CrossRef]

70. Ebrahimi-Khusfi, Z.; Taghizadeh-Mehrjardi, R.; Roustaei, F.; Ebrahimi-Khusfi, M.; Mosavi, A.H.; Heung, B.; Soleimani-Sardo, M.; Scholten, T. Determining the contribution of environmental factors in controlling dust pollution during cold and warm months of western Iran using different data mining algorithms and game theory. *Ecol. Indic.* **2021**, *132*, 108287. [CrossRef]

71. Mala, A.A.; Sherwani, A.F.H.; Younis, K.H.; Faraj, R.H.; Mosavi, A. Mechanical and fracture parameters of ultra-high performance fiber reinforcement concrete cured via steam and water: Optimization of binder content. *Materials* **2021**, *14*, 2016. [CrossRef]

72. Panahi, F.; Ehteram, M.; Ahmed, A.N.; Huang, Y.F.; Mosavi, A.; El-Shafie, A. Streamflow prediction with large climate indices using several hybrid multilayer perceptrons and copula Bayesian model averaging. *Ecol. Indic.* **2021**, *133*, 108285. [CrossRef]

73. Meiabadi, M.S.; Moradi, M.; Karamimoghadam, M.; Ardabili, S.; Bodaghi, M.; Shokri, M.; Mosavi, A.H. Modeling the producibility of 3D printing in polylactic acid using artificial neural networks and fused filament fabrication. *Polymers* **2021**, *13*, 3219. [CrossRef]

74. Peng, Y.; Ghahnaviyeh, M.B.; Ahmad, M.N.; Abdollahi, A.; Bagherzadeh, S.A.; Azimy, H.; Mosavi, A.; Karimipour, A. Analysis of the effect of roughness and concentration of Fe3O4/water nanofluid on the boiling heat transfer using the artificial neural network: An experimental and numerical study. *Int. J. Therm. Sci.* **2021**, *163*, 106863. [CrossRef]

75. Mousavi, S.M.; Ghasemi, M.; Dehghan Manshadi, M.; Mosavi, A. Deep learning for wave energy converter modeling using long short-term memory. *Mathematics* **2021**, *9*, 871. [CrossRef]

 *materials*

Article

Comparative Study of Experimental and Modeling of Fly Ash-Based Concrete

Kaffayatullah Khan [1],[*], Ayaz Ahmad [2], Muhammad Nasir Amin [1], Waqas Ahmad [3], Sohaib Nazar [3] and Abdullah Mohammad Abu Arab [1]

[1] Department of Civil and Environmental Engineering, College of Engineering, King Faisal University, Al-Ahsa 31982, Saudi Arabia; mgadir@kfu.edu.sa (M.N.A.); 219041496@student.kfu.edu.sa (A.M.A.A.)
[2] MaREI Centre, Ryan Institute and School of Engineering, College of Science and Engineering, National University of Ireland, H91 TK33 Galway, Ireland; a.ahmad8@nuigalway.ie
[3] Department of Civil Engineering, COMSATS University Islamabad, Abbottabad 22060, Pakistan; waqasahmad@cuiatd.edu.pk (W.A.); sohaibnazar@cuiatd.edu.pk (S.N.)
[*] Correspondence: kkhan@kfu.edu.sa

Abstract: The application of supplementary cementitious materials (SCMs) in concrete has been reported as the sustainable approach toward the appropriate development. This research aims to compare the result of compressive strength (C-S) obtained from the experimental method and results estimated by employing the various modeling techniques for the fly-ash-based concrete. Although this study covers two aspects, an experimental approach and modeling techniques for predictions, the emphasis of this research is on the application of modeling methods. The physical and chemical properties of the cement and fly ash, water absorption and specific gravity of the aggregate used, surface area of the cement, and gradation of the aggregate were analyzed in the laboratory. The four predictive machine learning (PML) algorithms, such as decision tree (DT), multi-linear perceptron (MLP), random forest (RF), and bagging regressor (BR), were investigated to anticipate the C-S of concrete. Results reveal that the RF model was observed more exact in investigating the C-S of concrete containing fly ash (FA), as opposed to other employed PML techniques. The high R2 value (0.96) for the RF model indicates the high precision level for forecasting the required output as compared to DT, MLP, and BR model R^2 results equal 0.88, 0.90, and 0.93, respectively. The statistical results and cross-validation (C-V) method also confirm the high predictive accuracy of the RF model. The highest contribution level of the cement towards the prediction was also reported in the sensitivity analysis and showed a 31.24% contribution. These PML methods can be effectively employed to anticipate the mechanical properties of concretes.

Keywords: concrete; fly ash; modeling; machine learning; compressive strength

Citation: Khan, K.; Ahmad, A.; Amin, M.N.; Ahmad, W.; Nazar, S.; Arab, A.M.A. Comparative Study of Experimental and Modeling of Fly Ash-Based Concrete. *Materials* **2022**, *15*, 3762. https://doi.org/10.3390/ma15113762

Academic Editor: Jorge Otero

Received: 26 April 2022
Accepted: 19 May 2022
Published: 24 May 2022

Publisher's Note: MDPI stays neutral with regard to jurisdictional claims in published maps and institutional affiliations.

Copyright: © 2022 by the authors. Licensee MDPI, Basel, Switzerland. This article is an open access article distributed under the terms and conditions of the Creative Commons Attribution (CC BY) license (https://creativecommons.org/licenses/by/4.0/).

1. Introduction

CO_2 emissions from industry, transportation, and services, and nitrogen and methane oxides from agriculture are significant greenhouse gases (GHGs) [1]. Worldwide worries about the environmental, economic, and social consequences of GHG emissions such as CO_2 have prompted the growth and deployment of a variety of CO_2 emission mitigation technologies and initiatives [2–7]. At this time, environmental sustainability has developed as a global objective for social interests [8–10]. Furthermore, ecological issues about CO_2 ejection from the Ordinary Portland Cement (OPC) manufacturing process have prompted past academics to look at the viability of other materials to substitute OPC during concrete production [11–13]. According to a study [14], the use of waste materials is desirable for the sustainability of the construction sector; however, another study [15] claims that the application of byproducts obtained from industries as a supplementary cementitious material (SCM) partly substitute OPC has substantially helped to achieve a more green environment. The growing demand for the strength properties along with the durability of concrete has necessitated the incorporation of a variety of industrial wastes with pozzolanic attributes into the OPC [16–21]. Additionally, these components used in OPC have a remarkable result in the microstructure alteration of cement pastes and the physio-mechanical parameters of concretes [19,22,23]. The application of waste products in concrete structures not only decreases ecological pollution but also improves the fresh and hardened properties of the selected concrete [22–27]. Due to these aspects, waste materials are frequently employed to improve the characteristics of concrete [28,29]. Nowadays, industrial wastes of various sorts and nanoparticles are employed in concrete [30,31]. A set of the waste materials frequently incorporated in concrete from the industries are ground granulated blast furnace slag, metakaolin, fly ash, and silica fume. However, nano industrial wastes which are frequently using in concrete are graphene, nano silica, titania–silica nanosphere, nano titanium, carbon nanotubes, and nano metakaolin.

FA is one of the most utilized SCM in concretes [32–36]. The FA obtained from coal incineration activities is not risky from the radiological fact [37]. Regrettably, it comprises trace levels of hazardous substances derived from coal-burning, including mercury, fluorine, and [38]. After burning, approximately 10–40% of chlorine and fluorine and 30–80% of mercury in coal are reported to retain in FA [39,40]. As a result, this industrial waste can be classified as a possibly hazardous substance in some instances. FA is an effective, very desirable waste for recycling purposes since concretes containing these supplements in proportions of up to 20% as OPC substitutes exhibit enhanced stability and fracture toughness [41–44], deterioration resistance [45], and tolerance to elevated temperatures [46]. Additionally, by utilizing FA, eco-friendly green material for civil engineering might be produced [47–51] and promote the development of a specific microstructure in concrete matrices, thereby facilitating the restriction of harmful elements [52]. Initially, the usage of FA in concrete enables the reduction of problematic disposal sites associated with this waste. It is worth noting that about 800 million tons of FA are generated annually on a global scale [53,54]. Due to the huge volume of combustion byproducts and their lack of usage, the necessity for dry or wet landfill sites to be constructed, maintained, and secured arises. It is a considerable environmental and public issue since the resulting contamination of the atmosphere has a detrimental effect on people's health and well-being and might contribute to the development of severe environmental infections. Dumping huge amounts of FA in landfilling is also detrimental, as they are extremely light and fine in dry conditions, making them easily dispersed by wind. Thus, the substitution of FA cement is an unambiguously environmentally acceptable alternative.

Moreover, it is necessary to introduce soft computing methods to accurately forecast the nature and performance/strength of materials. Artificial intelligence (AI) approaches are gaining more popularity in this aspect which are usually introduced to estimate the various characteristics of different materials [55–61]. Especially, the estimate of the mechanical characteristics of concrete is very important as it requires a lot of time, effort, and cost to have the experimental results. To minimize these parameters, numerous AI algorithms such as random forest (RF), multi-linear perception regression (MLP), artificial neural network (ANN), neuro-fuzzy regression, AdaBoost, bagging, and boosting are normally used for the estimate of concrete properties. Shariati et al. [62] research was based on the anticipation of concrete strength containing waste material (FA and furnace slag). The result reveals that the ANN approach shows a satisfactory prediction level for the compressive strength (C-S) of concrete. Han et al. [63] employed the RF algorithm for the anticipation of high-performance concrete and described that RF could be successfully employed for the forecast of C-S of concretes. Chaabene et al. [64] represent a comprehensive review of the number of PML approaches used for the prediction of the strength properties of concrete. They reported that ML models are more precise, adaptable, and can be retrained by incorporating the updated dataset.

This study describes the combined effect of experimental and soft computing predictive approaches for the concrete strength containing FA. A detailed investigation of the material used and mix ratios for preparing the concrete were carried out for the desired strength. The novelty of this research is to investigate the precision level of predictive algorithms (MLP, DT, BR, RF) employed in the experimental and data retrieved from literature for the strength property of FA-based concrete. The comparative study on the precision of employed algorithms towards the prediction of C-S would be beneficial for the scientists and researchers in the field of engineering to adopt the appropriate technique for the estimate of concrete's strength.

2. Materials and Methods

2.1. Materials

The materials utilized in this investigation were aggregates with a specific gravity of 2.79 and water absorption of 0.96% purchased from a local quarry, Ordinary Portland Cement Type I having a surface area of 380 m^2/kg, and class-F fly ash obtained from a nearby thermal power plant was introduced in the experimental work. The water absorption for the selected fine aggregate was noted as 2.32%, with its specific gravity of 2.65 obtained from the local source. As per the ASTM standard C494 superplastizers type A was used in the concrete during experimental work. Table 1 summarizes the physical properties and chemical composition of cementitious materials. As can be observed, cement has the highest specific gravity, as opposed to FA. Moreover, the amount of SiO_2, Fe_2O_3, and Al_2O_3 in FA is 77.9%, indicating that it is class-F FA. However, the fineness modulus FM of fine aggregate was noted as 2.65, while the result of fineness modulus for coarse aggregate and fine aggregate was calculated as 6.93, and 2.65, respectively.

Table 1. Physical properties and chemical composition of the cement and fly ash.

	Property/Composition	Cement	Fly Ash
	Initial setting time (minutes)	34	-
	Final setting time (minutes)	161	-
Physical properties	Standard consistency (%)	31.9	-
	Specific gravity	3.2	2.482
	Soundness (mm)	1	-
	Blaine fineness (m^2/kg)	2950	4300
	Silica as SiO_2 (%)	21.77	48.5
	Alumina as Al_2O_3 (%)	5.5	20.01
Chemical composition	Magnesium as MgO (%)	1.24	2.4
	Calcium as CaO (%)	63.3	16.45
	Iron as Fe_2O_3 (%)	4.6	8.5
	Sulphur as SO_3 (%)	1.91	1.72
	Loss of ignition (%)	1.68	2.42

2.2. Methods

In the laboratory, cylindrical specimens (100 mm diameter and 200 mm height) were made. Compaction was accomplished in two layers, with each layer receiving twenty blows, using a conventional 2.5 kg proctor hammer. This technique has been advocated over vibration and rodding. The number of random mixes was made with different mix ratios to obtain the maximum number of data points. Each batch was then subjected to curing for 7, 28, 56, and 90 days

2.3. Compressive Strength

The C-S of the FA-based concrete specimens was found using the ASTM C39/C 39M-99 standard [65]. The compressive axial load is applied to the specimens at a rate of 0.15 to 0.35 MPa/s until the failure. Concrete specimens were cured in water and then tested after 7, 28, 56, and 90 days. The maximum, minimum, and average C-S obtained from the experimental work in the laboratory were 60.90 MPa, 12.05 MPa, and 31.73 MPa, respectively.

2.4. Data Description

The 62 data points (mixes) were prepared from the experimental work in the laboratory, while 569 data points were retrieved from the literature [66,67] to have a maximum number of data samples for modeling. To run the selected models, a total of 631 data points with seven input parameters such as FA, water (W), cement (C), superplasticizers (SP), age, coarse aggregate (C-A), and fine aggregate (FA), with one output C-S were arranged in the tabulated form. The required dataset was then incorporated into the anaconda navigator software, in which the selected models were run one by one with the help of python coding. The result was obtained in the form of a coefficient of determination (R^2) value, which normally ranges from 0 to 1. The maximum R^2 value signifies the superior precision level of the employed method in forecasting desired outcome. In addition, the explanatory statistical analysis of the input parameters obtained from experiments and literature used in the study for the prediction (C-S) purpose can be seen in Table 2. The histograms indicating the relative frequency distribution in the percentage of each variable of the total dataset were developed using Jupyter Notebook (6.0.3) of the anaconda software, as depicted in Figure 1 and the units for each variable in the figure is kg/m^3, except age is days and strength in MPa. Moreover, the detailed schematic representation of this research is shown in Figure 2.

Table 2. Explanation of the statistical analysis for the input parameters.

Parameters	Cement (kg/m^3)	Fly Ash (kg/m^3)	Water (kg/m^3)	SP (kg/m^3)	C-A (kg/m^3)	FA (kg/m^3)	Age (days)
Mean values	282.13	77.29	180.95	5.45	1003.76	794.19	44.50
Standard deviation	94.88	61.91	17.97	5.28	72.84	68.18	58.66
Median of input	252.00	100.40	185.70	5.70	1006.40	794.90	28.00
Mode of input	213.50	0.00	192.00	0.00	968.00	613.00	28.00
Standard error	3.78	2.46	0.72	0.21	2.90	2.71	2.34
Range	405.30	200.10	88.00	28.20	324.00	351.00	364.00
Minimum values	134.70	0.00	140.00	0.00	801.00	594.00	1.00
Maximum values	540.00	200.10	228.00	28.20	1125.00	945.00	365.00

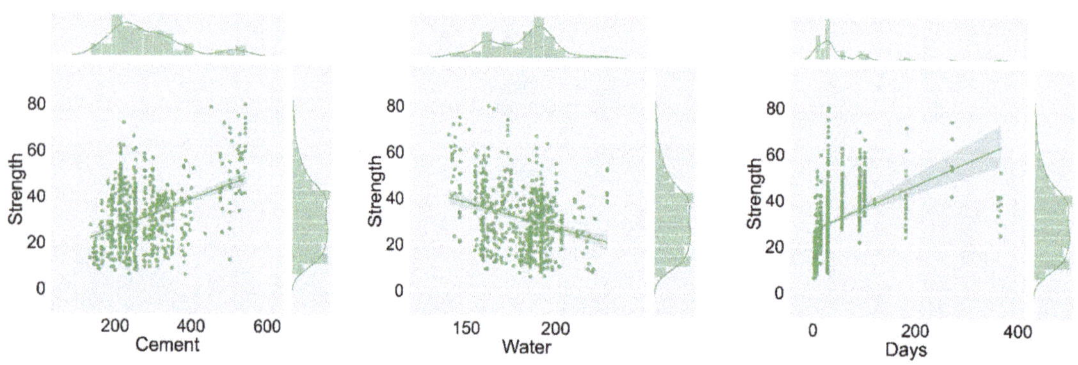

Figure 1. *Cont.*

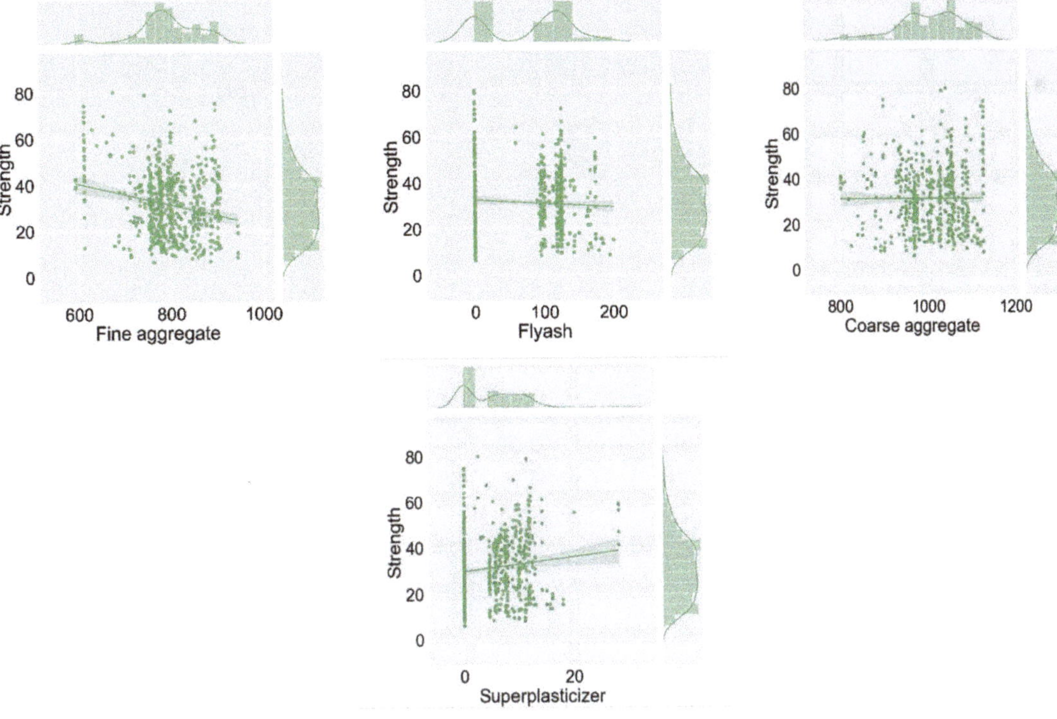

Figure 1. Reflection of histograms for inputs indicating the relative frequency distribution.

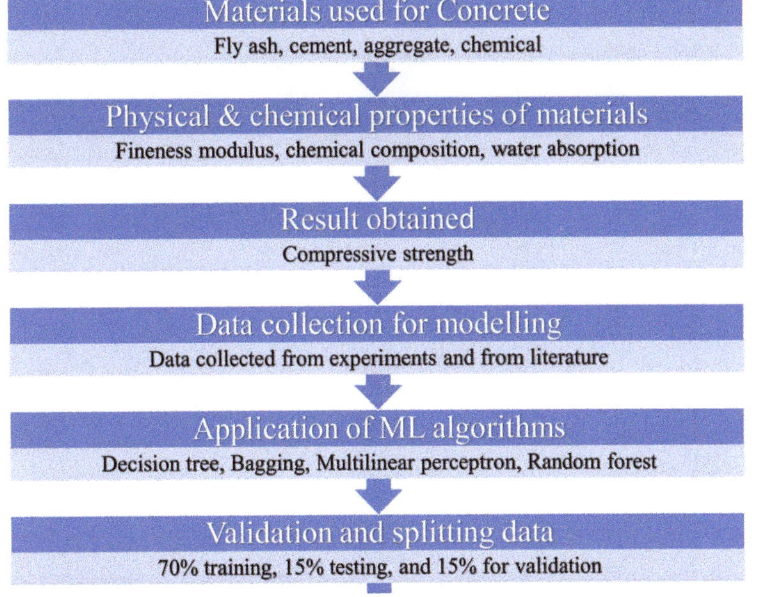

Figure 2. Flow chart of the research program indicating the step-by-step procedure.

3. Predictive Machine Learning (PML) Algorithms

3.1. Decision Tree

DT algorithms are well-recognized PML approaches that have been used for a variety of tasks, most notably classification. DTs are used to partition datasets in a nonparametric manner. Alternative data extracting methods include regression models, which depict variables' relations as cross-products. The DTs used in this research were chosen for their capacity to transform enormous, complex datasets into simple-to-understand yet knowledge-rich graphic presentations. More precisely, the resulting graphical tree image was deemed beneficial for rapidly elucidating the essential parameter value combinations that result in unacceptable product loss, which could then be turned into a set of rules. A DT employs a tree-like graph to describe a flowchart-like structure, with the "root" as the starting point. Each internal node of the tree corresponds to a test on a particular attribute or subset of attributes. Each branch from the node reflects the result of the test, while the final node represents a class label via a "leaf". A simple DT can be constructed manually. However, designing an algorithm that learns the tree from data is straightforward. As with other types of PML, supervised learning uses labeled samples to construct a classifier by computing the sequence of branch options. The flow chart of the DT model indicating the execution process for predicting the required outcome is shown in the Figure 3.

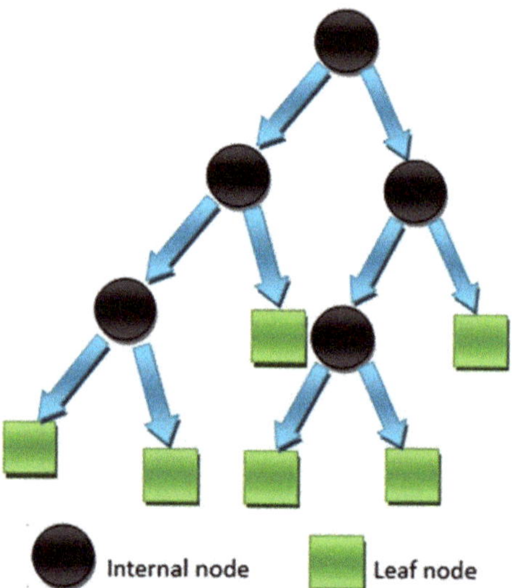

Figure 3. Execution process of the DT model [66].

3.2. MLP Algorithm

An MLP is a form of feedforward ANN that generates outputs based on a collection of inputs. Amongst the output and input layers, many layers of input nodes are linked through a targeted graph. Backpropagation is used to train the network in MLP. A MLP is a type of network (neural) that links many laps in a targeted graph, with signals traveling one way across the nodes. Except for the input nodes, each node has a nonlinear activation function, which is unique to it. MLPs are a type of supervised learning that makes use of backpropagation. Due to the number of laps of neurons in MLP, it is usually called a deep learning approach. MLP is commonly used in supervised learning applications and imputation pure science and parallel dispersed processing studies. Applications have machine translation, image perception, and speech realization. Initially, the algorithm selects predictors to employ during the regression phase to identify the variance inflation component (VIF). The VIF then evaluates the variance increase of an estimated regression coefficient due to collinearity. Finally, the algorithm eliminates variables with high VIFs in order to get the optimal forecasting solution as shown in the Figure 4.

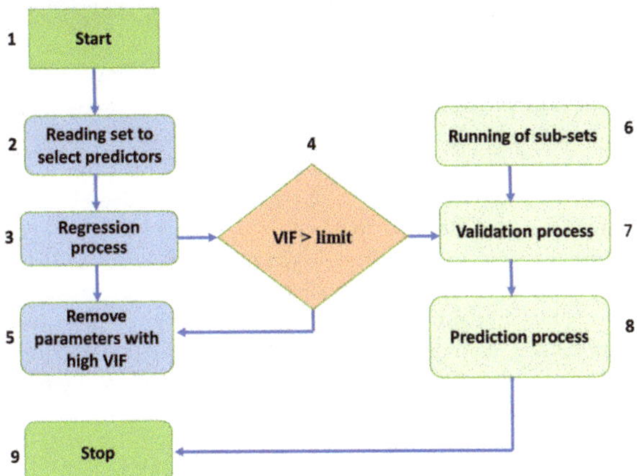

Figure 4. Flow chart of the MLP model showing the complete execution process.

3.3. Bagging Algorithm

BR, also known as bootstrap aggregation, is a technique for merging many editions of a predicted model. Every model is individually skilled and then averaged. The fundamental purpose of BR is to achieve a smaller deviation than any single model. Bootstrapping is the process of generating bootstrapped samples from a given dataset. The samples are generated by randomly picking and replacing data points. The resampled data have qualities that are unique from the original data in their entirety. It illustrates the data distribution and also tends to reserve divergence among bootstrapped samples, i.e., the data dispersal must remain together while maintaining distinction across bootstrapped samples. This helps to construct strong models. Furthermore, bootstrapping supports preventing the overfitting problem. When several training datasets are used to build the model, it becomes resistant to error creation and hence runs in a better manner with the test data, minimizing variation by creating a strong footing in the test set. Testing the model with numerous permutations guarantees that it is not partisan for an incorrect result. The flow chart of the bagging model can be seen in the Figure 5.

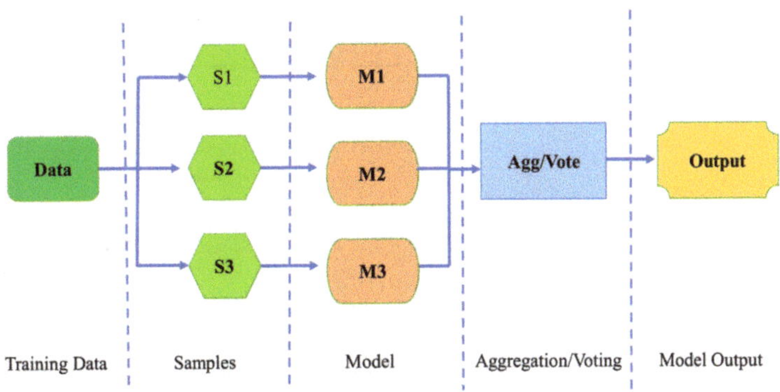

Figure 5. Flow chart of the bagging algorithm indicating the execution process.

3.4. Random Forest

An RF is a special kind of PML method that is utilized to deal with classification and relapse issues. It constructs the use of ensemble learning, a practice for settling complex problems through the application of various classifiers. An RF algorithm is made up of a huge number of decision trees. The RF approach creates a 'forest' that is trained using either backward regression or bootstrap aggregation. BR is an ensemble meta-algorithm that is used to improve the accuracy of PML systems. The RF technique creates the result based on the predictions of the DTs. Forecasting is accomplished by summing or scaling the output of distinct trees. Expanding the number of trees enhances the accuracy of the result. An RF algorithm solves the disadvantages of a deep learning system. It reduces overfitting and increases the accuracy of datasets. It makes predictions without needing the user to configure multiple packages (such as sci-kit-learn). A DT is composed of three components: decision nodes, leaf nodes, and root nodes. A DT technique partitions a training set into branches that subsequently split into additional branches. This method is continued till reaching a leaf node. It is not feasible to further segregate the leaf node. The nodes of the DT show the attributes that are used to anticipate the result. The decision nodes link the leaves together. The execution process of the RF model is depicted in the Figure 6.

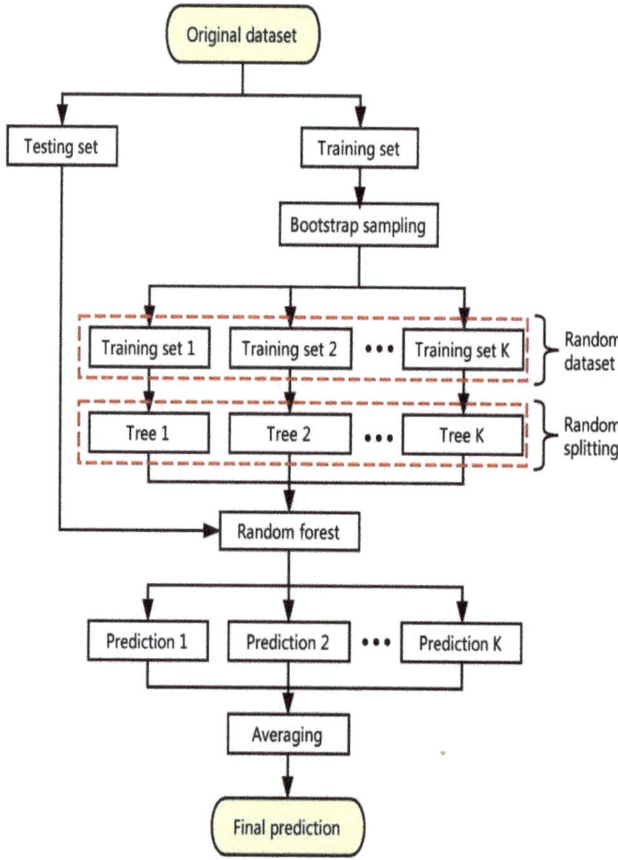

Figure 6. Predictive process of the RF model [63].

3.5. K-Fold Cross-Validation (C-V) Method

C-V is a statistical approach that is used to assess the prediction power of PML models. It is commonly applied in PML to match and select models for specific projecting modeling issues since it is simpler to understand and use and gives skill estimates that are typically less biased than those given by other approaches. C-V is a strategy for assessing PML models on a short sample of data. The method accepts a single parameter, k, which indicates how many groups a given data sample should be split into. As a result, the procedure is usually abbreviated as k-fold C-V. When an exact value for k is supplied, it may be used in place of k in the model's reference; for example, k = 10 becomes 10-fold C-V.

C-V is mostly utilized in applied PML to determine the skill of a PML model on formerly unknown data. That is, to assess the model's overall operation when employed to produce forecasts on data that were not used during the model's training. It is a popular method because it is simple to understand and offers a more accurate evaluation of model competency than other strategies, such as a simple train/test split. The general procedure is as follows: randomize the dataset, divide it into k distinct groups, treat one group as a reserve or test data collection, use the remaining groups as a source of training data on the training set, fit a model and evaluate it on the test set, keep the evaluation score and discard the model, and summarize the model's ability by examining a sample of model evaluation scores. Notably, each observation in the data sample is assigned to a unique group and remains assigned to that group throughout the process. This means that each sample is used just once in the hold outset and then used k times to train the model.

4. Result and Discussions

4.1. Decision Tree Model Outcome

The correlation amongst the experimental results and the findings found from the DT model (predicted) shows appreciable relation and gives the R^2 value equal to 0.88, as shown the Figure 7. However, Figure 8 depicts the spreading of errors from the predicted and experimental C-S results. This distribution ranges from 0 and gives the maximum value equal to 13.8 MPa, while the average result of this distribution was 3.09 MPa. In addition, 23.62% of the data were lying among 0 and 1 MPa, and 58.26% of the data were lying among 1 MPa and 5 MPa. However, only 18.11% of the error values were lying above 5 MPa.

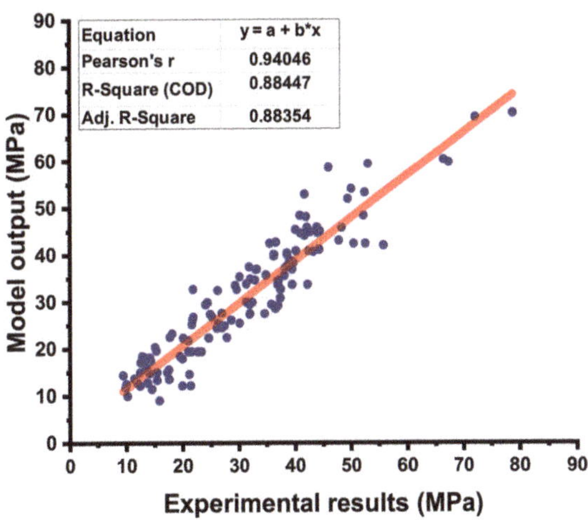

Figure 7. Correlation between the experimental C-S and projected C-S for the DT model.

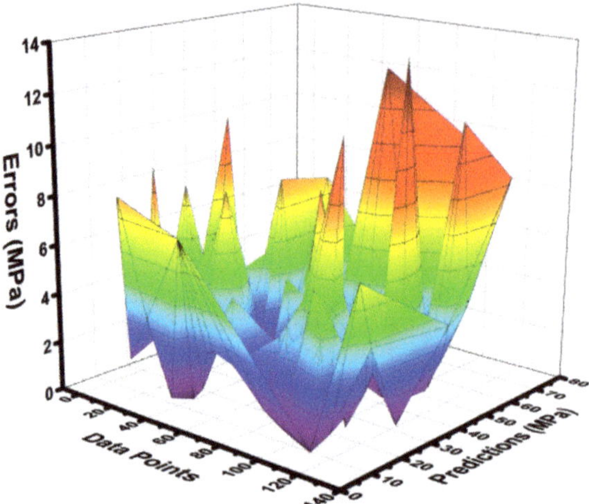

Figure 8. Difference between the experimental C-S and predicted C-S of the DT model.

4.2. MLP Model Outcome

The statistical result obtained from the MLP model between the experimental and predicted can be seen in Figure 9. The R^2 value equals 0.90 for the MLP model, showing a better predictive precision for C-S of concrete as opposed to the DT model. The difference (errors) between the experimental and forecasted C-S results for FA-based concrete are shown in Figure 10. This difference gives the maximum value equal to 15.22 MPa, while the minimum value was reported as 0.009 MPa, while this distribution shows the average value equals 3.74 MPa. Moreover, it was reported that 14.17% of data were lying up to 1 MPa, and 56.69% of data were lying among 1 MPa and 5 MPa. However, 29.13% of the data were lying above 5 MPa.

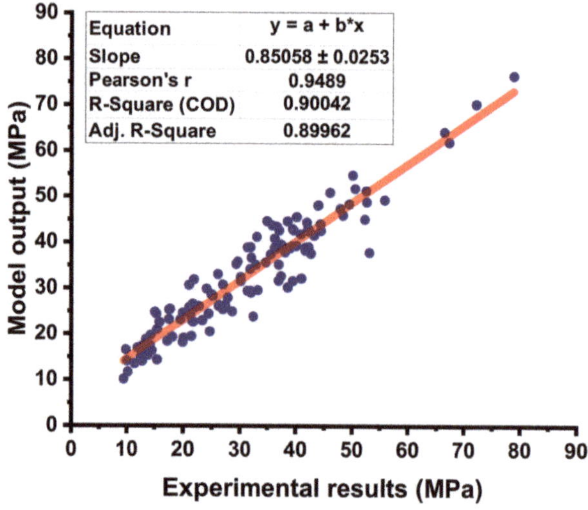

Figure 9. Correlation between the experimental C-S and the estimated C-S for the MLP model.

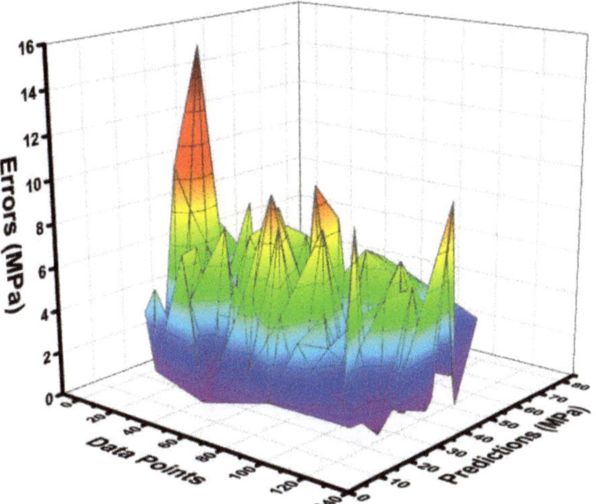

Figure 10. Difference between the experimental C-S and predicted C-S of the MLP model.

4.3. BR Model Outcome

The relationship between the experimental results of the C-S and the anticipated outputs of the concrete containing FA are shown in Figure 11. The results of the difference (errors) among the forecasted and experimental can be seen in Figure 12. The results of these differences give the highest, lowest, and average values of 9.01 MPa, 0.004 MPa, and 2.77 MPa, respectively. Moreover, 23.62% of the data were lying up to 1 MPa, 59.05% of data were found among 1 MPa, and 5 MPa, while only 17.32% of the data were lying above 5 MPa.

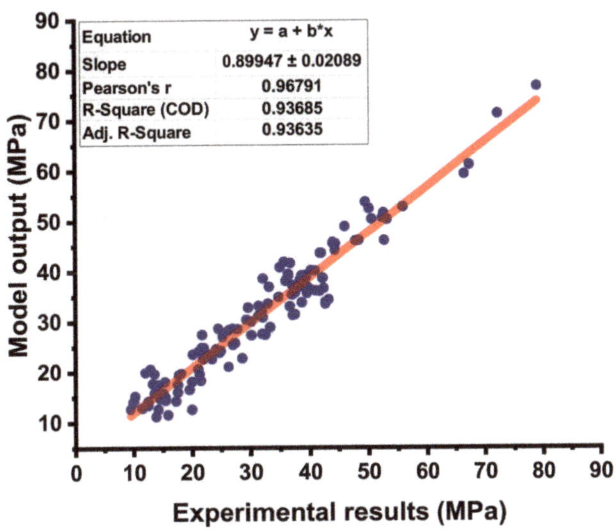

Figure 11. Correlation between the experimental C-S and projected C-S for the BR model.

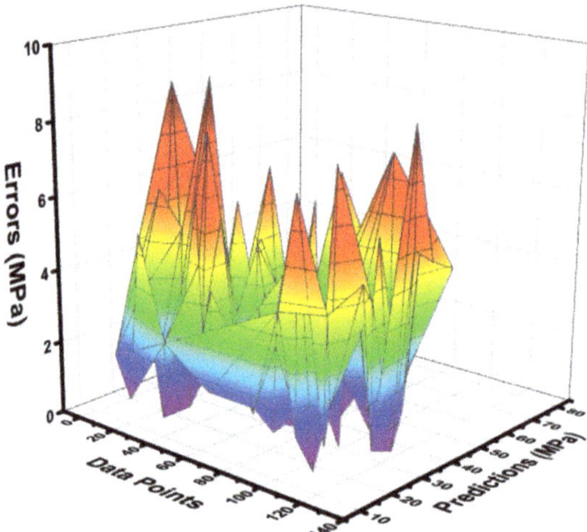

Figure 12. Difference between the experimental C-S and predicted C-S of the BR model.

4.4. RF Model Output

The statistical output for the RF model between the experimental C-S and predictive C-S of concrete containing FA is depicted in Figure 13. The RF model shows a much better predictive result when compared to other employed ML algorithms, as illustrated by the high R^2 value that equals 0.96. The errors distribution between the experimental C-S and forecasted C-S of concrete is shown in Figure 14. The RF model's error distribution gives the highest, lowest, and average values equal to 7.183 MPa, 0.056 MPa, and 2.170 MPa, respectively. Moreover, it was observed that 24.40% of the data were lying up to 1 Mpa, 67.71% of the data were lies among 1 MPa, and 5 MPa, while only 7.87% of the data were lying above 5 MPa.

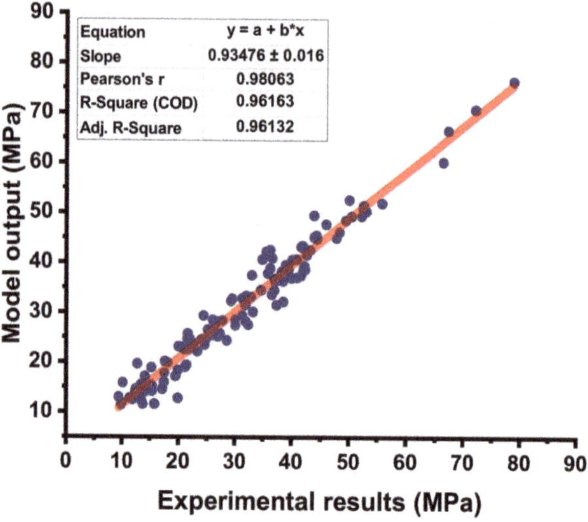

Figure 13. Correlation between the experimental C-S and projected C-S for the RF model.

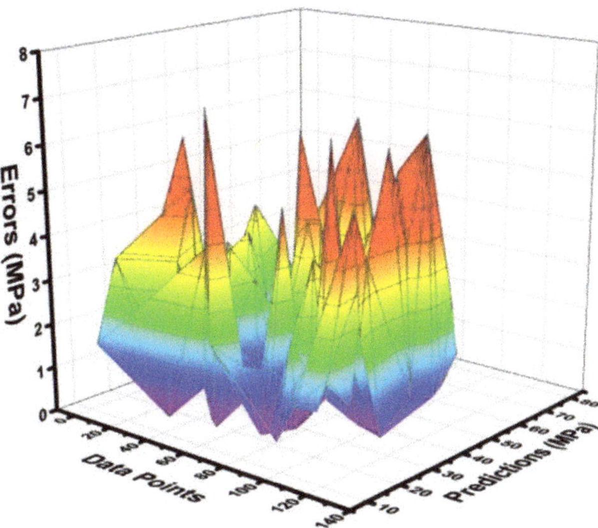

Figure 14. Difference between the experimental C-S and predicted C-S of the RF model.

4.5. K-Fold Outcome

C-V is a statistical approach that is used to analyze or approximate the factual performance of PML models in real-world situations. It is crucial to understand the effectiveness of the models that have been chosen. In order to accomplish this, a validation technique must be used to determine the level of correctness of the model's data. The k-fold validation test necessitates the randomization of the dataset as well as the division of the dataset into k-groups. According to the research detailed here, the data from experimental samples are separated into ten equal groups. It makes use of nine out of ten subsets, with the exception of one subset that is used for model validation purposes. The same approach used in this process is then replicated ten times in order to get the average precision of the ten replications carried out. It has been extensively established that the k-fold C-V approach accurately depicts the decision and correctness of the PML models, and this has been thoroughly confirmed.

The use of k-fold C-V might be employed to determine whether or not there is a bias or a variance reduction for the test set. As shown in Figure 15a–d, the outcomes of C-V are assessed using the R^2, the mean absolute error (MAE), the mean square error (MSE), and the root mean square error (RMSE). The RF model indicates the lower result of the proposed errors and high result of the R^2 as opposed to the other three employed models (BR, MLP, DT). RF shows the average value of R^2 equals 0.46, while the maximum and minimum values were equal to 0.88 and 0.07, respectively. The BR model's average R^2 value was noted as 0.63, and the highest and lowest value was reported as 0.87 and 0.25, respectively. Likewise, the average, least, and high value of R^2 for the MLP model was noted as 0.47, 0.07, and 0.88, respectively. However, the same result of the R^2 value for the DT model was reported as 0.57, 0.01, and 0.88, respectively.

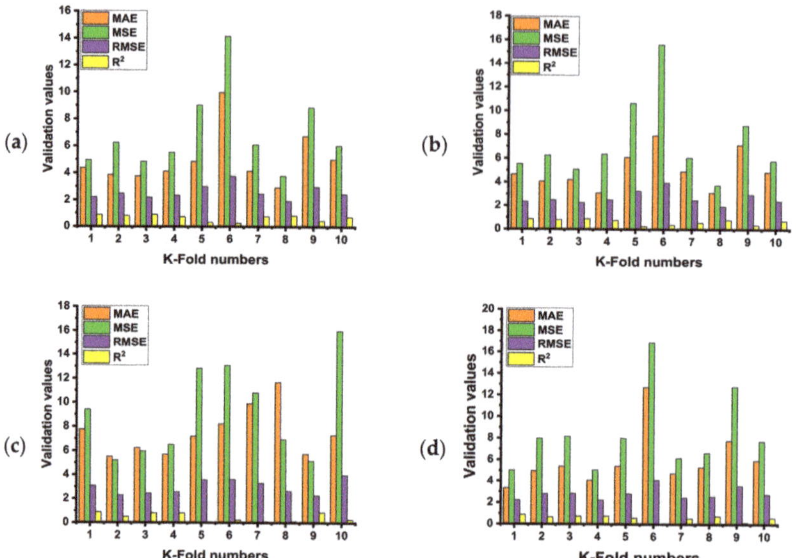

Figure 15. Statistical indicators of k-fold CV for the employed models; (**a**) RF model, (**b**) BR model, (**c**) MLP model, and (**d**) DT model.

5. Sensitivity Analysis (SA)

This analysis helps to find out the contribution level of each input factor employed for modeling to predict the C-S of FA-based concrete. It is also important to test the effect of each variable for the required outcome. SA reveals that the highest contribution towards the prediction of C-S was reported by cement and shows the 31.24 percent contribution, while the other variables contributed the least. The minimum contribution was reported by the superplasticizers, which contributed only 4.69 percent towards the anticipation of C-S of concrete, as shown in Figure 16.

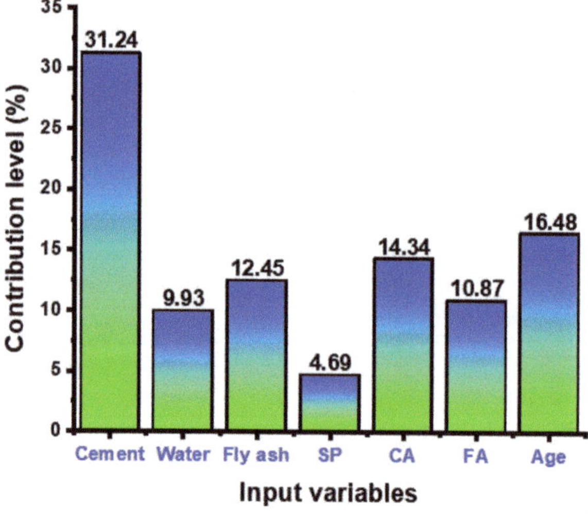

Figure 16. Parameter's effect on the strength property of FA-based concrete.

6. Discussion

This research described the comparative investigation of experimental results obtained in the laboratory and forecasted results acquired from the various modeling techniques for the C-S of concrete containing FA. It is the worth known fact that obtaining the strength of concrete must take a number of days (time), which is a time-consuming effort for researchers. To minimize time, effort on experiments, and cost, the application of such soft computing methods which can predict the desired strength initially are of great interest. The ML algorithms employed in this study also showed satisfactory outcomes when the experimental C-S result of the various mixes was compared with the forecasted C-S result. The comparison of four different types of ML approaches gives the anticipated result with a certain precision level based on the execution process of each approach. The RF ML technique gives the effective, precise result for C-S of FA-based concrete when compared to other employed ML algorithms (DT, MLP, and BR). The precision level of these models is normally evaluated from the R^2 value, which normally ranges from 0 to 1; the higher R^2 value of the model indicates a better precise result in terms of predictions. The high accuracy of the RF and BR is due to the execution process for the data and splitting of the model into the sub-models. The detailed information on the sub-models of RF and BR can be seen in Figure 17a,b, respectively. An RF is composed of a huge number of independent DTs that involve collaboration. Each tree in the RF produces a forecast for a class, and the class with the most choices becomes the model's prediction. The high accuracy of the RF model over the others has also been reported in the literature [68]. The applied statistical checks also give confirmation of high accuracy for the RF model. The lesser value of MAE, MSE, and RMSE shows that the R^2 value for the said model will be higher and vice versa.

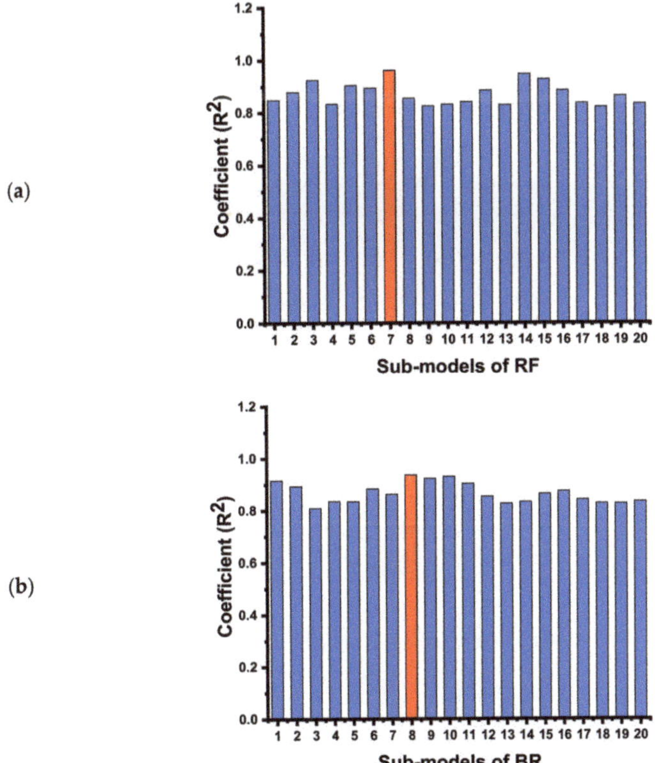

Figure 17. Coefficient of determination result of the 20 sub-models for; (a) RF, (b) BR models.

7. Conclusions

This research reported the comparative study of experimental C-S and the results from the various modeling approaches for concrete containing fly ash (FA). The 61 mixes were prepared in the laboratory with the random mix ratios to have the number of data points for further investigation in the modeling techniques. A similar database was also collected from the literature to make the database appreciable for modeling. The following conclusion can be drawn from the study.

- The RF model was more effective in predicting the C-S of concrete having FA as opposed to DT, MLP, and BR.
- RF gives the R^2 value equal to 0.96, which is the highest of the DT (0.88), MLP (0.90), and BR (0.93), indicating the highest precision level for forecasting the C-S of concrete.
- Statistical checks and the CV approach also validate the superior exactness level of the RF model as opposed to other employed models.
- The RF also gives a lesser result for the evaluated errors MAE (2.17 MPa), MSE (7.45 MPa), and RMSE (2.73 MPa) when compared with the error value of the DT, MLP, and BR. This lesser value of the error also confirms the high precision of the RF model.

Further studies can also be conducted using other supervised ML algorithms such as boosting regressor, Adaptive neuro-fuzzy inference system, and XGBoost technique to investigate their predictive performance. Furthermore, the experimental approach can also be enhanced to obtain the maximum number of data points to avoid overfitting the data. It is also recommended that the strain model can also be included in the study along with the use of supervised machine learning algorithms to strengthen the overall quality of research work. To compare the results with a database with restricted input parameters, the number of input variables might be expanded. The dimensions of the tested specimens, temperature, and humidity effects can also be considered to investigate the difference in the required outcome.

Author Contributions: K.K.: Funding acquisition, Formal analysis, Investigation, Supervision, Writing—Original draft. A.A.: Conceptualization, Software, Methodology, Validation, Supervision, Formal analysis, Investigation, Writing—Original draft. M.N.A.: Project administration, Resources, Visualization, Writing—Reviewing and editing. W.A.: Data curation, Resources, Formal analysis, Writing—Reviewing and editing. S.N.: Investigation, Validation, Writing—Reviewing and editing. A.M.A.A.: Data acquisition, Methodology, Writing—Reviewing and editing. All authors have read and agreed to the published version of the manuscript.

Funding: The study was supported by the Deanship of Scientific Research, Vice Presidency for Graduate Studies and Scientific Research, at King Faisal University, Al-Ahsa, Saudi Arabia. (Project No. GRANT646).

Institutional Review Board Statement: Not applicable.

Informed Consent Statement: Not applicable.

Data Availability Statement: The data used in this research has been properly cited and reported in the main text.

Acknowledgments: The authors acknowledge the Deanship of Scientific Research, Vice Presidency for Graduate Studies and Scientific Research, at King Faisal University, Al-Ahsa, Saudi Arabia. (Project No. GRANT646). The authors wish to express their gratitude for the financial support that has made this study possible.

Conflicts of Interest: The authors declare no conflict of interest.

References

1. Manan, Z.A.; Nawi, W.N.R.M.; Alwi, S.R.W.; Klemeš, J.J. Advances in Process Integration research for CO_2 emission reduction—A review. *J. Clean. Prod.* **2017**, *167*, 1–13. [CrossRef]
2. Fais, B.; Sabio, N.; Strachan, N.J.A.E. The critical role of the industrial sector in reaching long-term emission reduction, energy efficiency and renewable targets. *Appl. Energy* **2016**, *162*, 699–712. [CrossRef]

3. Rajamma, R.; Ball, R.J.; Tarelho, L.A.; Allen, G.C.; Labrincha, J.A.; Ferreira, V.M.J.J.o.h.m. Characterisation and use of biomass fly ash in cement-based materials. *J. Hazard. Mater.* **2009**, *172*, 1049–1060. [CrossRef] [PubMed]

4. Ahmad, W.; Ahmad, A.; Ostrowski, K.A.; Aslam, F.; Joyklad, P. A scientometric review of waste material utilization in concrete for sustainable construction. *Case Stud. Constr. Mater.* **2021**, *15*, e00683. [CrossRef]

5. Ahmad, W.; Ahmad, A.; Ostrowski, K.A.; Aslam, F.; Joyklad, P.; Zajdel, P. Sustainable approach of using sugarcane bagasse ash in cement-based composites: A systematic review. *Case Stud. Constr. Mater.* **2021**, *15*, e00698. [CrossRef]

6. Cao, M.; Mao, Y.; Khan, M.; Si, W.; Shen, S.J.C.; Materials, B. Different testing methods for assessing the synthetic fiber distribution in cement-based composites. *Constr. Build. Mater.* **2018**, *184*, 128–142. [CrossRef]

7. Arshad, S.; Sharif, M.B.; Irfan-ul-Hassan, M.; Khan, M.; Zhang, J.-L. Efficiency of supplementary cementitious materials and natural fiber on mechanical performance of concrete. *Arab. J. Sci. Eng.* **2020**, *45*, 8577–8589. [CrossRef]

8. Shi, X.; Collins, F.G.; Zhao, X.L.; Wang, Q.J.o.H.M. Mechanical properties and microstructure analysis of fly ash geopolymeric recycled concrete. *J. Hazard. Mater.* **2012**, *237*, 20–29. [CrossRef]

9. Zhang, B.; Ahmad, W.; Ahmad, A.; Aslam, F.; Joyklad, P. A scientometric analysis approach to analyze the present research on recycled aggregate concrete. *J. Build. Eng.* **2022**, *46*, 103679. [CrossRef]

10. Alyousef, R.; Ahmad, W.; Ahmad, A.; Aslam, F.; Joyklad, P.; Alabduljabbar, H. Potential use of recycled plastic and rubber aggregate in cementitious materials for sustainable construction: A review. *J. Clean. Prod.* **2021**, *329*, 129736. [CrossRef]

11. Embong, R.; Kusbiantoro, A.; Shafiq, N.; Nuruddin, M.F. Strength and microstructural properties of fly ash based geopolymer concrete containing high-calcium and water-absorptive aggregate. *J. Clean. Prod.* **2016**, *112*, 816–822. [CrossRef]

12. Yang, H.; Liu, L.; Yang, W.; Liu, H.; Ahmad, W.; Ahmad, A.; Aslam, F.; Joyklad, P. A comprehensive overview of geopolymer composites: A bibliometric analysis and literature review. *Case Stud. Constr. Mater.* **2022**, *16*, e00830. [CrossRef]

13. Li, X.; Qin, D.; Hu, Y.; Ahmad, W.; Ahmad, A.; Aslam, F.; Joyklad, P. A systematic review of waste materials in cement-based composites for construction applications. *J. Build. Eng.* **2021**, *45*, 103447. [CrossRef]

14. Łukowski, P.J.M. Polymer-cement composites containing waste perlite powder. *Materials* **2016**, *9*, 839. [CrossRef]

15. Li, M.; Zhu, X.; Mukherjee, A.; Huang, M.; Achal, V. Biomineralization in metakaolin modified cement mortar to improve its strength with lowered cement content. *J. Hazard. Mater.* **2017**, *329*, 178–184. [CrossRef]

16. Ercikdi, B.; Cihangir, F.; Kesimal, A.; Deveci, H.; Alp, İ. Utilization of industrial waste products as pozzolanic material in cemented paste backfill of high sulphide mill tailings. *J. Hazard. Mater.* **2009**, *168*, 848–856. [CrossRef]

17. Khan, M.; Ali, M. Improvement in concrete behavior with fly ash, silica-fume and coconut fibres. *Constr. Build. Mater.* **2019**, *203*, 174–187. [CrossRef]

18. Khan, M.; Cao, M.; Hussain, A.; Chu, S.H. Effect of silica-fume content on performance of CaCO3 whisker and basalt fiber at matrix interface in cement-based composites. *Constr. Build. Mater.* **2021**, *300*, 124046. [CrossRef]

19. Khan, M.; Rehman, A.; Ali, M. Efficiency of silica-fume content in plain and natural fiber reinforced concrete for concrete road. *Constr. Build. Mater.* **2020**, *244*, 118382. [CrossRef]

20. Xie, C.; Cao, M.; Guan, J.; Liu, Z.; Khan, M. Improvement of boundary effect model in multi-scale hybrid fibers reinforced cementitious composite and prediction of its structural failure behavior. *Compos. Part B Eng.* **2021**, *224*, 109219. [CrossRef]

21. Cao, M.; Khan, M.J.S.C. Effectiveness of multiscale hybrid fiber reinforced cementitious composites under single degree of freedom hydraulic shaking table. *Struct. Concr.* **2021**, *22*, 535–549. [CrossRef]

22. Siddique, R.; Jameel, A.; Singh, M.; Barnat-Hunek, D.; Aït-Mokhtar, A.; Belarbi, R.; Rajor, A.J.C.; Materials, B. Effect of bacteria on strength, permeation characteristics and micro-structure of silica fume concrete. *Constr. Build. Mater.* **2017**, *142*, 92–100. [CrossRef]

23. Hardjito, D.; Rangan, B.V. Development and Properties of Low-Calcium Fly Ash-Based Geopolymer Concrete. 2005. Available online: http://hdl.handle.net/20.500.11937/5594 (accessed on 22 May 2022).

24. Naik, T.R.; Ramme, B.W.; Kraus, R.N.; Siddique, R. Long-Term Performance of High-Volume Fly Ash. *ACI Mater. J.* **2003**, *100*, 150–155.

25. Mohamed, H.A. Effect of fly ash and silica fume on compressive strength of self-compacting concrete under different curing conditions. *Ain Shams Eng. J.* **2011**, *2*, 79–86. [CrossRef]

26. Qaidi, S.M.; Tayeh, B.A.; Isleem, H.F.; de Azevedo, A.R.; Ahmed, H.U.; Emad, W. Sustainable utilization of red mud waste (bauxite residue) and slag for the production of geopolymer composites: A review. *Case Stud. Constr. Mater.* **2022**, *16*, e00994. [CrossRef]

27. Khan, U.A.; Jahanzaib, H.M.; Khan, M.; Ali, M. Improving the tensile energy absorption of high strength natural fiber reinforced concrete with fly-ash for bridge girders. *Key Eng. Mater.* **2018**, *765*, 335–342. [CrossRef]

28. Chindaprasirt, P.; Jaturapitakkul, C.; Sinsiri, T. Effect of fly ash fineness on compressive strength and pore size of blended cement paste. *Cem. Concr. Compos.* **2005**, *27*, 425–428. [CrossRef]

29. Li, G.; Zhao, X. Properties of concrete incorporating fly ash and ground granulated blast-furnace slag. *Cem. Concr. Compos.* **2003**, *25*, 293–299. [CrossRef]

30. Starost, K.; Frijns, E.; Van Laer, J.; Faisal, N.; Egizabal, A.; Elizextea, C.; Blazquez, M.; Nelissen, I.; Njuguna, J. Assessment of nanoparticles release into the environment during drilling of carbon nanotubes/epoxy and carbon nanofibres/epoxy nanocomposites. *J. Hazard. Mater.* **2017**, *340*, 57–66. [CrossRef]

31. Niewiadomski, P.; Hoła, J.; Ćwirzeń, A. Study on properties of self-compacting concrete modified with nanoparticles. *Arch. Civ. Mech. Eng.* **2018**, *18*, 877–886. [CrossRef]

32. Li, G.; Zhou, C.; Ahmad, W.; Usanova, K.I.; Karelina, M.; Mohamed, A.M.; Khallaf, R. Fly Ash Application as Supplementary Cementitious Material: A Review. *Materials* **2022**, *15*, 2664. [CrossRef] [PubMed]

33. Berndt, M.L. Properties of sustainable concrete containing fly ash, slag and recycled concrete aggregate. *Constr. Build. Mater.* **2009**, *23*, 2606–2613. [CrossRef]

34. Celik, K.; Meral, C.; Mancio, M.; Mehta, P.K.; Monteiro, P.J.M. A comparative study of self-consolidating concretes incorporating high-volume natural pozzolan or high-volume fly ash. *Constr. Build. Mater.* **2014**, *67*, 14–19. [CrossRef]

35. Hefni, Y.; Abd El Zaher, Y.; Wahab, M.A. Influence of activation of fly ash on the mechanical properties of concrete. *Constr. Build. Mater.* **2018**, *172*, 728–734. [CrossRef]

36. Ibrahim, R.K.; Hamid, R.; Taha, M.R. Fire resistance of high-volume fly ash mortars with nanosilica addition. *Constr. Build. Mater.* **2012**, *36*, 779–786. [CrossRef]

37. Golewski, G.L. Studies of natural radioactivity of concrete with siliceous fly ash addition. *Cem. Wapno Beton* **2015**, *20*, 106.

38. Deng, S.; Shu, Y.; Li, S.; Tian, G.; Huang, J.; Zhang, F. Chemical forms of the fluorine, chlorine, oxygen and carbon in coal fly ash and their correlations with mercury retention. *J. Hazard. Mater.* **2016**, *301*, 400–406. [CrossRef]

39. Deng, S.; Zhang, C.; Liu, Y.; Cao, Q.; Xu, Y.; Wang, H.; Zhang, F. A full-scale field study on chlorine emission of pulverized coal-fired power plants in China. *Res. Environ. Sci.* **2014**, *27*, 127–133.

40. Wang, S.; Zhang, L.; Li, G.; Wu, Y.; Hao, J.; Pirrone, N.; Sprovieri, F.; Ancora, M.J.A.C. Mercury emission and speciation of coal-fired power plants in China. *Atmos. Chem. Phys.* **2010**, *10*, 1183–1192. [CrossRef]

41. Golewski, G. Improvement of fracture toughness of green concrete as a result of addition of coal fly ash. Characterization of fly ash microstructure. *Mater. Charact.* **2017**, *134*, 335–346. [CrossRef]

42. Golewski, G.; Sadowski, T.J.C.; Materials, B. An analysis of shear fracture toughness KIIc and microstructure in concretes containing fly-ash. *Constr. Build. Mater.* **2014**, *51*, 207–214. [CrossRef]

43. Sadowski, T.; Golewski, G. A failure analysis of concrete composites incorporating fly ash during torsional loading. *Compos. Struct.* **2018**, *183*, 527–535. [CrossRef]

44. Golewski, G.L.J.C.S. Effect of curing time on the fracture toughness of fly ash concrete composites. *Compos. Struct.* **2018**, *185*, 105–112. [CrossRef]

45. Chindaprasirt, P.; Rukzon, S.J.C.; Materials, B. Strength, porosity and corrosion resistance of ternary blend Portland cement, rice husk ash and fly ash mortar. *Constr. Build. Mater.* **2008**, *22*, 1601–1606. [CrossRef]

46. Nadeem, A.; Memon, S.A.; Lo, T.Y. The performance of fly ash and metakaolin concrete at elevated temperatures. *Constr. Build. Mater.* **2014**, *62*, 67–76. [CrossRef]

47. Lieberman, R.N.; Knop, Y.; Querol, X.; Moreno, N.; Muñoz-Quirós, C.; Mastai, Y.; Anker, Y.; Cohen, H. Environmental impact and potential use of coal fly ash and sub-economical quarry fine aggregates in concrete. *J. Hazard. Mater.* **2018**, *344*, 1043–1056. [CrossRef]

48. Khodair, Y.; Bommareddy, B. Self-consolidating concrete using recycled concrete aggregate and high volume of fly ash, and slag. *Constr. Build. Mater.* **2017**, *153*, 307–316. [CrossRef]

49. Mardani-Aghabaglou, A.; Sezer, G.İ.; Ramyar, K. Comparison of fly ash, silica fume and metakaolin from mechanical properties and durability performance of mortar mixtures view point. *Constr. Build. Mater.* **2014**, *70*, 17–25. [CrossRef]

50. Yao, Z.T.; Ji, X.S.; Sarker, P.K.; Tang, J.H.; Ge, L.Q.; Xia, M.S.; Xi, Y.Q. A comprehensive review on the applications of coal fly ash. *Earth-Sci. Rev.* **2015**, *141*, 105–121. [CrossRef]

51. Panda, L.; Dash, S. Characterization and utilization of coal fly ash: A review. *Emerg. Mater. Res.* **2020**, *9*, 921–934. [CrossRef]

52. Nedunuri, S.S.S.A.; Sertse, S.G.; Muhammad, S. Microstructural study of Portland cement partially replaced with fly ash, ground granulated blast furnace slag and silica fume as determined by pozzolanic activity. *Constr. Build. Mater.* **2020**, *238*, 117561. [CrossRef]

53. Belviso, C. State-of-the-art applications of fly ash from coal and biomass: A focus on zeolite synthesis processes and issues. *Prog. Energy Combust. Sci.* **2018**, *65*, 109–135. [CrossRef]

54. Almutairi, A.L.; Tayeh, B.A.; Adesina, A.; Isleem, H.F.; Zeyad, A.M. Potential applications of geopolymer concrete in construction: A review. *Case Stud. Constr. Mater.* **2021**, *15*, e00733. [CrossRef]

55. Ahmad, W.; Ahmad, A.; Ostrowski, K.A.; Aslam, F.; Joyklad, P.; Zajdel, P. Application of Advanced Machine Learning Approaches to Predict the Compressive Strength of Concrete Containing Supplementary Cementitious Materials. *Materials* **2021**, *14*, 5762. [CrossRef]

56. Xu, Y.; Ahmad, W.; Ahmad, A.; Ostrowski, K.A.; Dudek, M.; Aslam, F.; Joyklad, P. Computation of High-Performance Concrete Compressive Strength Using Standalone and Ensembled Machine Learning Techniques. *Materials* **2021**, *14*, 7034. [CrossRef]

57. Ahmad, A.; Ahmad, W.; Chaiyasarn, K.; Ostrowski, K.A.; Aslam, F.; Zajdel, P.; Joyklad, P. Prediction of geopolymer concrete compressive strength using novel machine learning algorithms. *Polymers* **2021**, *13*, 3389. [CrossRef]

58. Wang, Q.; Ahmad, W.; Ahmad, A.; Aslam, F.; Mohamed, A.; Vatin, N.I. Application of Soft Computing Techniques to Predict the Strength of Geopolymer Composites. *Polymers* **2022**, *14*, 1074. [CrossRef]

59. Isleem, H.F.; Tayeh, B.A.; Alaloul, W.S.; Musarat, M.A.; Raza, A.J.M. Artificial Neural Network (ANN) and Finite Element (FEM) Models for GFRP-Reinforced Concrete Columns under Axial Compression. *Materials* **2021**, *14*, 7172. [CrossRef]

60. Isleem, H.F.; Peng, F.; Tayeh, B.A. Confinement model for LRS FRP-confined concrete using conventional regression and artificial neural network techniques. *Compos. Struct.* **2022**, *279*, 114779. [CrossRef]

61. Isleem, H.F.; Wang, Z.; Wang, D. A new model for reinforced concrete columns strengthened with fibre-reinforced polymer. *Proc. Inst. Civ. Eng.-Struct. Build.* **2020**, *173*, 602–622. [CrossRef]

62. Shariati, M.; Mafipour, M.S.; Mehrabi, P.; Ahmadi, M.; Wakil, K.; Trung, N.T.; Toghroli, A.J.S.S.; Systems, A.I.J. Prediction of concrete strength in presence of furnace slag and fly ash using Hybrid ANN-GA (Artificial Neural Network-Genetic Algorithm). *Smart Struct. Syst. Int. J.* **2020**, *25*, 183–195.

63. Han, Q.; Gui, C.; Xu, J.; Lacidogna, G.J.C.; Materials, B. A generalized method to predict the compressive strength of high-performance concrete by improved random forest algorithm. *Constr. Build. Mater.* **2019**, *226*, 734–742. [CrossRef]

64. Chaabene, W.B.; Flah, M.; Nehdi, M.L.J.C.; Materials, B. Machine learning prediction of mechanical properties of concrete: Critical review. *Constr. Build. Mater.* **2020**, *260*, 119889. [CrossRef]

65. Concrete, A.I.C.C.o.; Aggregates, C. *Standard Test Method for Compressive Strength of Cylindrical Concrete Specimens*; ASTM International: West Conshohocken, PA, USA, 2014.

66. Song, H.; Ahmad, A.; Farooq, F.; Ostrowski, K.A.; Maślak, M.; Czarnecki, S.; Aslam, F.J.C.; Materials, B. Predicting the compressive strength of concrete with fly ash admixture using machine learning algorithms. *Constr. Build. Mater.* **2021**, *308*, 125021. [CrossRef]

67. Song, Y.; Zhao, J.; Ostrowski, K.A.; Javed, M.F.; Ahmad, A.; Khan, M.I.; Aslam, F.; Kinasz, R. Prediction of Compressive Strength of Fly-Ash-Based Concrete Using Ensemble and Non-Ensemble Supervised Machine-Learning Approaches. *Appl. Sci.* **2022**, *12*, 361. [CrossRef]

68. Chun, P.-j.; Ujike, I.; Mishima, K.; Kusumoto, M.; Okazaki, S.J.C.; Materials, B. Random forest-based evaluation technique for internal damage in reinforced concrete featuring multiple nondestructive testing results. *Constr. Build. Mater.* **2020**, *253*, 119238. [CrossRef]

Review

Fly Ash Application as Supplementary Cementitious Material: A Review

Guanlei Li [1], Chengke Zhou [1,*], Waqas Ahmad [2,*], Kseniia Iurevna Usanova [3], Maria Karelina [4], Abdeliazim Mustafa Mohamed [5,6] and Rana Khallaf [7]

1   Luohe Vocational Technology College, Luohe 462300, China; dayanta100@sina.com
2   Department of Civil Engineering, COMSATS University Islamabad, Abbottabad 22060, Pakistan
3   Peter the Great St. Petersburg Polytechnic University, 195291 St. Petersburg, Russia; usanova_kyu@spbstu.ru
4   Department of Machinery Parts and Theory of Mechanisms, Moscow Automobile and Road Construction University, 125319 Moscow, Russia; karelinamu@mail.ru
5   Department of Civil Engineering, College of Engineering in Al-Kharj, Prince Sattam bin Abdulaziz University, Al-Kharj 11942, Saudi Arabia; a.bilal@psau.edu.sa
6   Building & Construction Technology Department, Bayan College of Science and Technology, Khartoum 11115, Sudan
7   Structural Engineering and Construction Management, Faculty of Engineering and Technology, Future University in Egypt, New Cairo 11845, Egypt; rana.khallaf@fue.edu.eg
*   Correspondence: xinqiba001@sina.com (C.Z.); waqasahmad@cuiatd.edu.pk (W.A.)

Abstract: This study aimed to expand the knowledge on the application of the most common industrial byproduct, i.e., fly ash, as a supplementary cementitious material. The characteristics of cement-based composites containing fly ash as supplementary cementitious material were discussed. This research evaluated the mechanical, durability, and microstructural properties of FA-based concrete. Additionally, the various factors affecting the aforementioned properties are discussed, as well as the limitations associated with the use of FA in concrete. The addition of fly ash as supplementary cementitious material has a favorable impact on the material characteristics along with the environmental benefits; however, there is an optimum level of its inclusion (up to 20%) beyond which FA has a deleterious influence on the composite's performance. The evaluation of the literature identified potential solutions to the constraints and directed future research toward the application of FA in higher amounts. The delayed early strength development is one of the key downsides of FA use in cementitious composites. This can be overcome by chemical activation (alkali/sulphate) and the addition of nanomaterials, allowing for high-volume use of FA. By utilizing FA as an SCM, sustainable development may promote by lowering CO_2 emissions, conserving natural resources, managing waste effectively, reducing environmental pollution, and low hydration heat.

Keywords: cementitious composites; fly ash; supplementary cementitious material; mechanical properties; durability; microstructure

Citation: Li, G.; Zhou, C.; Ahmad, W.; Usanova, K.I.; Karelina, M.; Mohamed, A.M.; Khallaf, R. Fly Ash Application as Supplementary Cementitious Material: A Review. *Materials* **2022**, *15*, 2664. https://doi.org/10.3390/ma15072664

Academic Editor: Jorge Otero

Received: 1 March 2022
Accepted: 25 March 2022
Published: 5 April 2022

Publisher's Note: MDPI stays neutral with regard to jurisdictional claims in published maps and institutional affiliations.

Copyright: © 2022 by the authors. Licensee MDPI, Basel, Switzerland. This article is an open access article distributed under the terms and conditions of the Creative Commons Attribution (CC BY) license (https://creativecommons.org/licenses/by/4.0/).

1. Introduction

Industrial solid waste makes up a sizable portion of human-generated wastes, which come in a vast range of forms and are highly complex in nature [1–3]. Heavy metals are found in the majority of industrial wastes, such as red powder, metal cleaning, and radioactive waste [4–6]. Inappropriate handling of solid industrial waste can lead to leachate penetrating soil and groundwater, causing environmental irreversibility and endangering human health [7–9]; moreover, global warming and climate change are the two most serious environmental problems caused by CO_2 emissions [10–12]. The construction industry substantially impacts the environment, accounting for a significant portion of CO_2 emissions [13–15]. Each ton of cement produced emits about 0.8 tons of CO_2 [16–18], and cement production is increasing globally [19] due to the increasing demand for concrete [20–22]. Researchers worldwide are constantly on the lookout for new materials that can be utilized in place of, or in addition to, cement [23]. Since the last decade, the application of supplementary cementitious materials (SCMs) such as silica fume, fly ash (FA), slags, etc., as a cement replacement has been emphasized [24–26]. SCMs hydrate cement hydraulically or pozzolanically in pore solution [27–29]. Thus, utilizing industrial solid waste in construction as SCMs is an effective approach for eco-friendly construction [30]; it could reduce cement demand, reduce CO_2 emission, and solve waste management problems. From the various kinds of industrial byproducts that can be used as SCMs, the most common is FA.

FA is a byproduct of coal combustion that is accumulated at the top of boilers, particularly in coal-fired power plants [31,32]. The omitted mineral particles or mineral materials within the coal liquefy, evaporate, consolidate, or agglomerate during/after burning. By rapidly cooling in the post-combustion portion, sphere-shaped, amorphous FA grains are created due to surface tension force. When the entrapped volatile matter reaches a high temperature, it expands inside, forming a hollow cenosphere. Some residues may crystallize, while others may become glassy, reliant on the composition of residues and the heating/cooling circumstances [33]. FA is considered an SCM that is used in place of cement in cementitious materials [34,35]. FA increases workability, decreases the hydration heat, and thermal cracking in cementitious materials at initial ages, and improves the mechanical and durability characteristics of cementitious composites, mostly at later ages [36,37]. The application of FA is also being investigated in the manufacture of geopolymer concrete [38–40]; however, this study is limited to reviewing their utilization in cementitious composites.

The amorphous silica in FA reacts with the calcium hydroxide to form calcium-silicate-hydrate (CSH) [41]. FA's pozzolanic reaction boosts its utility not only in concrete but also in a variety of other construction applications [42]. Due to the pozzolanic process, the strength gain lasts significantly longer than with normal concrete [43]. Additionally, FA increases the workability of concrete by reducing bleeding [44]. FA has been shown to improve the long-term compressive strength (CS) of normal and recycled aggregate concrete [45]. Microstructural examination of FA samples following early curing reveals an abundance of un-hydrated spherical FA particles. Despite this, after a year of curing, the microstructure of FA samples appeared to be very compact, with no evidence of dehydrated FA particles [46]. FA requires a longer period of time to hydrate. As a result, during the initial phases of curing, low CS has been found. The strength development of FA, on the other hand, is dependent on its chemical and physical characteristics. It has been observed that FA with a fine particle size distribution had a better CS than FA with a coarse particle size distribution [47]. The binder causes the concrete to shrink during the hydration process, and excessive shrinkage can result in severe cracks in the concrete structure. FA is beneficial for shrinkage mitigation [48]. It has been noted that the use of large volume FA in concrete, specifically 50% replacement of cement with FA, resulted in a 30% reduction in shrinkage when compared to ordinary concrete [49].

The use of FA in low (<30%) and high (>30%) volume in concrete is a pioneering move that has already altered the worldwide concrete industry's approach. Mostly, FA is disposed of in landfills, which has had severe implications, the majority of its portion has been successfully utilized in the concrete industry for the last three to four decades. Additionally, the frequent generation of FA has compelled government officials and experts to develop a more dependable method of consuming it, while the application of FA to the development of sustainable concretes will almost certainly alter the future building industry [50–52]. Even though FA has been extensively investigated over the last few decades, experts have discovered some inconsistent results regarding the mechanical and durability characteristics of concrete. The chemical and physical features of FA have a major effect on the mechanical and durability characteristics of concrete. Additionally, the characteristics of FA vary depending on the source. This study focused on reviewing the mechanical, durability, and microstructural characteristics of FA-based concrete; the various factors influencing the aforesaid properties are highlighted, and limitations associated with the use of FA in concrete are described. Based on the review of the literature, possible solutions to the limitations are provided, and future research is directed to the application of FA in larger quantities.

2. Properties of Fly Ash

2.1. Physical and Chemical Properties

FA is a primary solid waste generated by coal-fired energy plants, and these plants are looking for economically viable ways to dispose of it. FA particles are generally spherical, solid/hollow in nature, mainly glassy (amorphous), with particle sizes varying from <1 μm to 150 μm [53–55]. The scanning electron microscopy (SEM) micrographs of the FA are shown in Figure 1. FA has a specific gravity of 2.1 to 3.0 [56] and a specific surface area of 170 to 1000 m^2/kg [57]. FA can range in color from tan to grey to black, based on the quantity of unburned carbon present [35,58]. Besides the environmental advantages of waste disposal and CO_2 reduction [59,60], FA increases workability [61], decreases the hydration heat and thermal cracking in concrete at the initial stage [62], and improves the performance of cementitious materials, especially at later stages [36,63]. Regardless of the advantages of FA, 100% application of FA is not possible due to a variety of reported limitations [64]. The ASTM categorizes FA into two categories: "C" and "F" [65]. FA classified as "Class F" is mostly generated by burning anthracite or bituminous coal containing SiO_2, Al_2O_3, and Fe_2O_3 concentrations greater than 70%. While "Class C" FA is generated by burning lignite or sub-bituminous coal that consists of 50% to 70% of the aforementioned chemicals [66]. Class F is a typical pozzolan and composed of silicate glass that has been modified with aluminum and iron [67]. CaO amount is less than 10% in "Class F" FA [68]; thus, to form CSH through pozzolanic reaction, $Ca(OH)_2$ formed during cement hydration is required; therefore, the chemical composition of FA performs a major part in determining its performance in cementitious composites [69]. The range of element oxide concentrations found in "Class F" and "Class C" FA has been listed in Table 1. As can be seen, there is a considerable difference in the element oxides contained within a single kind of FA, which might be ascribed to differences in source, processing conditions, and so on. There is a crucial need to utilize more FA in the construction materials due to the increase in FA production globally.

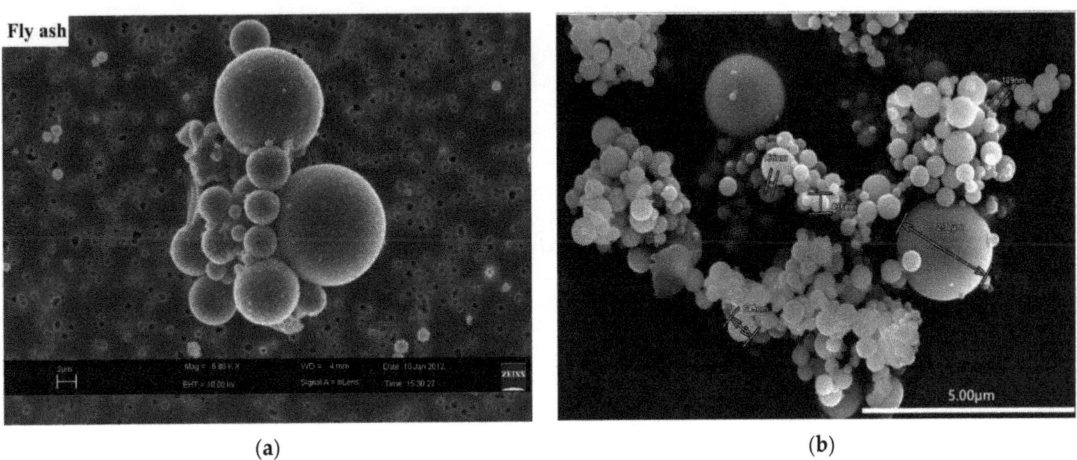

Figure 1. Micrograph of fly ash: (**a**) [70]; (**b**) [71].

Table 1. Element oxides range in fly ash types.

Element Oxides	Range of Element Oxides (%)	
	Class F	Class C
SiO_2	37.0–62.1	11.8–46.4
Al_2O_3	16.6–35.6	2.6–20.5
Fe_2O_3	2.6–21.2	1.4–15.6
CaO	0.5–14.0	15.1–54.8
MgO	0.3–5.2	0.1–6.7
SO_3	0.02–4.7	1.4–12.9
Na_2O	0.1–3.6	0.2–2.8
K_2O	0.1–4.1	0.3–9.3
TiO_2	0.5–2.6	0.6–1.0
P_2O_5	0.1–1.7	0.2–0.4
MnO	0.03–0.1	0.3–0.2
LOI	0.3–32.8	0.3–11.7

2.2. Cementing Efficiency and Pozzolanic Properties

Smith [72] proposed the notion of cementing factor (k) in order to develop a reasonable approach for incorporating FA into cement/concrete. Cementing efficiency can be employed to ascertain the overall quality, durability, and performance of composites. In general, FA has a low cementing efficiency at initial stages and acts more such as filler, but the pozzolanic feature turns out to be efficient at later ages, causing a significant increase in strength [73–75]. This clearly indicates that the pozzolanic reaction improves the cementing efficiency of FA with age. According to Smith [72], "the FA mass might be considered similar to the cement mass in terms of CS development." In other words, "k" is a factor that accounts for the variation among the contribution of cement to the development of a particular property and the contribution of mineral admixtures. CS tests are frequently used to measure this cementing efficiency due to their simplicity and repeatability. The "k" value of FA is determined by a variety of its intrinsic qualities, including physical properties such as particle size, distribution, and shape, as well as chemical composition [76]. Additionally, it was also reported that the "k" factor is dependent on other parameters such as the curing time, the concrete strength, and the FA type [77]; therefore, it was also discovered that the "k" value is dependent on external factors such as the water/cement ratio (w/c). They stated that for conventional FA, "k" is a function of the w/c and that the cementing efficiency of FA tends to decline as the w/c increases [78]. On the contrary, Smith [72] asserted that it is unaffected by the w/c.

Apart from cementing efficiency, pozzolanicity is another critical term in the context of FA concrete. Among the numerous favorable benefits of FA in cement/concrete, the pozzolanic effect is believed to be the most important [66]. The pozzolanic reaction is mostly dependent on the Al_2O_3 and SiO_2 content of FA and is stimulated by the Portlandite generated during cement hydration to generate a more hydrated gel. This gel plugs the capillary pores in the matrix, increasing its strength [79]. As a result, FA's reactivity is greatly dependent on its chemical properties; however, all pozzolanic materials are made of aluminosilicate glass that combines with Ca(OH)2 formed during hydration of cement to yield hydration products [80].

3. Properties of Composites Containing Fly Ash

3.1. Workability

FA has plasticizing properties that improve the workability of the composites [81]. Lee et al. [82] reported the subsequent factors as possible reasons for FA's plasticizing effect. Firstly, increased composite volume due to FA's lower density than cement. Secondly, FA decreases the flocculation of cement grains because of the dilution effect. Thirdly, the slower reaction rate of FA reduces hydration product growth at the initial time. Besides these causes, the spherical shape of FA grains facilitates the movement of nearby particles by the ball bearing effect, particularly at high replacement levels. Thus, FA can be a more cost-effective method with a low environmental effect to increase the workability than chemical superplasticizer [83]. Bentz et al. [84] also validated the positive impact of FA on the workability of the mix. The type of FA used also has a substantial impact on the workability of composites. According to Ponikiewski and Golaszewski [85], high calcium FA has a detrimental effect on workability, which adversely influences the mechanical strength of composites. The fresh state characteristics of a mix mostly depend on the flowability of cement paste, which is affected by a variety of aspects such as water-binder ratio (w/b), type, and quantity of SCM [86]. Conversely, some researchers reported a drop in the workability of mixes with FA addition at higher amounts [87]. The decrease in workability might be the high-water demand due to the smaller size and larger surface area of FA.

Lee et al. [82] highlighted the following aspects as possible explanations for FA's plasticizing action. To begin, increased paste volume due to FA's lower density than cement. Second, FA lowers the flocculation of cement particles due to the diluting effect. Thirdly, because of the FA's slower reaction rate as a result of the lowered development of hydration products at an initial age. In addition, the spherical shape of FA grains facilitates the movement of nearby fragments via the ball bearing effect, particularly at high replacement levels; thus, FA can be a more cost-effective method of increasing flowability than chemical superplasticizers [83]. Bentz et al. [84] also validated the favorable effect of FA on the fluidity of the mix. As previously observed [82], replacing cement with FA reduces yield stress due to the decreased density of FA and hence decreases the number of flocculated cement grain to cement particle contacts. The FA type utilized has also a considerable effect on the fresh properties. According to a study, high calcium FA has a detrimental effect on workability, which in turn has a negative effect on strength and durability [85]. The fresh state performance of a concrete is primarily dependent on the flowability of the cement paste, which is influenced by a variety of elements such as w/b, type, and dose of SCM [88].

Apart from these parameters, prior research has also demonstrated that the packing density of the cement-based composites also has a significant role in determining the cement paste's flowability, particularly at low w/b ratios [82,89]. Essentially, increased packing density results in decreased water demand, which results in increased water being released (after voids filling) to cover the solid fragments and lubricate the cement paste [90]; however, a higher specific surface area significantly increases the amount of solid surface area that can be covered with water [91–93]. These simultaneous actions of tiny fillers can enhance packing density while decreasing the quantity of surplus water per surface area; thus, to achieve a balance between the desired increase in packing density and the unwanted increase in surface area, a filler that is finer than cement but coarser than nano silica/silica fume is required [94]. This indicates that the fineness of the FA impacts the end material's properties. As a result, it has been concluded that some studies show lower water required for concrete workability due to the refined pores and spherical morphology of FA; others report a higher water requirement due to its increased surface area. This well-documented incompatibility between water demand and FA usage must be rectified.

3.2. Compressive Strength

Numerous tests are used to determine the concrete performance, but CS is often regarded as the most critical. CS tests provide a clear indication of the varied properties of concrete. The literature established that CS is related to a variety of mechanical and durability attributes directly or indirectly [95]. In other words, CS and the quality of concrete are inextricably related. FA's physical properties, particularly its size and shape, have a substantial effect on the performance of cement-based materials. Additionally, the chemical composition has been considered a base to ascertain the appropriateness of FA for use as SCM [96]. Thus, the hydration process of the FA-cement mix is strongly affected by the intrinsic characteristics of FA, for instance, crystalline structure, chemical and physical properties [97], as well as external factors such as w/b, replacement ratio, and curing temperature. FA fineness is a major factor [98] in controlling the appropriateness of FA in cementitious composites, as the FA grain size has a substantial impact on the performance of composites [99]. The packing and nucleation effects on the cement hydration are highly reliant on the particle size of the FA used [100]. Chindaprasirt et al. [100–102] conducted a thorough investigation to examine the effect of the fineness of FA on the composites' properties; they reported that using finer FA resulted in an increase in CS. It was discovered that coarse FA is less reactive and needs extra water, producing a more porous mortar. The detrimental effects of coarser FA are described as a cause of decreased strength. Numerous findings have indicated that the application of FA impairs the early-age strength development of composites [82,103–105]; however, FA generally enhances the strength and durability of composites over time, as it consumes the $Ca(OH)_2$ produced during the hydration of cement and makes secondary hydrates, for example, CSH [106].

The quantity of FA used as cement replacement in composites also affects their properties. The 28-days CS results of specimens at various replacement levels of FA have been provided in Table 2. In addition, the influence of FA replacement ratio, based on past studies, on 28-days CS of composites compared to the reference samples without FA has been shown in Figure 2. Barbuta et al. [107] observed a decrease in CS with the use of FA, and a higher quantity of FA as cement replacement resulted in greater loss of CS. The samples without fibers showed a decrease in CS by 11.3%, 30.4%, 24.8%, 33.7%, and 59.7%, with FA content of 10%, 15%, 20%, 30%, and 40%, respectively, related to the controlled sample. A comparable pattern was also noticed with the samples containing fibers. Gencel et al. [108] assessed the impact of FA on the properties of composites using 10%, 20%, and 30% FA in place of cement. A reduction in CS was observed with the use of FA, compared to the reference mix, also shown in the figure. The reduction in CS was more at higher FA contents. Huang et al. [109] studied the impact of two kinds of FA depending on loss on ignition (LOI) amount, i.e., low LOI (4.6%) and high LOI (7.8%) FA. The outcomes discovered that utilizing low LOI FA at lower proportions enhanced the CS. The maximum increase in CS was examined at 40% content of FA having 16.8% higher CS than that of the reference sample; however, at increased proportions of FA, the CS reduced, which may be ascribed to the finer grain size in low LOI FA, which made the microstructure denser and more compact. The CS reduced with the addition of high LOI FA was because of greater particle size and lower pozzolanic activity; it was also reported that the CS of composites containing higher contents of FA was improved at a later age (1 year) compared to the controlled sample because of the slow pozzolanic reaction.

The effect of FA addition has been investigated in self-compacting concrete (SCC). For example, substituting 35% FA for cement results in a 10% reduction, but substituting 55% FA leads to a 24% reduction compared to the control SCC mix [110]. Similarly, a reduction of approximately 46% and 35% have been seen in containing 50% FA mix, after 7 and 90 days of curing, respectively, when compared to a control SCC mix; moreover, at a 70% FA incorporation level in SCC, a severe reduction of approximately 63% and 47% was seen after 7 and 90 days of curing, respectively [111,112]. The presence of cement additives has a considerable effect on the CS of FA-based SCC. The addition of cement additives improves the performance of FA-based SCC mixtures at both low and high curing temperatures. Silica fume, metakaolin, and limestone filler have all been used previously to increase the CS of SCC mixes. At 90 days, a reduction of nearly 29%, 42%, and 15% was seen for a 50% level of FA with limestone filler (15%), metakaolin (20%), and silica fume (10%) in SCC, respectively [111,113,114].

Table 2. Compressive strength (28-days) of composites containing fly ash.

Reference	Replacement Ratio (%)	Compressive Strength (MPa)
Barbuta et al. [107], without fibers	0	33.4
	10	29.7
	15	23.3
	20	25.2
	30	22.2
	40	13.5
Barbuta et al. [107], with 0.25% and 30 mm long fibers	0	33.4
	10	31.8
	15	27.4
	20	25.4
	30	19.6
	40	14.4
Barbuta et al. [107], with 0.25% and 50 mm long fibers	0	33.4
	10	29.8
	15	24.1
	20	27.2
	30	20.9
	40	11.3
Gencel et al. [108], with 0% FSA	0	52.2
	10	44.7
	20	36.8
	30	29.6
Gencel et al. [108], with 25% FSA	0	50.7
	10	45.4
	20	37.8
	30	31.3
Gencel et al. [108], with 50% FSA	0	53.9
	10	45.8
	20	36.6
	30	29.6
Gencel et al. [108], with 75% FSA	0	55.6
	10	46.4
	20	37.3
	30	30.3
Paliwal and Maru [115]	0	26.4
	5	27.8
	10	29.4
	15	28.2
	20	27.5
Huang et al. [109], 24 MPa concrete and low LOI fly ash	0	25.0
	20	25.4
	40	25.6
	60	23.5
	80	20.9
Huang et al. [109], 35 MPa concrete and low LOI fly ash	0	34.5
	20	36.5
	40	40.3
	60	34.5
	80	30.0
Huang et al. [109], 35 MPa concrete and high LOI fly ash	0	34.5
	20	34.9
	40	34.1
	60	30.5
	80	25.2

FSA: ferrochromium slag aggregate, LOI: loss on ignition.

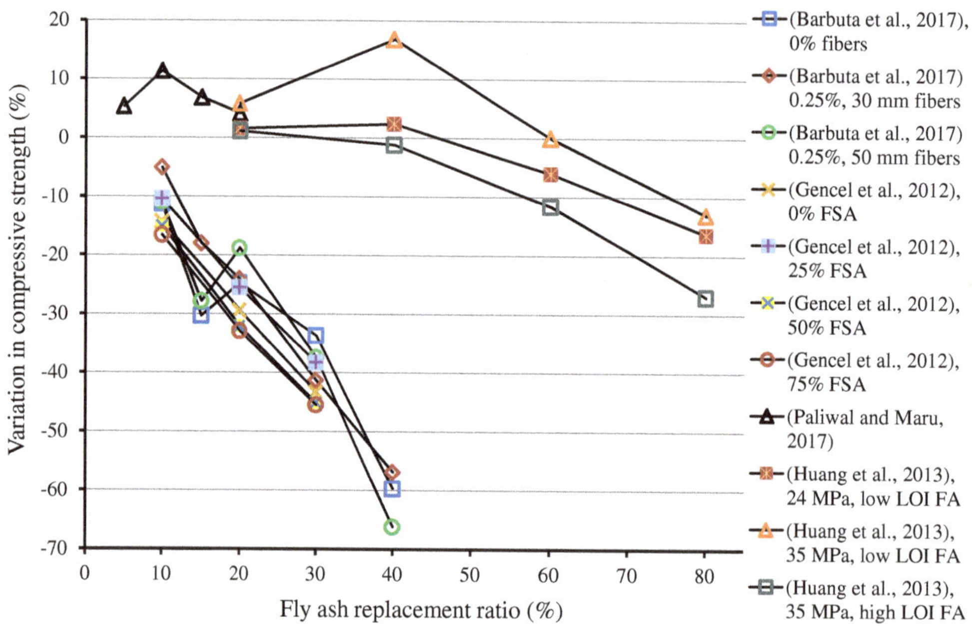

Figure 2. Influence of fly ash as cement replacement on 28-days compressive strength of composites. FSA: ferrochromium slag aggregate [107–109,115].

3.3. Split-Tensile Strength

Another essential mechanical characteristic of concrete is its tensile strength, which has a significant effect on the extent and size of cracking in concrete structures. Because concrete is weak in tension, it is critical to do a pre-evaluation of their split-tensile strength (STS) [116,117]. The use of FA in cementitious composites has a detrimental effect on STS. The 28-days STS results of composites containing FA as SCM are displayed in Table 3. Figure 3 is generated on the data acquired from the literature depicting the variation in 28-days STS due to the replacement of cement by FA. Mostly, a reduction in STS is observed, especially at higher replacement ratios. From the experimental data performed by Barbuta et al. [107], the samples without fibers showed a decrease in STS by 8.1%, 8.1%, 48.3%, 29.6%, and 48.3% with FA content of 10%, 15%, 20%, 30%, and 40%, respectively, as compared to the control sample. The sample containing fibers (0.25% and 50 mm long) and 10% FA showed 12.8% higher STS when compared to the control sample; however, with the further addition of FA, STS was reduced. Gencel et al. [108] studied the combined effect of FA as SCM and ferrochromium slag as an aggregate replacement on STS of composites. They also reported decreasing trend with the addition of FA. The STS of specimens without ferrochromium slag was reduced by 9.7%, 20.7%, and 30.2%, with FA content of 10%, 20%, and 30%, respectively, compared to the sample without FA. A similar pattern of decreasing STS with FA addition was also noted in specimens containing ferrochromium slag as aggregate replacement.

Table 3. Split-tensile strength (28-days) of composites containing fly ash.

Reference	Fly Ash Replacement Ratio (%)	Split-Tensile Strength (MPa)
Barbuta et al. [107], without fibers	0	1.72
	10	1.58
	15	1.58
	20	0.89
	30	1.21
	40	0.89
Barbuta et al. [107], with 0.25% and 30 mm long fibers	0	1.72
	10	1.51
	15	1.37
	20	1.05
	30	1.71
	40	1.02
Barbuta et al. [107], with 0.25% and 50 mm long fibers	0	1.72
	10	1.94
	15	1.45
	20	0.87
	30	1.82
	40	0.85
Gencel et al. [108], with 0% FSA	0	5.20
	10	4.70
	20	4.12
	30	3.63
Gencel et al. [108], with 25% FSA	0	5.31
	10	4.74
	20	4.17
	30	3.66
Gencel et al. [108], with 50% FSA	0	5.24
	10	4.78
	20	4.19
	30	3.78
Gencel et al. [108], with 75% FSA	0	5.30
	10	4.83
	20	4.22
	30	3.70

FSA: ferrochromium slag aggregate.

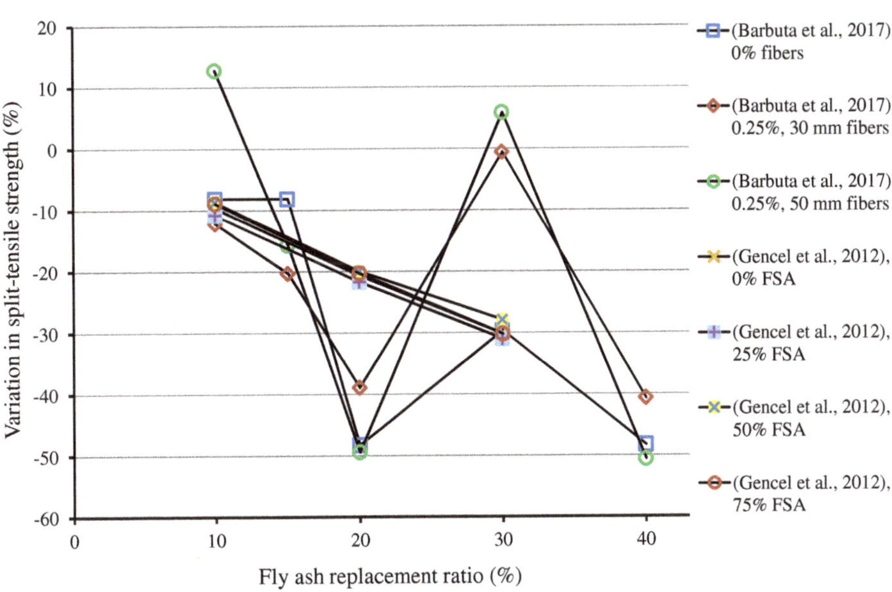

Figure 3. Influence of fly ash as cement replacement on 28-days split-tensile strength of composites. FSA: ferrochromium slag aggregate [107,108].

STS was increased at higher curing ages in comparison to lower curing ages, as examined in most previous findings. According to published reports, FA interacts with calcium ions from $Ca(OH)_2$ to CSH, the binder phase. Due to the lack of CSH and $Ca(OH)_2$ in FA-containing concrete, it is unable to build early age strength [118,119]; moreover, the addition of FA reduces the STS of SCC mixtures due to its intrinsic propensity to reduce water [112,120]; however, in FA-based mixes, a considerable increase in STS has been seen with an increase in curing time, despite the presence of minor decrements. Due to the extensive research conducted to date on the effect of curing on FA-based mixes, similar improvements have been noticed in various studies [121].

3.4. Flexural Strength

The review of the past studies revealed that using FA at lower replacement levels can improve the flexural strength (FS) of composites, as shown in Table 4. Figure 4 is generated, indicating the percentage variation in 28-days FS of specimens at various replacement levels of FA. The improvement of 20.33% was observed in the FS when 10% FA was used SCM, while further increase in FA content decreased the FS compared with the reference specimen [107]. The results of Barbuta et al. [107] of specimens with 30 mm long fibers exhibited improvement in FS of 10.4%, 0%, 18.7%, and 8.2% with FA content of 10%, 15%, 20%, and 30%, respectively. While FS reduced by 8.2% at 40% replacement of FA. The specimens containing 50 mm long fibers showed 10.4%, 14.3%, 36.8%, 12.6%, and 7.7% increase in FS when 10%, 15%, 20%, 30%, and 40% cement was replaced by FA, respectively. Hence, it resulted that using a higher amount of FA has a negative influence on FS [107,108], as shown in Figure 4. Paliwal and Maru [115] noted maximum FS at 10% FA content as cement replacement. It can be concluded that the size, type, chemical composition, and content of FA used in cementitious composites have distinct effects on their mechanical properties. The finer particle size improves, while coarser particle size reduces the strength of composites [109]. Additionally, lower content of FA improves while higher FA content reduces the strength of composites [107–109]; hence, finer FA and a lower replacement ratio are preferable.

Table 4. Flexural strength (28-days) of composites containing fly ash.

	Fly Ash	
Reference	**Replacement Ratio (%)**	**Flexural Strength (MPa)**
	0	1.82
	10	2.19
Barbuta et al. [107],	15	1.62
without fibers	20	1.53
	30	1.62
	40	1.04
	0	1.82
	10	2.01
Barbuta et al. [107], with	15	1.82
0.25% and 30 mm long fibers	20	2.16
	30	1.97
	40	1.67
	0	1.82
	10	2.01
Barbuta et al. [107], with	15	2.08
0.25% and 50 mm long fibers	20	2.49
	30	2.05
	40	1.96

Table 4. *Cont.*

Reference	Fly Ash	
	Replacement Ratio (%)	Flexural Strength (MPa)
Paliwal and Maru [115]	0	3.48
	5	3.88
	10	4.44
	15	4.14
	20	3.62
Huang et al. [109], 35 MPa concrete and low LOI fly ash	0	5.1
	20	5.3
	40	5.2
	60	5
	80	3.7
Huang et al. [109], 35 MPa concrete and high LOI fly ash	0	5.1
	20	5
	40	5.1
	60	4.5
	80	3.2

LOI: loss on ignition.

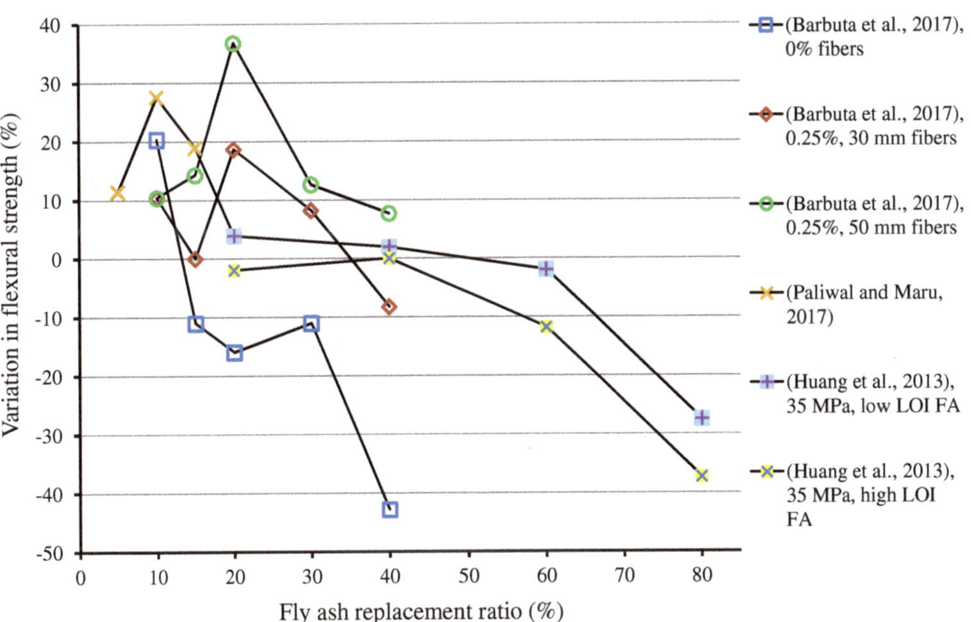

Figure 4. Influence of fly ash as cement replacement on 28-days flexural strength of composites [107,109,115].

3.5. Durability

3.5.1. Chloride Penetration

The addition of FA also enhances the durability performance of cementitious composites. Saha [122] investigated the durability properties of concrete containing FA at various replacement levels and curing ages of 28 and 90 days. The results of the chloride ion penetration test revealed a decrease in penetration depth with FA addition (see Figure 5). At 28-days of curing, the chloride ion penetration depth reduced by around 18%, 39%, 52%, and 61% at FA content of 10%, 20%, 30%, and 40%, respectively. After

180 days, chloride penetration decreased marginally for all samples. The incorporation of 10%, 20%, 30%, and 40% FA content resulted in chloride ion penetration reduction of about 7%, 27%, 48%, and 53%, respectively, compared to the control mix. While the volume of the paste remains constant for mixes, the penetration of chloride ions into the matrix is determined by two fundamental factors, including the interlinking pores of the matrix and the free hydroxyl ion in the pore solution. Due to the finer particle size of FA, it may have minimized the interconnecting spaces and decreased the chloride ion penetration. FA can help concrete perform better over time in terms of CS, STS, FS, porosity, chloride penetration, creep, capillary absorption, drying shrinkage, surface scaling, and sulphate attack. Mainly, FA enhanced CS marginally but greatly increased the long-term STS and FS of concretes [123–125]. Class F FA in concrete provided more CS and chloride penetration resistance than Class C FA, and the maximum long-term CS was achieved for a FA concrete (67% Class F FA) at the age of seven years, along with exceptional surface scaling resistance [126]. Even when exposed to a sea environment for five years, FA concrete demonstrated strength development. Additionally, utilizing FA in concrete can help prevent chloride permeability and rusting of embedded steel bars [127]. All these long-term advantages can be ascribed to the pozzolanic nature of FA, which improves the amount of CSH, causing cross-linking hydrates at the molecular level and a compact and crack-free microstructure, thereby enhancing durability [125].

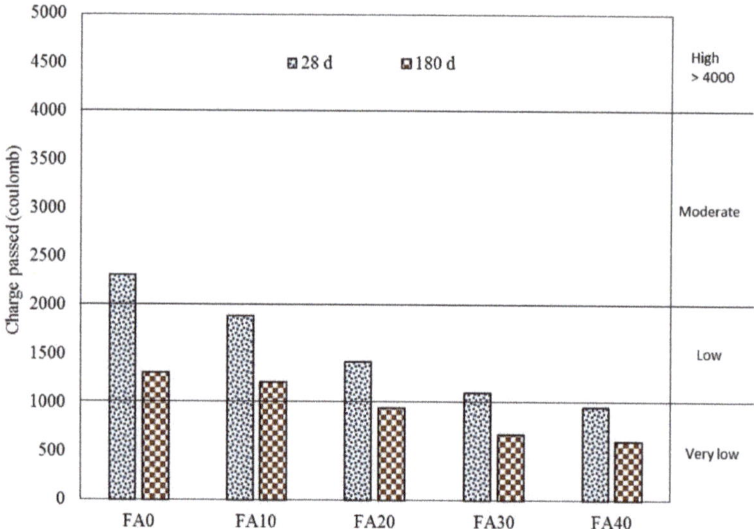

Figure 5. Chloride penetration of fly ash based-concrete at 28 and 180 days of curing [122].

3.5.2. Shrinkage

One of the main causes of concrete cracking is the strains caused by shrinkage. While the stresses created by restricted shrinkage have no effect on the structure's integrity, they do raise the likelihood of durability issues [128,129]. While drying shrinkage occurs as a result of the concrete losing water, autogenous shrinkage occurs as a result of a variation in macroscopic volume when no moisture is transported to the adjacent environment. As a result, composites' shrinkage must be considered cumulative, taking into account both drying and autogenous deformations. According to reports, volume variation because of shrinkage can frequently be addressed utilizing fillers such as FA [130]. A previous study found that SCM-based composites displayed more drying shrinkage than conventional cement-based composites [131]. Particularly, mixes comprising FA shrink more during the drying process than mixtures, including micro silica and slag cement. The pore structure of a concrete mixture containing SCMs such as FA, micro silica, and slag cement is more refined than that of a concrete mixture comprising only cement. As a result, these mixtures have a greater number of smaller capillary spaces, and thus, the water removal from these pores might result in increased drying shrinkage [94]. Another study also noted that composites with a greater proportion of SCMs have a finer pore structure, which may enhance free shrinkage proportionately [132]. Specifically, the loss caused by autogenous shrinkage can be considerably decreased by the inclusion of FA in composites [133]. Both Class C and Class F FA are thought to be beneficial for minimizing drying shrinkage [134]. The inclusion of FA reduces shrinkage by densifying the mix and preventing internal moisture evaporation [135]. Another reason for the limited shrinkage documented in the literature is the existence of un-hydrated FA grains in the matrix, which act as fine aggregates [136]; however, a few studies have found that FA with a smaller particle size than cement increases autogenous shrinkage [137]. A small-sized FA reduces the space between particles, which reduces the pore size in the paste and, as a result, capillary pressure increases in the paste while consuming water in the hydration process. While some research finds decreased shrinkage as a result of FA addition, a few others report an increase in shrinkage properties as a result of FA incorporation; thus, the influence of varying quantities of distinct FA types on the shrinkage of cement-based materials must be investigated to identify how shrinkage is reduced in FA concrete.

3.5.3. Sulfate Resistance

The effect of FA on the resistance of mortar and concrete to sulphate attack has been widely studied. The significant range in performance of FA cement blends is due to the variety of FA kinds and compositions, as well as changes in mix proportions and construction techniques. In general, low Ca FA is more resistant to sulphate than high Ca FA because it can consume more $Ca(OH)_2$ from the hydrated cement paste, generating more sulfate-resistant CSH without incorporating additional reactive phases present in high Ca FA that can accelerate sulfate-induced deterioration, while high Ca FA can hydrate independently during the generation of additional $Ca(OH)_2$, hence accelerating sulfate-induced deterioration [138]. Apart from changes in calcium concentration, the amount of oxides in FA, including silica, alumina, and iron, as well as their amorphous and crystalline forms, has been demonstrated to have a substantial effect on their sulphate attack efficacy. FA containing less than 5% CaO is anticipated to have no reactive alumina and hence would not react with external sulphates to create expansive ettringite crystals [139]. Most of the studies reported an increase in sulphate resistance of the concrete with FA addition [140–142].

3.5.4. Water Absorption

Water absorption (WA) is a feature of cementitious materials that are directly associated with its durability or long-term behavior. The existence of pores, cracks, and fissures in the matrix increases WA, which influences the mechanical and other durability aspects. In general, an increase in WA associated with an increase in FA indicates an increase in the volume of accessible pores [114,143]. Pitroda et al. [144] concluded that when 10% FA is replaced with cement, the WA of concrete decreases. Additionally, they discovered an increasing trend in WA as the level of cement replaced by FA increased by more than 10%. The WA of FA-incorporated concrete, on the other hand, was found to be greater than those of conventional concrete. In contrast, Hatungimana et al. [145] noticed a reduction in the WA of concrete with FA addition, as depicted in Figure 6. They reported that WA values increased as the amount of FA substitution increased, probably because the 28-day curing period was insufficient to complete the pozzolanic reaction; however, at 10% and 20% FA content, the WA of the samples was reduced by 14.6% and 12.2%, respectively, compared to the control mix. Whereas at 30% FA content, the FA concrete sample exhibited a comparable WA capacity to that of the control mix. Finally, the results indicated that FA could be employed as an SCM with some prudent engineering judgments [144].

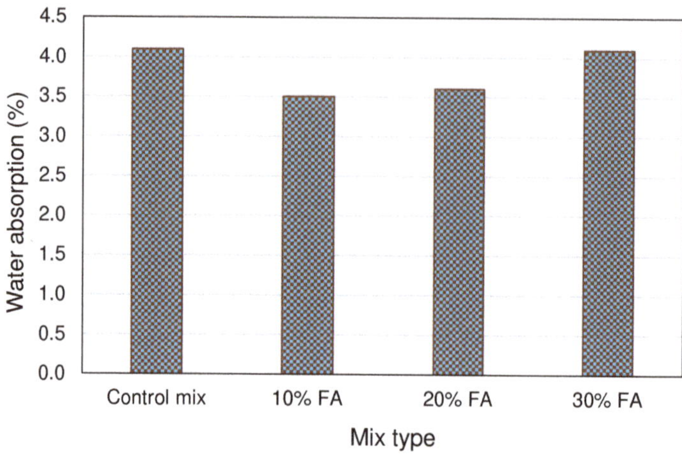

Figure 6. Water absorption of concrete at various contents of fly ash [145].

3.6. Microstructure

Gunasekara et al. [146] performed SEM evaluation to explore the microstructure of composites without FA and composites with FA. The SEM analysis of samples was carried out at the age of 28-days. A dense, compacted, and uniformly distributed matrix was observed for the sample without FA, while cracked, porous, and partially reacted FA grains were observed in the sample containing FA. These observations are consistent with the reduced mechanical properties of FA-based composites, as discussed earlier. Ahad et al. [147] also studied and compared the microstructure of a reference mix without FA and a mix containing 30% FA, as depicted in Figure 7. Crack was observed in the matrix of the reference mix (Figure 7a). This may be due to the high heat of cement hydration resulting in the micro-cracks in the matrix. Though voids in the matrix are less and a denser and compacted matrix can be observed. Micro-cracks are not observed in the matrix of composite containing FA because FA addition reduced the heat of hydration; however, more voids and partially reacted FA particles are observed resulting in less dense matrix (Figure 7b). This also supports the detrimental effects of FA on cementitious composites.

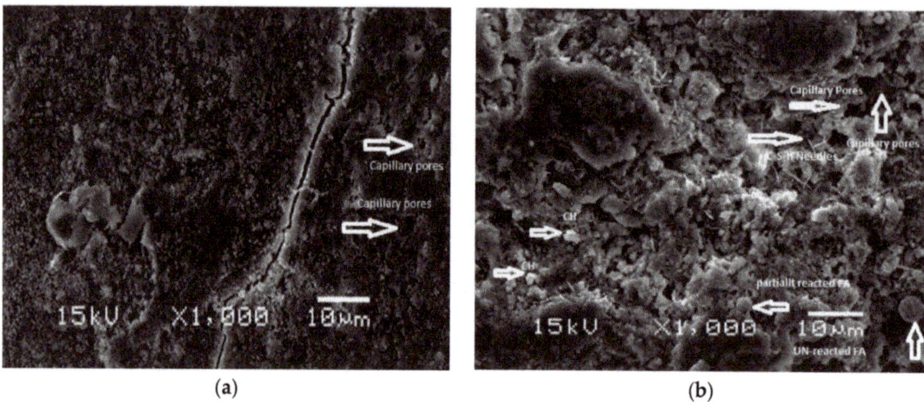

(a) (b)

Figure 7. Microstructure of samples: (**a**) Without fly ash; (**b**) With 30% fly ash [147].

Saha [122] performed a microstructural study of FA concrete, and their micrographs are shown in Figure 8. Figure 8a illustrates the microstructural images of cracked cementitious material containing 40% FA as an SCM at the age of 28 days. Ettringite needles initiate to form in the voids of the binder matrix and on the surface of the FA. Smooth spherical FA grains are also visible, indicating that the FA has been hydrated during its first phase. FA has a somewhat spherical shape, and the existence of spherical grains in the microstructure of the matrix at the age of 28 days suggests that the FA grains have not reacted with the cement during the early hydration phase. Due to exposure to a harsh environment, the spherical form of FA rapidly spoils in cement mix and is substituted with ettringite needles [148,149]. This corroborates the idea that FA retards concrete hydration. Figure 8b illustrates the microstructure of FA concrete after 180 days of curing. The spherical grains were substituted by ettringite as a result of the pozzolanic reaction of FA. The voids between the aggregates are densely packed with ettringite needles. Additionally, the ettringite needles are longer, fill up the gaps in the cement mix; therefore, the ettringite needles fill the spaces between the aggregates by the pozzolanic reaction of FA. As a result, FA concrete produces a denser binder matrix than normal concrete [122].

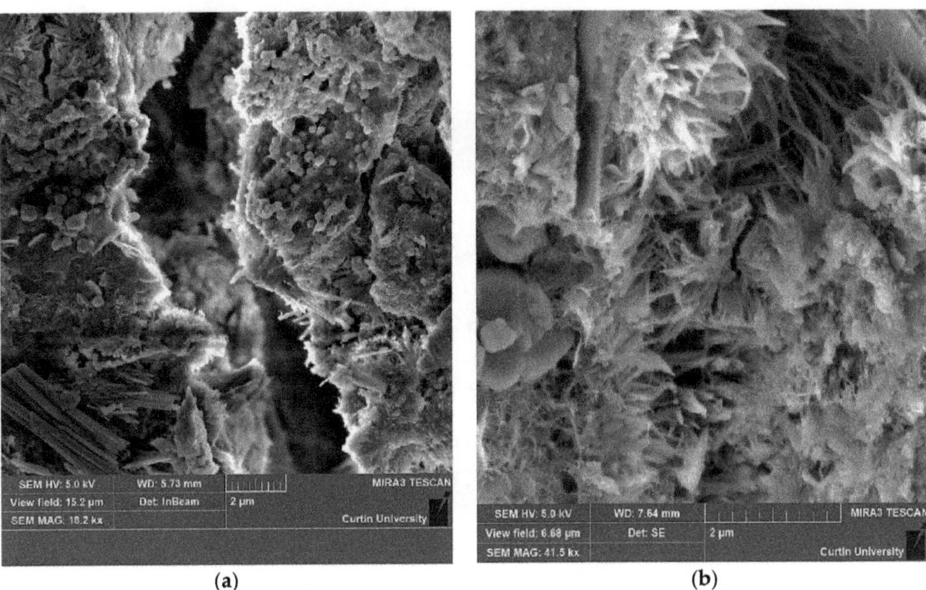

(a)　　　　(b)

Figure 8. SEM micrographs of composites containing 40% fly ash as a biner at: (a) 28-days; (b) 180-days [122].

4. Discussions

The challenges correlated with the manufacture and application of cement are well known. The growing need for concrete, and therefore cement, poses a severe danger to both the environment and human life. In this context, scholars are concentrating on the use of SCMs that can substitute cement in the manufacture of concrete, encouraging eco-friendly development. This study examined the usage of the most common industrial byproducts, i.e., FA, in cementitious composites as SCMs. This study highlighted and discussed the most critical sections, including the properties of FA, the characteristics of composites containing FA as SCMs, i.e., workability, compressive, split-tensile, and flexural strength, durability, and microstructural properties. Table 5 has been prepared to summarize the various parameters examined in this study for FA use as SCM. As can be noticed from the table, FA utilization contributes to construction sustainability. In addition to the sustainable benefits, FA has the further advantage of low heat of hydration. FA is a pozzolanic material and is utilized as SCM in cement-based composites. The use of

FA as cement replacement in lower replacement ratios (up to 20%) has a positive effect on the mechanical and durability properties of composites, while at higher replacement ratios, it has a negative effect; moreover, the size of FA alters the properties of composites. The smaller size FA is preferable, as it has a more positive impact on the performance of composites.

Table 5. Comparison of various aspects of utilizing silica fume and fly ash in cementitious materials.

Aspect	Detail
Sustainability	Reduction in CO_2 emission
	Effective waste management
	Decreases environmental pollution
	Preserve natural resources
	Cost-effective
	Low heat of hydration
Influence on material properties	Inconsistent influence on workability
	Enhances the performance of composites when used in lower replacement ratios (up to 20%)
	Negative influence on material's performance at higher proportions
Limitations	Utilization at higher replacement levels is not preferable
	Low early strength development

FA consumption has increased in the concrete industry because of its benefits, which include reduced hydration heat and increased durability; however, due to the slow pozzolanic action, its contribution to strength begins only at a later age. Attempts have been undertaken to address this well-documented FA deficit using a variety of approaches. Chemical activation is one of these ways and can be accomplished using either alkali or sulphate. In alkali activation, the glass phases of FA are broken down to expedite the reaction at an early stage [150], whereas, in sulphate activation, sulphate combines with the aluminum oxide in the glass phase of FA to form ettringite [151]. In each of these instances, strength is developed at a young age [152]. Alkali activation of FA is a physicochemical method that converts powdered ash to a material with excellent cementitious characteristics [153,154], developing high mechanical strength and exceptional bonding to reinforcing bars [155]. The use of nanoparticles in FA-based cementitious composites to accelerate its early strength gain is becoming more common as a result of its benefits. The nanoparticles serve as nuclei for the cement, accelerating hydration and densifying the microstructure and interfacial transition zone, hence decreasing permeability [156]. Additionally, the combination of FA and nanomaterials enables the hydration product to be tightly bound, which is a critical element in accelerating the pozzolanic process since it compensates for the poor initial strength growth [157–159]; hence, the strength of the FA-based cementitious composites can be increased by various methods, including those covered above. For successful strength enhancement via alkali activation or nanoparticle addition, knowledge of the properties of FA is required. Ca/Si and Ca/Al ratios are regarded as critical factors in the formation of CSH gel in the case of nano addition and alumino silicate gel in the case of alkali activation, respectively. Using either of these approaches, it is possible to replace up to 60% of cement with FA without sacrificing strength or durability [94]. For high-volume FA concrete, a ternary blend of cement, FA, and nanomaterials can be advised.

5. Conclusions

The present study aimed to review the different aspects of the fly ash (FA) application as supplementary cementitious material (SCM) in cement-based materials. The influence of the FA characteristics of the mechanical, durability, and microstructural properties of the material is discussed. The various limitations of the FA use in higher proportions, and their potential solutions are described. This study reached the following conclusions:

- The influence of FA incorporation on the workability of fresh concrete was found to be inconsistent. Some studies reported an increase in the workability because of

FA spherical shape, increased volume of the mix due to lower density of FA, and slower development of hydration products due to FA addition; however, some studies found a reduction in the workability of the fresh mix due to the smaller size and larger surface area of FA.

- Numerous studies have demonstrated that the application of FA inhibits composites' early-age strength development; however, FA mostly improves the strength of concrete over time by consuming the $Ca(OH)_2$ produced during cement hydration and producing secondary hydrates such as CSH; moreover, the mechanical strength of the composites improves when FA is incorporated in lower concentrations (up to 20%). In addition, finer particle size FA enhances while coarser particle size FA reduces the mechanical strength of composites.
- The resistance of composites to chloride penetration increases with the addition of FA, especially at later ages. The finer particle size FA decreases the interconnecting voids in the matrix, and pozzolanic action further improves the microstructure, resulting in increased resistance to chloride penetration.
- There is a contradiction regarding the influence of FA on the shrinkage of cementitious composites. FA incorporation may increase shrinkage due to the creation of a higher amount of small capillary spaces, which facilitates the evaporation of water, causing higher shrinkage. In contrast, due to finer particle size FA addition, the density and compactness of the mix increases, which prevents the internal moisture evaporation, causing reduced shrinkage.
- The influence of FA on sulphate resistance of composite is determined by the FA type. Generally, low calcium FA (Class C) exhibits more resistance to sulphate resistance than high calcium FA (Class F). This is because low calcium FA consumes more $Ca(OH)_2$ to form CSH, compared to the high calcium FA.
- The incorporation of FA as SCM up to 20% content is beneficial to composites in terms of reducing water absorption; however, at higher replacement levels, the water absorption capacity increases.
- The slow early strength development is one of the major drawbacks of FA use in cementitious composites; however, this can be overcome by chemical activation (alkali/sulphate) and/or adding nanomaterials, and this can facilitate high volume usage of FA.
- Using FA as SCM will promote sustainable development due to reduction in CO_2 emissions, preservation of natural resources, effective waste management, reduction in environmental pollution, and low heat of hydration.

6. Future Recommendations

After reviewing the different aspects of the FA application as SCM, this study suggests the following future research directions:

- Despite the numerous stated procedures for using FA in large quantities, 100% utilization of FA has not been accomplished due to the existence of some grey areas highlighted in this study that must be addressed in the future to increase FA's usage as SCM. Future studies should be directed on resolving the stated disparities in shrinkage, water demand, and faster curing.
- While some attributes of structural performance, such as flexural and shear resistance, have been covered previously, additional research is necessary to support the case for FA concrete usage in reinforced concrete structures. These may include the bond strength of steel rebars, the examination of beam-column junctions, and seismic design.
- Since FA reactivity is based on multiple factors, trying to control one property may deteriorate other properties. For example, while decreasing the w/b in high volume FA concrete increases early/late age strength, the mix results in early-age cracking. Thus, it is critical to consider the combined effect of multiple parameters in order to maximize the benefits of FA and cement in an optimum planned mix.

- Presently, there is a growing tendency toward the production of geopolymers that include 100% FA [160]; however, to promote the applicability of geopolymer concrete, certain shortcomings such as curing regime, availability of activators, efflorescence, and alkali-silica reaction [39] must be addressed. In this respect, it is reported that designed FA concrete, which can replace up to 60% of cement, is a superior alternative in terms of strength and durability.
- Based on the comprehensive examination of FA as SCM, it is proposed that further fly ash classifications be added to the existing ASTM classifications.

Author Contributions: G.L.: conceptualization, resources, methodology, investigation, validation, project administration, writing-original draft; C.Z.: data curation, formal analysis, supervision, investigation, visualization, writing, reviewing, and editing; W.A.: conceptualization, data curation, methodology, software, supervision, writing—original draft; K.I.U.: funding acquisition, validation, formal analysis, writing, reviewing, and editing; M.K.: resources, visualization, writing, reviewing, and editing; A.M.M.: data curation, formal analysis, writing, reviewing, and editing; R.K.: methodology, investigation, writing, reviewing, and editing. All authors have read and agreed to the published version of the manuscript.

Funding: The research is partially funded by the Ministry of Science and Higher Education of the Russian Federation under the strategic academic leadership program 'Priority 2030' (Agreement 075-15-2021-1333 dated 30.09.2021).

Institutional Review Board Statement: Not applicable.

Informed Consent Statement: Not applicable.

Data Availability Statement: Not applicable.

Acknowledgments: This research is supported by COMSATS University Islamabad, Abbottabad Campus.

Conflicts of Interest: The authors declare no conflict of interest.

References

1. Kishore, K.; Gupta, N. Application of domestic & industrial waste materials in concrete: A review. *Mater. Today Proc.* **2020**, *26*, 2926–2931.
2. Tang, P.; Brouwers, H.J.H. The durability and environmental properties of self-compacting concrete incorporating cold bonded lightweight aggregates produced from combined industrial solid wastes. *Constr. Build. Mater.* **2018**, *167*, 271–285. [CrossRef]
3. Ren, C.; Wang, W.; Li, G. Preparation of high-performance cementitious materials from industrial solid waste. *Constr. Build. Mater.* **2017**, *152*, 39–47. [CrossRef]
4. Brooks, S.J.; Escudero-Oñate, C.; Lillicrap, A.D. An ecotoxicological assessment of mine tailings from three Norwegian mines. *Chemosphere* **2019**, *233*, 818–827. [CrossRef]
5. Park, I.; Tabelin, C.B.; Jeon, S.; Li, X.; Seno, K.; Ito, M.; Hiroyoshi, N. A review of recent strategies for acid mine drainage prevention and mine tailings recycling. *Chemosphere* **2019**, *219*, 588–606. [CrossRef]
6. Ahmaruzzaman, M. Industrial wastes as low-cost potential adsorbents for the treatment of wastewater laden with heavy metals. *Adv. Colloid Interface Sci.* **2011**, *166*, 36–59. [CrossRef]
7. Friedlander, L.R.; Weisbrod, N.; Garb, Y.J. Climatic and soil-mineralogical controls on the mobility of trace metal contamination released by informal electronic waste (e-waste) processing. *Chemosphere* **2019**, *232*, 130–139. [CrossRef]
8. Walls, M.; Palmer, K. Upstream pollution, downstream waste disposal, and the design of comprehensive environmental policies. *J. Environ. Econ. Manag.* **2001**, *41*, 94–108. [CrossRef]
9. Awomeso, J.A.; Taiwo, A.M.; Gbadebo, A.M.; Arimoro, A.O. Waste disposal and pollution management in urban areas: A workable remedy for the environment in developing countries. *Am. J. Environ. Sci.* **2010**, *6*, 26–32. [CrossRef]
10. Wilberforce, T.; Baroutaji, A.; Soudan, B.; Al-Alami, A.H.; Olabi, A.G. Outlook of carbon capture technology and challenges. *Sci. Total Environ.* **2019**, *657*, 56–72. [CrossRef]
11. Mastali, M.; Abdollahnejad, Z.; Pacheco-Torgal, F. Carbon dioxide sequestration on fly ash/waste glassalkali-based mortars with recycled aggregates: Compressive strength, hydration products, carbon footprint, and cost analysis. In *Carbon Dioxide Sequestration in Cementitious Construction Materials*; Elsevier: Amsterdam, The Netherlands, 2018; pp. 299–348.
12. Koneswaran, G.; Nierenberg, D. Global farm animal production and global warming: Impacting and mitigating climate change. *Environ. Health Perspect.* **2008**, *116*, 578–582. [CrossRef]
13. Drissi, S.; Ling, T.-C.; Mo, K.H.; Eddhahak, A. A review of microencapsulated and composite phase change materials: Alteration of strength and thermal properties of cement-based materials. *Renew. Sustain. Energy Rev.* **2019**, *110*, 467–484. [CrossRef]

14. Khan, M.; Ali, M. Improvement in concrete behavior with fly ash, silica-fume and coconut fibres. *Constr. Build. Mater.* **2019**, *203*, 174–187. [CrossRef]

15. Khan, M.; Rehman, A.; Ali, M. Efficiency of silica-fume content in plain and natural fiber reinforced concrete for concrete road. *Constr. Build. Mater.* **2020**, *244*, 118382. [CrossRef]

16. Meng, Y.; Ling, T.-C.; Mo, K.H.; Tian, W. Enhancement of high temperature performance of cement blocks via CO_2 curing. *Sci. Total Environ.* **2019**, *671*, 827–837. [CrossRef]

17. Kumar, V.P.; Prasad, D.R. Influence of supplementary cementitious materials on strength and durability characteristics of concrete. *Adv. Concr. Constr.* **2019**, *7*, 75.

18. Alyousef, R.; Ahmad, W.; Ahmad, A.; Aslam, F.; Joyklad, P.; Alabduljabbar, H. Potential use of recycled plastic and rubber aggregate in cementitious materials for sustainable construction: A review. *J. Clean. Prod.* **2021**, *329*, 129736. [CrossRef]

19. Cao, M.; Mao, Y.; Khan, M.; Si, W.; Shen, S. Different testing methods for assessing the synthetic fiber distribution in cement-based composites. *Constr. Build. Mater.* **2018**, *184*, 128–142. [CrossRef]

20. Xie, C.; Cao, M.; Khan, M.; Yin, H.; Guan, J. Review on different testing methods and factors affecting fracture properties of fiber reinforced cementitious composites. *Constr. Build. Mater.* **2020**, *273*, 121766. [CrossRef]

21. Khan, M.; Cao, M.; Xie, C.; Ali, M. Efficiency of basalt fiber length and content on mechanical and microstructural properties of hybrid fiber concrete. *Fatigue Fract. Eng. Mater. Struct.* **2021**, *44*, 2135–2152. [CrossRef]

22. Li, L.; Khan, M.; Bai, C.; Shi, K. Uniaxial Tensile Behavior, Flexural Properties, Empirical Calculation and Microstructure of Multi-Scale Fiber Reinforced Cement-Based Material at Elevated Temperature. *Materials* **2021**, *14*, 1827. [CrossRef] [PubMed]

23. Schneider, M.; Romer, M.; Tschudin, M.; Bolio, H. Sustainable cement production—Present and future. *Cem. Concr. Res.* **2011**, *41*, 642–650. [CrossRef]

24. Ali, K.; Qureshi, M.I.; Saleem, S.; Khan, S.U. Effect of waste electronic plastic and silica fume on mechanical properties and thermal performance of concrete. *Constr. Build. Mater.* **2021**, *285*, 122952. [CrossRef]

25. Adil, G.; Kevern, J.T.; Mann, D. Influence of silica fume on mechanical and durability of pervious concrete. *Constr. Build. Mater.* **2020**, *247*, 118453. [CrossRef]

26. Ahmad, W.; Ahmad, A.; Ostrowski, K.A.; Aslam, F.; Joyklad, P.; Zajdel, P. Application of Advanced Machine Learning Approaches to Predict the Compressive Strength of Concrete Containing Supplementary Cementitious Materials. *Materials* **2021**, *14*, 5762. [CrossRef]

27. Federico, L. Waste Glass—A Supplementary Cementitious Material. Ph.D. Thesis, McMaster University, Hamilton, ON, Canada, 2013.

28. Shanmugasundaram, N.; Praveenkumar, S. Influence of supplementary cementitious materials, curing conditions and mixing ratios on fresh and mechanical properties of engineered cementitious composites—A review. *Constr. Build. Mater.* **2021**, *309*, 125038. [CrossRef]

29. Gupta, S.; Chaudhary, S. State of the art review on Supplementary Cementitious Materials in India—I: An overview of legal perspective, governing organizations, and development patterns. *J. Clean. Prod.* **2020**, *261*, 121203. [CrossRef]

30. Ahmad, W.; Ahmad, A.; Ostrowski, K.A.; Aslam, F.; Joyklad, P.; Zajdel, P. Sustainable approach of using sugarcane bagasse ash in cement-based composites: A systematic review. *Case Stud. Constr. Mater.* **2021**, *15*, e00698. [CrossRef]

31. Provis, J.L.; Palomo, A.; Shi, C. Advances in understanding alkali-activated materials. *Cem. Concr. Res.* **2015**, *78*, 110–125. [CrossRef]

32. Usanova, K.; Barabanshchikov, Y.G. Cold-Bonded fly ash aggregate concrete. *Mag. Civ. Eng.* **2020**, *3*, 95.

33. Kutchko, B.G.; Kim, A.G. Fly ash characterization by SEM–EDS. *Fuel* **2006**, *85*, 2537–2544. [CrossRef]

34. Sua-iam, G.; Makul, N. Rheological and mechanical properties of cement–fly ash self-consolidating concrete incorporating high volumes of alumina-based material as fine aggregate. *Constr. Build. Mater.* **2015**, *95*, 736–747. [CrossRef]

35. Amran, M.; Fediuk, R.; Murali, G.; Avudaiappan, S.; Ozbakkaloglu, T.; Vatin, N.; Karelina, M.; Klyuev, S.; Gholampour, A. Fly ash-based eco-efficient concretes: A comprehensive review of the short-term properties. *Materials* **2021**, *14*, 4264. [CrossRef] [PubMed]

36. Şahmaran, M.; Li, V.C. Durability properties of micro-cracked ECC containing high volumes fly ash. *Cem. Concr. Res.* **2009**, *39*, 1033–1043. [CrossRef]

37. Uthaman, S.; Vishwakarma, V.; George, R.P.; Ramachandran, D.; Kumari, K.; Preetha, R.; Premila, M.; Rajaraman, R.; Mudali, U.K.; Amarendra, G. Enhancement of strength and durability of fly ash concrete in seawater environments: Synergistic effect of nanoparticles. *Constr. Build. Mater.* **2018**, *187*, 448–459. [CrossRef]

38. Azimi, E.A.; Abdullah, M.M.A.B.; Vizureanu, P.; Salleh, M.A.A.M.; Sandu, A.V.; Chaiprapa, J.; Yoriya, S.; Hussin, K.; Aziz, I.H. Strength development and elemental distribution of dolomite/fly ash geopolymer composite under elevated temperature. *Materials* **2020**, *13*, 1015. [CrossRef]

39. Yang, H.; Liu, L.; Yang, W.; Liu, H.; Ahmad, W.; Ahmad, A.; Aslam, F.; Joyklad, P. A comprehensive overview of geopolymer composites: A bibliometric analysis and literature review. *Case Stud. Constr. Mater.* **2022**, *16*, e00830. [CrossRef]

40. Li, X.; Qin, D.; Hu, Y.; Ahmad, W.; Ahmad, A.; Aslam, F.; Joyklad, P. A systematic review of waste materials in cement-based composites for construction applications. *J. Build. Eng.* **2021**, *45*, 103447. [CrossRef]

41. John, E.; Matschei, T.; Stephan, D. Nucleation seeding with calcium silicate hydrate—A review. *Cem. Concr. Res.* **2018**, *113*, 74–85. [CrossRef]

42. González, A.; Navia, R.; Moreno, N. Fly ashes from coal and petroleum coke combustion: Current and innovative potential applications. *Waste Manag. Res.* **2009**, *27*, 976–987. [CrossRef]

43. Saha, A.K.; Sarker, P.K. Sustainable use of ferronickel slag fine aggregate and fly ash in structural concrete: Mechanical properties and leaching study. *J. Clean. Prod.* **2017**, *162*, 438–448. [CrossRef]

44. Yao, Z.T.; Ji, X.S.; Sarker, P.K.; Tang, J.H.; Ge, L.Q.; Xia, M.S.; Xi, Y.Q. A comprehensive review on the applications of coal fly ash. *Earth-Sci. Rev.* **2015**, *141*, 105–121. [CrossRef]

45. Berndt, M.L. Properties of sustainable concrete containing fly ash, slag and recycled concrete aggregate. *Constr. Build. Mater.* **2009**, *23*, 2606–2613. [CrossRef]

46. Li, G.; Zhao, X. Properties of concrete incorporating fly ash and ground granulated blast-furnace slag. *Cem. Concr. Compos.* **2003**, *25*, 293–299. [CrossRef]

47. Chindaprasirt, P.; Jaturapitakkul, C.; Sinsiri, T. Effect of fly ash fineness on compressive strength and pore size of blended cement paste. *Cem. Concr. Compos.* **2005**, *27*, 425–428. [CrossRef]

48. Lee, H.K.; Lee, K.M.; Kim, B.G. Autogenous shrinkage of high-performance concrete containing fly ash. *Mag. Concr. Res.* **2003**, *55*, 507–515. [CrossRef]

49. Atiş, C.D. High-Volume fly ash concrete with high strength and low drying shrinkage. *J. Mater. Civ. Eng.* **2003**, *15*, 153–156. [CrossRef]

50. Baite, E.; Messan, A.; Hannawi, K.; Tsobnang, F.; Prince, W. Physical and transfer properties of mortar containing coal bottom ash aggregates from Tefereyre (Niger). *Constr. Build. Mater.* **2016**, *125*, 919–926. [CrossRef]

51. Siddique, R. Utilization of industrial byproducts in concrete. *Procedia Eng.* **2014**, *95*, 335–347. [CrossRef]

52. Kim, H.-K. Utilization of sieved and ground coal bottom ash powders as a coarse binder in high-strength mortar to improve workability. *Constr. Build. Mater.* **2015**, *91*, 57–64. [CrossRef]

53. Thomas, M. *Supplementary Cementing Materials in Concrete*; CRC Press: Boca Raton, FL, USA, 2013.

54. Ma, J.; Wang, D.; Zhao, S.; Duan, P.; Yang, S. Influence of particle morphology of ground fly ash on the fluidity and strength of cement paste. *Materials* **2021**, *14*, 283. [CrossRef] [PubMed]

55. Bicer, A. Effect of fly ash particle size on thermal and mechanical properties of fly ash-cement composites. *Therm. Sci. Eng. Prog.* **2018**, *8*, 78–82. [CrossRef]

56. Panda, L.; Dash, S. Characterization and utilization of coal fly ash: A review. *Emerg. Mater. Res.* **2020**, *9*, 921–934. [CrossRef]

57. Song, M.; Kaewmee, P.; Takahashi, F. Mechanism analysis of modification of coal fly ash on high strength biodegradable porous composites synthesis. In Proceedings of the 29th Annual Conference of Japan Society of Material Cycles and Waste Management, Nagoya, Japan, 12–15 September 2018; p. 591.

58. Ahmaruzzaman, M. A review on the utilization of fly ash. *Prog. Energy Combust. Sci.* **2010**, *36*, 327–363. [CrossRef]

59. Dananjayan, R.R.T.; Kandasamy, P.; Andimuthu, R. Direct mineral carbonation of coal fly ash for CO_2 sequestration. *J. Clean. Prod.* **2016**, *112*, 4173–4182. [CrossRef]

60. Ukwattage, N.L.; Ranjith, P.; Yellishetty, M.; Bui, H.H.; Xu, T. A laboratory-scale study of the aqueous mineral carbonation of coal fly ash for CO_2 sequestration. *J. Clean. Prod.* **2015**, *103*, 665–674. [CrossRef]

61. Xie, J.; Wang, J.; Rao, R.; Wang, C.; Fang, C. Effects of combined usage of GGBS and fly ash on workability and mechanical properties of alkali activated geopolymer concrete with recycled aggregate. *Compos. Part B Eng.* **2019**, *164*, 179–190. [CrossRef]

62. Zhao, Z.; Wang, K.; Lange, D.A.; Zhou, H.; Wang, W.; Zhu, D. Creep and thermal cracking of ultra-high volume fly ash mass concrete at early age. *Cem. Concr. Compos.* **2019**, *99*, 191–202. [CrossRef]

63. Hefni, Y.; Abd El Zaher, Y.; Wahab, M.A. Influence of activation of fly ash on the mechanical properties of concrete. *Constr. Build. Mater.* **2018**, *172*, 728–734. [CrossRef]

64. Vargas, J.; Halog, A. Effective carbon emission reductions from using upgraded fly ash in the cement industry. *J. Clean. Prod.* **2015**, *103*, 948–959. [CrossRef]

65. American Society for Testing and Materials. *ASTM C 618-Standard Specification for Coal Fly Ash and Raw or Calcined Natural Pozzolan for Use in Concrete*; ASTM International: West Conshohocken, PA, USA, 2012.

66. Siddique, R. Performance characteristics of high-volume Class F fly ash concrete. *Cem. Concr. Res.* **2004**, *34*, 487–493. [CrossRef]

67. Committee, R.-S. Siliceous byproducts for use in concrete. *Mater. Struct* **1988**, *21*, 69–80.

68. Atiş, C.D.; Görür, E.B.; Karahan, O.; Bilim, C.; İlkentapar, S.; Luga, E. Very high strength (120 MPa) class F fly ash geopolymer mortar activated at different NaOH amount, heat curing temperature and heat curing duration. *Constr. Build. Mater.* **2015**, *96*, 673–678. [CrossRef]

69. Shehata, M.H.; Thomas, M.D. The effect of fly ash composition on the expansion of concrete due to alkali–silica reaction. *Cem. Concr. Res.* **2000**, *30*, 1063–1072. [CrossRef]

70. Belviso, C.; Giannossa, L.C.; Huertas, F.J.; Lettino, A.; Mangone, A.; Fiore, S. Synthesis of zeolites at low temperatures in fly ash-kaolinite mixtures. *Microporous Mesoporous Mater.* **2015**, *212*, 35–47. [CrossRef]

71. Sun, J.; Zhang, Z.; Hou, G. Utilization of fly ash microsphere powder as a mineral admixture of cement: Effects on early hydration and microstructure at different curing temperatures. *Powder Technol.* **2020**, *375*, 262–270. [CrossRef]

72. Smith, I.A. The design of fly-ash concretes. *Proc. Inst. Civ. Eng.* **1967**, *36*, 769–790. [CrossRef]

73. Bui, P.T.; Ogawa, Y.; Kawai, K. Long-Term pozzolanic reaction of fly ash in hardened cement-based paste internally activated by natural injection of saturated $Ca(OH)_2$ solution. *Mater. Struct.* **2018**, *51*, 144. [CrossRef]

74. Yu, Z.; Ni, C.; Tang, M.; Shen, X. Relationship between water permeability and pore structure of Portland cement paste blended with fly ash. *Constr. Build. Mater.* **2018**, *175*, 458–466. [CrossRef]
75. Nedunuri, S.S.S.A.; Sertse, S.G.; Muhammad, S. Microstructural study of Portland cement partially replaced with fly ash, ground granulated blast furnace slag and silica fume as determined by pozzolanic activity. *Constr. Build. Mater.* **2020**, *238*, 117561. [CrossRef]
76. Babu, K.G.; Rao, G.S.N. Efficiency of fly ash in concrete with age. *Cem. Concr. Res.* **1996**, *26*, 465–474. [CrossRef]
77. Gopalan, M.K.; Hague, M.N. Design of flyash concrete. *Cem. Concr. Res.* **1985**, *15*, 694–702. [CrossRef]
78. Bijen, J.; Van Selst, R. Cement equivalence factors for fly ash. *Cem. Concr. Res.* **1993**, *23*, 1029–1039. [CrossRef]
79. Cao, C.; Sun, W.; Qin, H. The analysis on strength and fly ash effect of roller-compacted concrete with high volume fly ash. *Cem. Concr. Res.* **2000**, *30*, 71–75. [CrossRef]
80. Zeng, Q.; Li, K.; Fen-chong, T.; Dangla, P. Determination of cement hydration and pozzolanic reaction extents for fly-ash cement pastes. *Constr. Build. Mater.* **2012**, *27*, 560–569. [CrossRef]
81. Kim, J.H.; Noemi, N.; Shah, S.P. Effect of powder materials on the rheology and formwork pressure of self-consolidating concrete. *Cem. Concr. Compos.* **2012**, *34*, 746–753. [CrossRef]
82. Lee, C.Y.; Lee, H.K.; Lee, K.M. Strength and microstructural characteristics of chemically activated fly ash–cement systems. *Cem. Concr. Res.* **2003**, *33*, 425–431. [CrossRef]
83. Kashani, A.; San Nicolas, R.; Qiao, G.G.; van Deventer, J.S.; Provis, J.L. Modelling the yield stress of ternary cement–slag–fly ash pastes based on particle size distribution. *Powder Technol.* **2014**, *266*, 203–209. [CrossRef]
84. Bentz, D.P.; Ferraris, C.F.; Galler, M.A.; Hansen, A.S.; Guynn, J.M. Influence of particle size distributions on yield stress and viscosity of cement–fly ash pastes. *Cem. Concr. Res.* **2012**, *42*, 404–409. [CrossRef]
85. Ponikiewski, T.; Gołaszewski, J. The influence of high-calcium fly ash on the properties of fresh and hardened self-compacting concrete and high performance self-compacting concrete. *J. Clean. Prod.* **2014**, *72*, 212–221. [CrossRef]
86. ACI Committee 234. *ACI PRC-234-06: Guide for the Use of Silica Fume in Concrete*; ACI: Farmington Hills, MI, USA, 2006.
87. Chandra Sekar, G.; Ch, H.K.; Manikanta, V.; Simhachalam, M. Effect of fly ash on mechanical properties of light weight vermiculite concrete. *Int. J. Innov. Res. Sci. Eng. Technol.* **2016**, *5*, 4106–4112.
88. Aldred, J.M.; Holland, T.C.; Morgan, D.R.; Roy, D.M.; Bury, M.A.; Hooton, R.D.; Olek, J.; Scali, M.J.; Detwiler, R.J.; Jaber, T.M. *Guide for the Use of Silica Fume in Concrete*; ACI–American Concrete Institute–Committee: Farmington Hills, MI, USA, 2006; Volume 234.
89. Kwan, A.K.H.; Wong, H.H.C. Packing density of cementitious materials: Part 2—Packing and flow of OPC + PFA + CSF. *Mater. Struct.* **2008**, *41*, 773–784. [CrossRef]
90. Nanthagopalan, P.; Haist, M.; Santhanam, M.; Müller, H.S. Investigation on the influence of granular packing on the flow properties of cementitious suspensions. *Cem. Concr. Compos.* **2008**, *30*, 763–768. [CrossRef]
91. Fung, W.W.S.; Kwan, A.K.H. Role of water film thickness in rheology of CSF mortar. *Cem. Concr. Compos.* **2010**, *32*, 255–264. [CrossRef]
92. Nochaiya, T.; Wongkeo, W.; Chaipanich, A. Utilization of fly ash with silica fume and properties of Portland cement–fly ash–silica fume concrete. *Fuel* **2010**, *89*, 768–774. [CrossRef]
93. Wong, H.H.C.; Kwan, A.K.H. Rheology of cement paste: Role of excess water to solid surface area ratio. *J. Mater. Civ. Eng.* **2008**, *20*, 189–197. [CrossRef]
94. Hemalatha, T.; Ramaswamy, A. A review on fly ash characteristics—Towards promoting high volume utilization in developing sustainable concrete. *J. Clean. Prod.* **2017**, *147*, 546–559. [CrossRef]
95. Singh, N.; Kumar, P.; Goyal, P. Reviewing the behaviour of high volume fly ash based self compacting concrete. *J. Build. Eng.* **2019**, *26*, 100882. [CrossRef]
96. Blissett, R.; Rowson, N. A review of the multi-component utilisation of coal fly ash. *Fuel* **2012**, *97*, 1–23. [CrossRef]
97. Durdziński, P.T.; Dunant, C.F.; Haha, M.B.; Scrivener, K.L. A new quantification method based on SEM-EDS to assess fly ash composition and study the reaction of its individual components in hydrating cement paste. *Cem. Concr. Res.* **2015**, *73*, 111–122. [CrossRef]
98. Slanička, Š. The influence of fly ash fineness on the strength of concrete. *Cem. Concr. Res.* **1991**, *21*, 285–296. [CrossRef]
99. Erdoğdu, K.; Türker, P. Effects of fly ash particle size on strength of Portland cement fly ash mortars. *Cem. Concr. Res.* **1998**, *28*, 1217–1222. [CrossRef]
100. Chindaprasirt, P.; Jaturapitakkul, C.; Sinsiri, T. Effect of fly ash fineness on microstructure of blended cement paste. *Constr. Build. Mater.* **2007**, *21*, 1534–1541. [CrossRef]
101. Chindaprasirt, P.; Chotithanorm, C.; Cao, H.; Sirivivatnanon, V. Influence of fly ash fineness on the chloride penetration of concrete. *Constr. Build. Mater.* **2007**, *21*, 356–361. [CrossRef]
102. Chindaprasirt, P.; Homwuttiwong, S.; Sirivivatnanon, V. Influence of fly ash fineness on strength, drying shrinkage and sulfate resistance of blended cement mortar. *Cem. Concr. Res.* **2004**, *34*, 1087–1092. [CrossRef]
103. Harwalkar, A.B.; Awanti, S. Laboratory and field investigations on high-volume fly ash concrete for rigid pavement. *Transp. Res. Rec.* **2014**, *2441*, 121–127. [CrossRef]
104. Khan, U.A.; Jahanzaib, H.M.; Khan, M.; Ali, M. Improving the Tensile Energy Absorption of High Strength Natural Fiber Reinforced Concrete with Fly-Ash for Bridge Girders. *Key Eng. Mater.* **2018**, *765*, 335–342. [CrossRef]

105. Yao, Y.; Gong, J.K.; Cui, Z. Anti-corrosion performance and microstructure analysis on a marine concrete utilizing coal combustion byproducts and blast furnace slag. *Clean Technol. Environ. Policy* **2014**, *16*, 545–554. [CrossRef]
106. Kocak, Y.; Nas, S. The effect of using fly ash on the strength and hydration characteristics of blended cements. *Constr. Build. Mater.* **2014**, *73*, 25–32. [CrossRef]
107. Barbuta, M.; Bucur, R.; Serbanoiu, A.A.; Scutarasu, S.; Burlacu, A. Combined effect of fly ash and fibers on properties of cement concrete. *Procedia Eng.* **2017**, *181*, 280–284. [CrossRef]
108. Gencel, O.; Koksal, F.; Ozel, C.; Brostow, W. Combined effects of fly ash and waste ferrochromium on properties of concrete. *Constr. Build. Mater.* **2012**, *29*, 633–640. [CrossRef]
109. Huang, C.-H.; Lin, S.-K.; Chang, C.-S.; Chen, H.-J. Mix proportions and mechanical properties of concrete containing very high-volume of Class F fly ash. *Constr. Build. Mater.* **2013**, *46*, 71–78. [CrossRef]
110. Şahmaran, M.; Keskin, S.B.; Ozerkan, G.; Yaman, I.O. Self-Healing of mechanically-loaded self consolidating concretes with high volumes of fly ash. *Cem. Concr. Compos.* **2008**, *30*, 872–879. [CrossRef]
111. Celik, K.; Meral, C.; Mancio, M.; Mehta, P.K.; Monteiro, P.J.M. A comparative study of self-consolidating concretes incorporating high-volume natural pozzolan or high-volume fly ash. *Constr. Build. Mater.* **2014**, *67*, 14–19. [CrossRef]
112. Şahmaran, M.; Yaman, İ.Ö.; Tokyay, M. Transport and mechanical properties of self consolidating concrete with high volume fly ash. *Cem. Concr. Compos.* **2009**, *31*, 99–106. [CrossRef]
113. Anjos, M.A.S.; Camões, A.; Jesus, C. Eco-Efficient self-compacting concrete with reduced Portland cement content and high volume of fly ash and metakaolin. *Key Eng. Mater.* **2015**, *634*, 172–181. [CrossRef]
114. Wongkeo, W.; Thongsanitgarn, P.; Ngamjarurojana, A.; Chaipanich, A. Compressive strength and chloride resistance of self-compacting concrete containing high level fly ash and silica fume. *Mater. Des.* **2014**, *64*, 261–269. [CrossRef]
115. Paliwal, G.; Maru, S. Effect of fly ash and plastic waste on mechanical and durability properties of concrete. *Adv. Concr. Constr.* **2017**, *5*, 575–586.
116. Ibrahim, M.H.W.; Hamzah, A.F.; Jamaluddin, N.; Ramadhansyah, P.J.; Fadzil, A.M. Split tensile strength on self-compacting concrete containing coal bottom ash. *Procedia-Soc. Behav. Sci.* **2015**, *195*, 2280–2289. [CrossRef]
117. Yoshitake, I.; Komure, H.; Nassif, A.Y.; Fukumoto, S. Tensile properties of high volume fly-ash (HVFA) concrete with limestone aggregate. *Constr. Build. Mater.* **2013**, *49*, 101–109. [CrossRef]
118. Singh, N.; Singh, S.P. Carbonation and electrical resistance of self compacting concrete made with recycled concrete aggregates and metakaolin. *Constr. Build. Mater.* **2016**, *121*, 400–409. [CrossRef]
119. Singh, N.; Singh, S.P. Carbonation resistance and microstructural analysis of low and high volume fly ash self compacting concrete containing recycled concrete aggregates. *Constr. Build. Mater.* **2016**, *127*, 828–842. [CrossRef]
120. Liu, M. Self-Compacting concrete with different levels of pulverized fuel ash. *Constr. Build. Mater.* **2010**, *24*, 1245–1252. [CrossRef]
121. Kapoor, K.; Singh, S.P.; Singh, B. Durability of self-compacting concrete made with Recycled Concrete Aggregates and mineral admixtures. *Constr. Build. Mater.* **2016**, *128*, 67–76. [CrossRef]
122. Saha, A.K. Effect of class F fly ash on the durability properties of concrete. *Sustain. Environ. Res.* **2018**, *28*, 25–31. [CrossRef]
123. Gonen, T.; Yazicioglu, S. The influence of mineral admixtures on the short and long-term performance of concrete. *Build. Environ.* **2007**, *42*, 3080–3085. [CrossRef]
124. Li, G. Properties of high-volume fly ash concrete incorporating nano-SiO_2. *Cem. Concr. Res.* **2004**, *34*, 1043–1049. [CrossRef]
125. Thomas, M.D.A. *Optimizing the Use of Fly Ash in Concrete*; Portland Cement Association: Skokie, IL, USA, 2007; Volume 5420.
126. Naik, T.R.; Ramme, B.W.; Kraus, R.N.; Siddique, R. Long-Term Performance of High-Volume Fly Ash. *ACI Mater. J.* **2003**, *100*, 150–155.
127. Kwon, S.-J.; Lee, H.-S.; Karthick, S.; Saraswathy, V.; Yang, H.-M. Long-Term corrosion performance of blended cement concrete in the marine environment—A real-time study. *Constr. Build. Mater.* **2017**, *154*, 349–360. [CrossRef]
128. Al-Saleh, S.A.; Al-Zaid, R.Z. Effects of drying conditions, admixtures and specimen size on shrinkage strains. *Cem. Concr. Res.* **2006**, *36*, 1985–1991. [CrossRef]
129. Holt, E.; Leivo, M. Cracking risks associated with early age shrinkage. *Cem. Concr. Compos.* **2004**, *26*, 521–530. [CrossRef]
130. Rao, G.A. Long-Term drying shrinkage of mortar—Influence of silica fume and size of fine aggregate. *Cem. Concr. Res.* **2001**, *31*, 171–175. [CrossRef]
131. Mokarem, D.W.; Weyers, R.E.; Lane, D.S. Development of a shrinkage performance specifications and prediction model analysis for supplemental cementitious material concrete mixtures. *Cem. Concr. Res.* **2005**, *35*, 918–925. [CrossRef]
132. Güneyisi, E.; Gesoğlu, M.; Özbay, E. Strength and drying shrinkage properties of self-compacting concretes incorporating multi-system blended mineral admixtures. *Constr. Build. Mater.* **2010**, *24*, 1878–1887. [CrossRef]
133. Akkaya, Y.; Ouyang, C.; Shah, S.P. Effect of supplementary cementitious materials on shrinkage and crack development in concrete. *Cem. Concr. Compos.* **2007**, *29*, 117–123. [CrossRef]
134. Zhao, H.; Sun, W.; Wu, X.; Gao, B. The properties of the self-compacting concrete with fly ash and ground granulated blast furnace slag mineral admixtures. *J. Clean. Prod.* **2015**, *95*, 66–74. [CrossRef]
135. Şahmaran, M.; Yaman, Ö.; Tokyay, M. Development of high-volume low-lime and high-lime fly-ash-incorporated self-consolidating concrete. *Mag. Concr. Res.* **2007**, *59*, 97–106. [CrossRef]
136. Zhang, M.H. Microstructure, crack propagation, and mechanical properties of cement pastes containing high volumes of fly ashes. *Cem. Concr. Res.* **1995**, *25*, 1165–1178. [CrossRef]

137. Wang, K.; Shah, S.P.; Phuaksuk, P. Plastic shrinkage cracking in concrete materials—Influence of fly ash and fibers. *Mater. J.* **2001**, *98*, 458–464.
138. Mardani-Aghabaglou, A.; Sezer, G.İ.; Ramyar, K. Comparison of fly ash, silica fume and metakaolin from mechanical properties and durability performance of mortar mixtures view point. *Constr. Build. Mater.* **2014**, *70*, 17–25. [CrossRef]
139. Baghabra, O.S.; Maslehuddin, M.; Saadi, M.M. Effect of magnesium sulfate and sodium sulfate on the durability performance of plain and blended cements. *Mater. J.* **1995**, *92*, 15–24.
140. Sumer, M. Compressive strength and sulfate resistance properties of concretes containing Class F and Class C fly ashes. *Constr. Build. Mater.* **2012**, *34*, 531–536. [CrossRef]
141. Ramyar, K.; Inan, G. Sodium sulfate attack on plain and blended cements. *Build. Environ.* **2007**, *42*, 1368–1372. [CrossRef]
142. Mirvalad, S.; Nokken, M. Minimum SCM requirements in mixtures containing limestone cement to control thaumasite sulfate attack. *Constr. Build. Mater.* **2015**, *84*, 19–29.
143. Khodair, Y.; Bommareddy, B. Self-Consolidating concrete using recycled concrete aggregate and high volume of fly ash, and slag. *Constr. Build. Mater.* **2017**, *153*, 307–316. [CrossRef]
144. Pitroda, J.; Umrigar, D.F.S.; Principal, B.; Anand, G.I. Evaluation of sorptivity and water absorption of concrete with partial replacement of cement by thermal industry waste (Fly Ash). *Int. J. Eng. Innov. Technol. (IJEIT)* **2013**, *2*, 245–249.
145. Hatungimana, D.; Taşköprü, C.; İçhedef, M.; Saç, M.M.; Yazıcı, Ş. Compressive strength, water absorption, water sorptivity and surface radon exhalation rate of silica fume and fly ash based mortar. *J. Build. Eng.* **2019**, *23*, 369–376. [CrossRef]
146. Gunasekara, C.; Zhou, Z.; Law, D.W.; Sofi, M.; Setunge, S.; Mendis, P. Microstructure and strength development of quaternary blend high-volume fly ash concrete. *J. Mater. Sci.* **2020**, *55*, 6441–6456. [CrossRef]
147. Ahad, M.Z.; Ashraf, M.; Kumar, R.; Ullah, M. Thermal, Physico-Chemical, and Mechanical Behaviour of Mass Concrete with Hybrid Blends of Bentonite and Fly Ash. *Materials* **2019**, *12*, 60.
148. Saha, A.K.; Sarker, P.K. Expansion due to alkali-silica reaction of ferronickel slag fine aggregate in OPC and blended cement mortars. *Constr. Build. Mater.* **2016**, *123*, 135–142. [CrossRef]
149. Ibrahim, R.K.; Hamid, R.; Taha, M.R. Fire resistance of high-volume fly ash mortars with nanosilica addition. *Constr. Build. Mater.* **2012**, *36*, 779–786.
150. Poon, C.S.; Kou, S.C.; Lam, L.; Lin, Z.S. Activation of fly ash/cement systems using calcium sulfate anhydrite ($CaSO_4$). *Cem. Concr. Res.* **2001**, *31*, 873–881.
151. Min, Y.; Jueshi, Q.; Ying, P. Activation of fly ash–lime systems using calcined phosphogypsum. *Constr. Build. Mater.* **2008**, *22*, 1004–1008. [CrossRef]
152. Aimin, X.; Sarkar, S.L. Microstructural study of gypsum activated fly ash hydration in cement paste. *Cem. Concr. Res.* **1991**, *21*, 1137–1147.
153. Fernández-Jiménez, A.; Palomo, A. Characterisation of fly ashes. Potential reactivity as alkaline cements. *Fuel* **2003**, *82*, 2259–2265. [CrossRef]
154. Palomo, A.; Alonso, S.; Fernandez-Jiménez, A.; Sobrados, I.; Sanz, J. Alkaline activation of fly ashes: NMR study of the reaction products. *J. Am. Ceram. Soc.* **2004**, *87*, 1141–1145.
155. Fernandez-Jimenez, A.M.; Palomo, A.; Lopez-Hombrados, C. Engineering properties of alkali-activated fly ash concrete. *ACI Mater. J.* **2006**, *103*, 106.
156. Sanchez, F.; Sobolev, K. Nanotechnology in concrete—A review. *Constr. Build. Mater.* **2010**, *24*, 2060–2071.
157. Shaikh, F.U.A.; Supit, S.W.M.; Sarker, P.K. A study on the effect of nano silica on compressive strength of high volume fly ash mortars and concretes. *Mater. Des.* **2014**, *60*, 433–442. [CrossRef]
158. Supit, S.W.M.; Shaikh, F.U.A.; Sarker, P.K. Effect of nano silica and ultrafine fly ash on compressive strength of high volume fly ash mortar. *Appl. Mech. Mater.* **2013**, *368–370*, 1061–1065. [CrossRef]
159. Supit, S.W.M.; Shaikh, F.U.A. Durability properties of high volume fly ash concrete containing nano-silica. *Mater. Struct.* **2015**, *48*, 2431–2445.
160. Meesala, C.R.; Verma, N.K.; Kumar, S. Critical review on fly-ash based geopolymer concrete. *Struct. Concr.* **2020**, *21*, 1013–1028. [CrossRef]

MDPI

St. Alban-Anlage 66

4052 Basel

Switzerland

Tel. +41 61 683 77 34

Fax +41 61 302 89 18

www.mdpi.com

Materials Editorial Office

E-mail: materials@mdpi.com

www.mdpi.com/journal/materials

www.ingramcontent.com/pod-product-compliance
Lightning Source LLC
LaVergne TN
LVHW071241170726
843515LV00006BA/1294